History and Philosophy of the Social Sciences

Yafeng Shan
Editor

History and Philosophy of the Social Sciences

Springer

Editor
Yafeng Shan
Center for Philosophy of Science and Division of Humanities
The Hong Kong University of Science and Technology
Hong Kong, Hong Kong

ISBN 978-3-032-08166-7 ISBN 978-3-032-08167-4 (eBook)
https://doi.org/10.1007/978-3-032-08167-4

This Springer imprint is published by the registered company Springer Nature Switzerland AG
The registered company address is: Gewerbestrasse 11, 6330 Cham, Switzerland

Acknowledgements

Even 6 years ago I was not very interested in the philosophy of social science, because it seemed to me that there are no intrinsic differences between natural science and social science. Issues in the philosophy of social science can be easily addressed if similar issues are well addressed in the general philosophy of science. My interest in the philosophy of social science started when I worked with Jon Williamson on a project on Evidential Pluralism in the social sciences at the University of Kent. I was looking for some interesting case studies in the social sciences. The more I read the work of social science, the more I was fascinated by its plurality and complexity. Gradually I came to realise and appreciate the autonomy and significance of the philosophy of social science. Meanwhile, I was surprised to notice that an integrated HPS approach is rarely employed in the philosophy of social science. My reflections on these issues eventually led to the idea of this volume. At the time, I was travelling around the Balkan peninsula. It took longer than I expected to complete the volume. I would like to thank my friends and students for their invaluable support, especially Yiwen Zhan, Xiaotao Liu, Qinyi Wang, and Qiyue Zhang. I am very grateful to all the authors and referees for their hard work on this volume and to Christopher Wilby, Solomon George, and the team at Springer for their work in bringing the book to fruition. I am also indebted to my wife, Zifei Li, for her boundless support and companionship. The volume is dedicated to Thomas Hylland Eriksen, who contributed a wonderful chapter but unfortunately cannot see it in print.

Contents

Contributors

Ivan Boldyrev Radboud University, Nijmegen, The Netherlands

Michel Bourdeau Centre National de la Recherche Scientifique, Paris, France

Paul Carls Université de Montréal, Montreal, QC, Canada

Sharon Crasnow Norco College, Norco, CA, USA

Thomas Hylland Eriksen University of Oslo, Oslo, Norway

Thomas Kemple University of British Columbia, Vancouver, BC, Canada

Basit Bilal Koshul Lahore University of Management Sciences, Lahore, Pakistan

Pauline Launay Caen University Hospital, Caen, France

Stephan Moebius Universität Graz, Graz, Austria

Eric S. Nelson The Hong Kong University of Science and Technology, Hong Kong, Hong Kong

William Outhwaite Newcastle University, Newcastle, UK

Olli Pyyhtinen Tampere University, Tampere, Finland

Cherry Schrecker Université Grenoble Alpes, Grenoble, France

Yafeng Shan Center for Philosophy of Science and Division of Humanities, The Hong Kong University of Science and Technology, Hong Kong, Hong Kong

Stephen Turner University of South Florida, Tampa, FL, USA

Thomas Uebel University of Manchester, Manchester, UK

Jeroen Van Bouwel Ghent University, Ghent, Belgium

Petri Ylikoski Sociology, University of Helsinki, Helsinki, Finland

Chapter 1
Let's Do HPSS!

Yafeng Shan

Abstract Given that HPS is now an active and vibrant programme in contemporary philosophy of science, it is a bit surprising that there is no similar programmes in the philosophy of social science. The chapter makes the case for a new direction of research, History and Philosophy of the Social Sciences (HPSS), especially by promoting a historical approach to the philosophy of the social sciences. It shows that HPSS is fruitful and sheds new light on many central issues on the nature, development, and methodologies of the social sciences. I argue that there are three obvious benefits of doing HPSS. First, an in-depth analysis of the classic work on the nature and methodology of the social sciences helps inform contemporary social science research. Second, careful historical case studies will help reassess the important debates in the philosophy of the social sciences as well as the methodological disputes between social scientists. Third, a careful historical examination of the origins and development of the social sciences sheds light on the distinctive features and methodologies of the social sciences.

Keywords History and philosophy of social science · History and philosophy of science · HPSS · HPS

1.1 HPSS: A Manifesto

In the 1950s, more and more philosophers of science recognised the significance of the history of science and paid serious attention to historical cases in the philosophical examination of science. Such a turn, often called the historical turn (Bird 2008), ultimately led to the emergence of History and Philosophy of Science (aka HPS). Many HPS departments and journals have been founded since then, while HPS

Y. Shan (✉)
Center for Philosophy of Science and Division of Humanities, The Hong Kong University of Science and Technology, Hong Kong, Hong Kong
e-mail: hmyfshan@ust.hk

Y. Shan (ed.), *History and Philosophy of the Social Sciences*,
https://doi.org/10.1007/978-3-032-08167-4_1

conferences and workshops are regularly organised. HPS, as an academic discipline, has been controversial in many aspects (e.g. Giere 1973; Burian 1977; Laudan 1989; Pinnick and Gale 2000; Steinle and Burian 2002; Schickore 2011; Arabatzis and Schickore 2012; Shan 2020). That said, there is no doubt that HPS is now an active and vibrant programme among philosophers of science as well as some historians of sciences, inspiring and promoting the development of the historical approach to the philosophy of science.

There are three main approaches to HPS: History of Philosophy of Science (aka HOPOS), Integrated History and Philosophy of Science (aka integrated HPS or &HPS), and Philosophical History of Science (aka PHS). HOPOS is primarily concerned with the history of the philosophy of science and the relationship between the history of modern philosophy and the history of modern science. HOPOS practitioners contend that a careful study of the history of philosophy of science 'provides further ground for hope that a new kind of productive relationship between [history of science and philosophy of science] may now be possible' (Friedman 2008, 127). Integrated HPS aims at doing good philosophy of science and good history of science at the same time, based on the motivation that history of science and philosophy of science 'have a special affinity and one can effectively advance both simultaneously' (Arabatzis and Howard 2015, 1). PHS is 'to reconstruct particular historical episodes or to address historiographical questions by engaging with philosophical issues about, for example, experimentation or conceptual change' (Arabatzis 2017, 69–70).

However, there is still no similar programmes in the philosophy of social science. A historical approach has yet to be systemically developed within the philosophy of the social sciences. As Stephen Turner points out, 'There is a kind of allergy in philosophy of social science to historical issues'. I argue that History of Philosophy of Social Science (HPSS) is a promising programme and HOPOS, integrated HPS, and PHS can be fruitfully applied to HPSS. First, a careful examination of the history of the philosophy of social science will inform and inspire our philosophical enquiry about the nature, development, and methodology of the social sciences. Second, aiming at doing philosophy of social science and history of social science is not only possible but also beneficial for both disciplines. For example, careful historical case studies will help reassess the important debates in the philosophy of the social sciences as well as the methodological disputes between social scientists. Third, a philosophical reconstruction of the origins and development of the social sciences will shed light on the distinctive features and methodologies of the social sciences. Questions like where social sciences come from and how social sciences develop are key to understand the questions like what is special about social science and what differs the social sciences from natural sciences. Thus, there can be corresponding approaches to HOPOS, integrated HPS, and PHS in HPSS: History of Philosophy of Social Science (HOPOSS), Integrated History and Philosophy of Social Science (integrated HPSS or &HPSS), and Philosophical History of Social Science (PHSS).

Accordingly, this volume is structured by two parts. Part I is a collection of the work of HOPOSS by revisiting the work of some important figures in the philosophy of social science, such as August Comte, Wilhelm Dilthey, Georg Simmel,

Émile Durkheim, Max Weber, logical empiricists, and Thomas Kuhn, whose views greatly influenced the origins and development of the social sciences. Part II consists of the work of PHSS by examining the historical development of different socials sciences and their key concepts, including anthropology, geography, economics, political science, and sociology. This volume aims to make the case for a new direction of research, History and Philosophy of the Social Sciences (HPSS), especially by promoting a historical approach to the philosophy of the social sciences. It will show that HPSS is fruitful and sheds new light on many central issues on the nature, development, and methodologies of the social sciences.

1.2 Doing HOPOSS: Revisiting Pioneers and Classics

A natural starting point of doing HOPOSS is to revisit the seminal work on the nature and methodology of the social sciences. It is evident that an in-depth analysis of the classic work on the nature and methodology of the social sciences may help inform contemporary issues in the philosophy of social science.

Comte is widely credited as the father of social science. However, Michel Bourdeau, in Chap. 2, argues that Comte should be treated as rather a synthesis of previous endeavours than a starting point of the origins of social science. He carefully examines the historical and intellectual context of Comte, especially the influence of Condorcet, Montesquieu, and Cabanis. Bourdeau also shows how conjectural history and political economy, as two immediate predecessors of sociology, informed Comte's two research programmes: the science of man and moral and political sciences. This sheds light on the profound influence of the Scottish Enlightenment on Comte's intellectual path, which has been overlooked.

Eric Nelson, in Chap. 3, examines the work of Dilthey, another pioneer in social science, by providing a reconstruction of Dilthey's critique of historical reason. He identifies two prerequisites of Dilthey's account of the human sciences: the individual and normativity. For Dilthey, the human sciences aim at understanding individuals and encompass propositions about perceived historical realities, abstracted and theoretically constructed uniformities and regularities, and normative rules and value judgments. Nelson argues that an important implication of Dilthey's philosophy of the human sciences cannot mainly consist of a priori axiologies or conceptual analysis of games and rules. Thus, the philosophical significance of Dilthey's work is his emphasis on the immanent interconnections between the empirical, social systematic, and normative dimensions of the human world and its scientific investigation.

In Chap. 4, Olli Pyyhtinen reassesses the work of Simmel and argues that the significance of Simmel's work on sociology cannot be fully appreciated if they are read independent of his philosophical concerns. He shows that Simmel's approach to sociology is distinctive and unique, compared with the work of his contemporaries (especially Durkheim, Weber, and Karl Marx), in the manner that Simmel places social relation in the centre of his sociological thought as the ultimate unit of analysis. Pyythitnen argues that Simmel's approach can be applied not only to the human world but also to the more-than-human world and help us understand the affective, multispecies, technological, and planetary processes which humans are entangled with today.

Along with Weber and Marx, Durkheim is regarded as a founding father of sociology. However, there is a puzzle for historians and philosophers of social science. Despite his historical significance, Durkheim's work is poorly received in many fields of contemporary social science and harshly treated by prominent philosophers, sociologists, and political scientists. Paul Carls, in Chap. 5, argues that the reception of Durkheim's work was due to some misrepresentations and misunderstandings of some sociological and philosophical concepts related to Durkheim's social ontology and sociological method. Carls argues that these misrepresentations construct a caricature of Durkheim, which unfortunately led to a number of missed opportunities of exploring Durkheim's legacy in a more fruitful way.

In Chap. 6, Basit Bilal Koshul examines Weber's contribution to the philosophy of science. He compares Weber's reflections on the methodology of the social sciences with John Dewey's on the emergence of modern science. Koshul argues that Weber bridges the divide between the hard and soft sciences by showing that scientific enquiry is one type of cultural activities. He concludes that Weber's philosophical reflections take one step further from Dewey's.

In contrast to Comte, Dilthey, Simmel, Durkheim, and Weber, logical empiricism and Kuhn are not widely examined in the context of the philosophy of social science. In Chap. 7, Thomas Uebel makes a reassessment of the significance of logical empiricism in the history of philosophy of social science by scrutinising Otto Neurath's work. He argues that the view of logical empiricism on social science cannot be simply characterised as unduly positivist in a reductionist sense. Uebel shows that Neurath's encyclopaedic account of the unity of science does not require reductive hierarchy of homogeneous laws with physics at its base or methodological monism.

William Outhwaite, in Chap. 8, examines Kuhn's view on social science and its influence on the development of the social sciences. Outhwaite argues that although Kuhn's knowledge of social science was uneven and his own practice of it left him unsatisfied, his paradigm-based account of scientific change profoundly influenced social scientists' reflections on the nature and development of the social sciences. He also argues that Kuhn's holistic approach contributed to shape structural Marxism and the concept of paradigm became an important conceptual tool for social scientists.

1.3 Doing PHSS: Scrutinising the Development and Methodologies of the Social Sciences

Pauline Launay and Cherry Schrecker, in Chap. 9, offer an in-depth analysis of the methodological development of French sociology by focussing on the issue of the relationship between the researcher and their research object. They scrutinise the influence of Durkheim's and Weber's work and interinfluence between French sociology and sociology overseas.

In Chap. 10, Stephen Turner reviews the development of American sociology. He argues that the issues and methodologies of American sociology evolved around two core motivations: the aspiration of sociologists to be scientific and the reform movements. Turner points out that the aim of sociology becoming a science like the physical sciences vanished while labour statistics survive as an important form of bureaucratic social research.

Stephan Moebius, in Chap. 11, outlines the history of German sociology from 1842 to 1945. He shows that German sociology began with the social question, addressed by Marx, Friedrich Engels, Lorenz von Stein, Robert von Mohl, and others. Moebius also shows that the focus of German sociology shifted from the social question to the cultural question, influenced by the work of Simmel and Weber, during the German Reich. And sociology in Germany began institutionalisation as an academic discipline during the Weimar Republic. Moebius concludes with a summary of defining features of sociology.

In Chap. 12, Thomas Hylland Eriksen examines the development of anthropological theory and methodology by identifying three major breaks. The first break is mainly a conceptual shift from evolutionism about cultural change to cultural relativism, led by Franz Boas. The second break is more about methodology, associated with Bronislaw Malinowski, in which the ethnographic method was established as the paradigmatic approach to social and cultural anthropology. The third break, also known as 'the Writing Cultural turn', shifted anthropologists' attention to more fundamental questions about the validity of anthropological concepts, methods, and theoretical constructions. Eriksen argues that all these historical breaks show that anthropologists are a community of disagreement and anthropology develops through internal friction based on some fundamental, shared principles of a theoretical, epistemological, and methodological nature.

Ivan Boldyrev, in Chap. 13, examines the origins of market design and its major conceptual and policy achievements. He argues that the development of market design challenges the traditional concepts of economic models as epistemic artifacts to solely understand the world. Boldyrev also argues that market design models serve as instruments for intervention. He urges historians and philosophers of economics to explore research challenges to market design in greater depth, which will ultimately illuminate important but overlooked issues for stakeholders and society at large,

In Chap. 14, Sharon Crasnow offers a historical and philosophical analysis to show in what sense political science should be understood as a science of causal inference. She reviews the debate over the scientific status of political science in the first part of the twentieth century in the USA. Crasnow argues that the aim of political science is the production of causal knowledge and causal knowledge in political science is context-dependent.

Petri Ylikoski, in Chap. 15, examines two important origins of mechanism-based thinking in the social sciences. He shows that Rom Harré's realist philosophy of natural science, as the source of critical realism about mechanisms, provides a metaphysical theory of causation, which anticipated many later developments. Ylikoski also shows that Jon Elster develops the idea of toolbox of mechanisms as a basis of knowledge in the social sciences. He argues that these two lines of theorising mechanisms are largely independent of each other and have very different sources of inspiration. Ylikoski concludes that a good understanding of heterogeneous origins of mechanism-based thinking helps to explain the plurality and diversity of mechanism talks in contemporary philosophy of social science.

In Chap. 16, Jeroen van Bouwel reviews the history of the debate over methodological individualism and holism. Following Lars Udenn's tripartite analysis of the debate, van Bouwel discusses some important distinctions and limitations of stances in the debate. He also examines two relevant discussions: one is about microfoundations, the other economics imperialism. Van Bouwel suggests that when discussing methodological individualism and holism, philosophers ought to carefully distinguish and spell out the variety of positions that can be evaluated along multiple dimensions.

1.4 More to Do: Remarks and Prospects

In a nutshell, the volume argues for HPSS as a promising direction in the philosophy of the social sciences. It maintains that HPSS, as a currently under-employed resource in the philosophy of social science, contains many proposals and suggestions that can be usefully brought into contemporary debates. The volume not only shows the benefits of HPSS but also provides some examples of how to do HPSS. The chapters in Part I are good cases of how HOPOSS can contribute to a viable programme of HPSS, while Part II provides good cases of PHSS. Having said that, I wish to highlight that the volume itself is merely an attempt to promote the manifesto. There is much more to do for HPSS. Compared with HOPOSS and PHSS, integrated HPSS is still underdeveloped. The history of social science is not short of fascinating episodes for philosophising. In addition, many more works of philosophers of social science (e.g. Marx, Herbert Spencer, Alfred Schütz, Karl Popper, and Peter Winch) are worthy of examination. Let's do HPSS!

References

Arabatzis, Theodore. 2017. What's in It for the Historian of Science? Reflections on the Value of Philosophy of Science for History of Science. *International Studies in the Philosophy of Science* 31 (1): 69–82.

Arabatzis, Theodore, and Don Howard. 2015. Introduction: Integrated History and Philosophy of Science in Practice. *Studies in History and Philosophy of Science* 50 (1): 1–3.

Arabatzis, Theodore, and Jutta Schickore. 2012. Introduction: Ways of Integrating History and Philosophy of Science. *Perspectives on Science* 20 (4): 395–408.

Bird, Alexander. 2008. The Historical Turn in the Philosophy of Science. In *The Routledge Companion to Philosophy of Science*, ed. Stathis Psillos and Martin Curd, 67–77. London and New York: Routledge.

Burian, Richard M. 1977. More than a Marriage of Convenience: On the Inextricability of History and Philosophy of Science. *Philosophy of Science* 44 (1): 1–42.

Friedman, Michael. 2008. History and Philosophy of Science in a New Key. *Isis* 99 (1): 125–134.

Giere, Ronald N. 1973. History and Philosophy of Science: Intimate Relationship or Marriage of Convenience? *The British Journal for the Philosophy of Science* 24 (3): 282–297.

Laudan, Larry. 1989. Thoughts on HPS: 20 Years Later. *Studies in History and Philosophy of Science* 20 (1): 9–13.

Pinnick, Cassandra, and George Gale. 2000. Philosophy of Science and History of Science: A Troubling Interaction. *Journal for General Philosophy of Science* 31 (1): 109–125.

Schickore, Jutta. 2011. More Thoughts on HPS: Another 20 Years Later. *Perspectives on Science* 19 (4): 453–481.

Shan, Yafeng. 2020. Doing Integrated History and Philosophy of Science: A Case Study of the Origin of Genetics. In *Boston Studies in the Philosophy and History of Science*, 1st ed. Cham: Springer. https://doi.org/10.1007/978-3-030-50617-9.

Steinle, Friedrich, and Richard M. Burian. 2002. Introduction: History of Science and Philosophy of Science. *Perspectives on Science* 10 (4): 391–397.

Part I
Philosophy and the Social Sciences

Chapter 2
Auguste Comte and the Origins of Social Science

Michel Bourdeau

Abstract Comte's role in the history of sociology remains a controversial issue. To assess his claim of being its founder, this chapter briefly reviews the previous attempts to establish a social science and shows how Comte synthesized them. The first part presents his own account. As a matter of fact, Comte, given his strong historical perspective, wrote extensively about this topic in 1822 and 1839. He focused on Montesquieu and Condorcet, highlighting their positive contributions and limitations, and on physiological Ideology, which aimed to derive sociology from biology. Although this initial approach is useful, it is also somehow one-sided: Comte, for instance did not mention Saint-Simon, despite having been his secretary for seven years.

The second part offers a new perspective by exploring two research programs widely discussed at that time: the science of man and the moral and political sciences. This approach provides new insights into Condorcet and the Ideologues. However, both programs ultimately proved to be dead ends. In fact, the actual roots of social science trace back to political economy and conjectural history, two fields that developed significantly during the eighteenth century, especially in Scotland. Comte's debt to the Scottish Enlightenment has been overlooked, but is extensive. Regarding political economy, he is usually remembered for his critical stance, but we thereby forget that he began reading Adam Smith and J.B. Say, and that sociology emerged from recognizing the limitations of political economy. As for conjectural history, this perspective helps explain the central role of social dynamics in Comte's sociology: Hume, he said, guided him toward discovering the law of the three stages.

Keywords Sociology · Science of man · Moral and political sciences · Ideology · Conjectural history · Scottish enlightenment · Cabanis · Auguste Comte · Condorcet · Henry Saint-Simon

M. Bourdeau (✉)
Centre National de la Recherche Scientifique, Paris, France
e-mail: michel.bourdeau@ehess.fr

Y. Shan (ed.), *History and Philosophy of the Social Sciences*,
https://doi.org/10.1007/978-3-032-08167-4_2

2.1 Introduction

The sociological approach to sciences requires them to be associated with institutions (teaching chairs, societies, journals). From this point of view, sociology was born at the end of the nineteenth century, with Durkheim and Weber and, even if Comte coined the word, he would at best be its godfather, not its father. His main work, the *Cours de philosophie positive* (1830–1842), remains one of the classics of the philosophy of science; but it is too often forgotten that the *Cours* had two aims: a general one, to systematise all knowledge, and a special one, to found sociology. And, whatever judgement one makes of the result, it cannot be denied that he was the first to elaborate a full fledged theory of social phenomena. In what follows, he will therefore be considered as the founder of social science. However, it does not seem useful here to give an account of Comte's sociology, as it is not difficult to find good presentations of it, which the reader is invited to consult.[1] Instead, in order to focus on the origins of social sciences, Comte will be taken not as a starting point but rather as a synthesis of previous endeavours.

Comte has given his own account of this story but, as it is questionable, this chapter will consist of two parts, of unequal length. We will begin by sketching the way Comte, in the *Plan des travaux scientifiques nécessaires pour réorganiser la société* (1822), and then in the 47th lesson of the *Cours* (1839), described the attempts made before him to constitute social science. We will then focus on how these origins appear to us today. Faced with the rapid development of the natural sciences, Pope had reminded us that "The proper study of mankind is man", and the eighteenth century had devised various research programs (science of man, moral and political sciences) which, together with several academic disciplines in the process of being constituted, such as conjectural history and political economy, were to lead to the gradual identification of the specific nature of the social.

Of course, there is no question of giving an exhaustive picture of the history, or rather the prehistory, of social science in the eighteenth century, as it has already been done.[2] The aim is simply to set out the main milestones, distinguishing between those that lead to Comte and those that are dead ends.

Two more comments. These origins are largely French, in particular because the upheaval produced by the Revolution was accompanied by a profusion of research; which also means that, at this stage, there was a close relationship between sociology and politics, which lasted well into the nineteenth century, social science being associated with a social art which was to guide politicians.

---

[1] See in particular Evans-Pritchard (1970), Gane (2006), Guillin (2018). Aron (1965) remains without doubt the best presentation.

[2] See for instance Gusdorf (1976) and Olson (1993).

2.2 Comte's Own Account

Comte began his career as Saint-Simon's secretary. Following the lock out of the *École Polytechnique* in April 1816, he had to look for a job, and meeting the *Père Simon* in the summer of 1817 was a windfall for him. The former aristocrat, author of the *Introduction aux travaux scientifiques du XIXe siècle* (1810) and the *Mémoire sur la science de l'homme* (1813), had become a liberal and introduced his secretary to the world of industry. As early as 1819, Comte was writing a booklet on positive politics, which was to include a survey of the previous attempts. Among the manuscripts published after his death were an examination of the previous attempts to base social science on physiology, and a review of Condorcet's political works (Comte 1970, 473–490), which was largely taken up in what he always referred to as his "fundamental booklet", the *Plan des travaux scientifiques nécessaires pour réorganiser la société* (1822–1824).

From the *Plan* to the *Cours*, some changes were introduced. Thus, in 1825, Comte insisted on the twofold need which the constitution of what was still called *social physics* had to meet: on the scientific level, it had to complement the system of knowledge by making positive the knowledge of social phenomena; on the political level, it had to indicate what the reorganisation that was to end the Revolution should consist of (Comte 1998, 166–170).

2.2.1 *The* "Fundamental Booklet"

Montesquieu and Condorcet While Comte, in the 47th lesson of the *Cours*, entitled *Principal philosophical attempts to constitute a social science,* goes back to Aristotle, the *Plan* begins with Montesquieu, who was prized for being the first to recognise that the social world is subject to laws. Moreover, his theory of climates, conceived as a study of the social influence of local causes, is undeniably positive. Unfortunately, he greatly overestimated its importance. More generally, as chemistry and biology were still insufficiently developed, he lacked the elements needed to constitute a social science. In short, his enterprise was premature.

As to Condorcet, "He was the first to see clearly that civilization is subjected to a progressive course all of whose steps are rigorously connected to each other according to natural laws, which can be unveiled by philosophical observation of the past. For each era these laws determine in a wholly positive manner the improvements which the social state is called to experience, whether in its parts or in its overall shape" (Comte 1998, 116).

History of sociology does not usually pay much attention to Condorcet, but at that time, the man whom Comte always considered his immediate forerunner was a mandatory reference, not only in France, but also abroad (Winch 1996). However, the praise was followed by harsh criticisms: Condorcet conceived well, but he executed very badly (Macherey 2013, 233–248; Pickering 1993, 48–53).

Condorcet's great merit, indeed his only one, in his *Esquisse d'un tableau historique des progrès de l'esprit humain* (1795), was to clearly identify the idea of progress and to make it the primary object of a social science. Admittedly, a step had already been taken with the theory of perfectibility put forward by Turgot, whose *protégé* he had been. But the notion remained equivocal: *perfectible man* could refer to both the individual man and the species, and the discovery of progress goes with the decision to consider mankind as a single people (*l'humanité comme un seul peuple*). The idea that there is an unbreakable link between sociology and progress has become largely foreign to us, but it explains the primacy given by Comte to social dynamics: the founding act of sociology is the discovery of the law of three states, and in the *Cours,* dynamics will occupy seven chapters, some of them several hundred pages long, as opposed to a short lesson of a few dozen pages for statics. But progress is not just a fact, it is also a task. If Comte congratulates Condorcet on having ended his work with a picture of the future, it is because the march of civilisation is not over. Sociology's link to history thus guarantees its link to politics. In other words, Condorcet's project consisted in "founding politics on history" (Comte 1876, 527): By observing the past we can predict the future and act accordingly. Condorcet was a key player in the French Revolution, which cost him his life. But for the next generation, now that the *Ancien Régime* had been abolished, the political objective set for social science was to help complete the Revolution.

As remarkable as Condorcet's project was, the result was disappointing. First of all, the division into ten epochs is unsatisfactory. Above all, the *Esquisse* suffers from a "fundamental contradiction, [...] in view of the immense improvement which mankind was said to have reached at the end of the eighteenth century, compared with the eminently retrograde influence that the author almost constantly attributes, throughout the past, to all the doctrines, all the institutions, all the powers that were actually preponderant" (Comte 2012, 128–129). Condorcet had not been able to free himself of the critical prejudices peculiar to the eighteenth century and, in this respect, Comte acknowledged Montesquieu's "profound philosophical superiority" (Comte 2012, p. 129).

Three Dead Ends Besides the two previous attempts which, if unsuccessful, were at least heading in the right direction, there were also three others, which received much attention at that time but led to dead ends.

The first one also comes from Condorcet. As a matter of fact, the *Esquisse* is not very representative of his work, which was much more concerned with what he called *mathématique sociale*, the science of applying the calculus of probabilities to moral and political sciences. Condorcet had written some famous essays, such as the *Essai sur l'application de l'analyse à la probabilité des décisions rendues à la pluralité des voix* (1785), in which he proved what is now called Arrow's impossibility theorem; and today that comes under the heading of social choice theory rather than sociology proper. Comte's criticism stems from a radical hostility to the calculus of probabilities (Comte 2012, 231n), which also led him to turn his back on the introduction of statistics.[3]

The second attempt came from Cabanis, the leader of the Ideologues. In order to develop moral and political sciences, the Ideologues proposed that social physics should be seen as a "branch of physiology, that is to say of the study of man" (Comte 1998, 132); in this broad sense, social physics became a "physiology of the species" (Comte 1998, 134), one of the "two great sections of total physiology" (Comte 1998, 137). This is why, although Comte agrees with Cabanis that the study of moral phenomena is a matter for physiology, he insists on "the absolute necessity of separating the study of the collective phenomena of the human race from that of the individual phenomena" (*Ibid.*). Moreover, man is not the only animal to live in society. What distinguishes him from other social animals is historicity, the action of one generation on another; this is also why, of the two major axes around which social life is organised, synchronic solidarity and diachronic continuity, Comte will always give primacy to continuity.

A third attempt was to derive the basis of social science directly from Ideology. Briefly mentioned in 1819 (Comte 1970, 479–480) but absent in 1822, because it was too obviously unsatisfactory, it was presented by Tracy in the fourth volume of his *Eléments d'idéologie*, entitled *Traité de la volonté et de ses effets* (1815), which the author was to republish in 1823 under the title *Traité d'économie politique*.

2.2.2 *The* Cours *and the Criticism of Political Economy*

When the subject was taken up again in the *Cours* in 1839, Montesquieu and Condorcet were of course still there but, since social science had now been established, there was no longer any need to mention the last three attempts, which had in the meantime been forgotten, and their examination was deferred to the passages in which the points of doctrine that concerned them more particularly were set out.[4] On the other hand, the development of political economy made it crucial to show that, contrary to its claims, it can not be seen as a sound basis for the constitution of social science.

To understand Comte's position, it is important to see that he does not deny economics some merits. By representing "the various human interests as necessarily interdependent, and therefore amenable to a stable fundamental conciliation, [… it has dispelled] the disastrous and immoral prejudice which, either between individuals or between peoples, represented the improvement of the material condition of some as only being able to result from a corresponding deterioration in others; […]

[3] Comte, who hiterto had used *physique sociale*, coined *sociology* after Quetelet took over the term in *Sur l'homme et le développement de ses facultés, ou Essai de physique sociale,* a book "in which it is at most a question of mere statistics" (Comte 2012, 30n). See also Porter (1986).

[4] For social mathematics, see Comte (2012, 190–191 and 230–231). For the reduction to biology, see Comte (2012, 218–219).

a great service, which true social science will have to carefully preserve and improve" (Comte 2012, 135–136).

Nevertheless, the criticism was radical. First of all, on the theoretical level, political economy has none of the characteristics of a science: no consensus between those who practice it, no progress from one generation to the next. Instead, there are endless discussions about utility, production and value. However, the fundamental mistake lies elsewhere, in the ignorance of the specific nature of social phenomena. Whereas "future social physics [...] can only be properly founded by embracing, from a single great philosophical viewpoint, the rational whole of all the various social aspects" (Comte 2012, 137), economists do quite the opposite: they begin by isolating their object, by positing it as "entirely distinct and independent", after which "they propose their way of proceeding as the type according to which all social theories must finally be regenerated" (Comte 2012, 131). Moreover, *homo œconomicus* is only a fiction, for "man does not solely or even mainly act on the basis of calculations, and in the second place he is not always or even normally capable of calculating accurately" (Comte 1998, 221). In Comte's terminology, political economy is metaphysical.

This judgment is confirmed from a practical point of view. For Comte, metaphysics is characterised by several attributes: it is abstract, but also critical and dissolving (Bourdeau 2006, 93–115). In this respect, by contributing to the destruction of the industrial system of the *Ancien Régime*, political economy was a factor of progress; but this goal has been achieved and from now on it will only "systematise anarchy": "For having, more or less imperfectly, observed, in a few particular cases of very secondary importance, the natural tendency of human societies towards a certain necessary order, this so-called science has very viciously concluded the fundamental uselessness of any special institution, directly intended to regulate this spontaneous coordination, instead of seeing in it only the primary source of the possibility of such organisation" (Comte 2012, 135), which is tantamount to granting that we are powerless in the face of the crises that are bound to arise, instead of making use of the room for manoeuvre available to us. And so,

> the general spirit of political economy, for anyone who has properly appreciated it [...], leads essentially today to erect as a universal dogma the necessary absence of any regulatory intervention whatsoever, as constituting, by the nature of the subject, the most suitable means of supporting the spontaneous development of society; in such a way that, on each serious occasion that successively arises, this doctrine is usually only able to respond to the most urgent practical needs by the vain and uniform reproduction of this systematic negation. [...] This vain and irrational disposition to admit only that degree of order which establishes itself, is obviously equivalent, in social practice, to a sort of solemn resignation given by this so-called science with regard to every slightly serious difficulty which industrial development comes to raise (Comte 2012, 135–136).

Finally, less as an attempt to establish social science than as an indication of such a need, Comte mentions the revival of history, "because it is obvious [that] historical science, properly conceived, and political science, rationally treated, coincide, in general, of necessity" (Comte 2012, 138). Special mention should be made of the German school of historical law, as well as Kant's *Idea of a Universal History from a Cosmopolitical Point of View* (1784), which had made the strongest impression on Comte: "it is prodigious for its time and, if I had known it six or seven years earlier, it would have saved me a lot of trouble".[5]

2.3 A Fresh Look at the Origins: An Omission

Thanks to Comte, we got a first look at the main attemps made before him to constitute social science, but his account is one-sided and needs to be supplemented in two ways. Firstly, there is no mention of Saint-Simon. Comte's injustice towards his mentor is obvious and it is impossible to dismiss the issue. Moreover, recent studies have shed new light on the origins of social science. A more in depth study is required, where we shall meet some of the authors already mentionned.

With regard to Saint-Simon, two questions need to be distinguished, even if they are not unrelated. The first concerns what Comte owes to the man whose secretary he was for seven years. The other concerns the exact nature of Saint-Simon's contribution to sociology. As the first does not concern us directly, we will limit ourselves to two remarks. First of all, the question is still open between those who, like Durkheim or Hayek, only want to see Comte as a Saint-Simonian and those who emphasise his originality and tend to minimise his debt.[6] The existence of this debt is undeniable, starting with the focus on industry, or the necessary separation between a temporal power and a spiritual power, which should be entrusted to scientists. However, dividing up what belongs to one and what belongs to the other is singularly complicated by the question of attributions. For several years, Comte did not sign his writings, which were therefore published under the name of Saint-Simon. "He was justified in later life for feeling frustrated that the letters in *L'Organisateur* had not been properly attributed to him; historians even today base much of their exposition of Saint-Simon's theory of history on Comte's '*Sommaire appréciation de l'ensemble du passé moderne*'" (Pickering 1993, 175). It is therefore unfortunate that the recent edition of Saint-Simon keeps perpetuating this mistake and continues to attribute to him texts written by Comte.

As to Saint-Simon's contribution to social science, the question is quite controversial. Asking who should be credited with founding sociology, Durkheim did not hesitate to answer "It is to Saint-Simon that we must, in all fairness, attribute the honour commonly attributed to Comte" (Durkheim 1928, 148). But that is open to question, for just as Comte, in order better to conceal his debt to Saint-Simon, appealed to Condorcet, so Durkheim, in order better to conceal his debt to Comte, appealed to Saint-Simon. Furthermore, it is of the utmost importance to make a clear distinction between Saint-Simon and the Saint-Simonians. When we speak of

[5] Comte (1973, 143). On Comte's relationship with Herder, Kant and Hegel, see Pickering (1993, 275–305).

[6] See Bourdeau (2016, 277–282). Pickering (1993, 175) sums up the situation: Comte "undeniably took many of Saint-Simon's ideas, but he gave them a new and more extensive interpretation".

Saint-Simonism, what we are usually referring to is not Saint-Simon so much as his disciples, who had the singular distinction of not having known *le Père Simon*, and acknowledged that it was in Comte's 1824 pamphlet that they had discovered the master's doctrine (Bazard 1831, 373–375). However, it was the latter who were influential, perhaps more on socialism than on sociology.

If we look at Saint-Simon himself, we find a clear desire to develop a science of man, a clear awareness that social phenomena are not reducible to biology, the discovery of the world of industry and a critical stance toward political economy. But Saint-Simon was a very versatile thinker. In just over 20 years, he went through no less than five successive phases: scientism, pacifism, industrialism, socialism and moralism. He was an alert mind who knew how to awaken people's minds, but anyone who has read up on his works knows how difficult it is to pick out a clear and elaborated doctrine. After 1815, his publications consisted mainly of occasional writings, in which practical concerns tended to dominate. Georges Gurvitch aptly summed up Saint-Simon's contribution: he "raises with astonishing acuity most of the issues that agitated sociological thought in the nineteenth and twentieth centuries, but his answers are poorly developed and sometimes even barely sketched out".[7]

Bonald should also be mentioned. The author of the *Théorie du pouvoir* (1796) certainly did not see himself as a sociologist, and he is not known to have exerted any considerable direct influence on sociology. However, he did exert an undeniable influence on a number of sociologists, including Comte.[8]

2.4 A Fresh Look at the Origins: A Broader Historical Context

In addition to this long-acknowledged omission, Comte's account needs to be set in a broader historical context. The first occurrence of *science sociale* dates back to 1789, in a writing by Abbé Sieyès. Not only do the French seem to have played a key role in the genesis of the idea of social science, but this coincidence with the revolutionary epoch explains why, in its early days, the term was so closely linked to political concerns. In what follows, we will begin by examining the two research programs within which the idea of social science emerged, one of which focused on the idea of a science of man, and the other on the idea of moral and political sciences. The development of the latter program is closely associated with the Ideologues, a group around Cabanis in the second class of the *Institut national des*

[7] Saint-Simon (1965, 7). Gurvitch is also careful to temper Engels's "exaggerated enthusiasm" (45n). See also Frick (1983).

[8] See Spaemann (1959) and Karsenti (2003). Bonald is usually associated with de Maistre, whom Comte mentions much more often, but this is because of his treatise *Du pape*, on spiritual power, a topic which belongs much more to politics than to sociology proper. The distinction, fundamental for Comte after 1848, between the three forms of society, domestic (the family), political (the city), and religious (the Church), is very explicitly stated in Bonald.

sciences et des arts, founded in 1795 to replace the academies abolished by the Revolution. However, none of these programs produced the expected results and it is in another direction that we must look for the first sketches of a social science. In fact, at the same time, two academic disciplines pointed directly to comtean sociology; and—whether it be conjectural history, which leads to social dynamics, or political economy, which leads to social statics—, this time, those we meet are mainly representatives of the Scottish Enlightenment.

Each time, we will be witnessing the emergence of the social, as distinct from the political or the economic. During the eighteenth century, the word *society* took on a new meaning. Until then, it had been used to designate virtually any type of voluntary association (commercial society, learned society, such as the *Royal Society*). The members of a political body thus formed a society, the civil society. All that had to be done was to delete the adjective, and we had a new object, society, which called for the creation of a new branch of knowledge. Arguably, this was Comte's aim, and if he is indeed the founder of social science, it is because his work lies at the confluence of these various attempts to define the social. In what follows, the topics or authors discussed will not be studied for themselves. It is simply a matter of assigning them a place in the origins of social science, without even being able to describe in detail their own contribution.

2.4.1 Two Research Programs

Science of Man There is no need to dwell at length on the idea of science of man because, while it is the framework within which social science is embedded, it is also a vague project, with a fundamental ambiguity: what exactly are we interested in, the individual man or the human species? In the first case, we distinguish between physical man, the study of which comes under biology, and moral man, the study of which points in the direction of psychology rather than sociology. So, it is easy to understand why Comte, in a statement that has given rise to many misunderstandings, came to present science of man as having no object: "Man indeed, as an individual, cannot properly be said to exist, except in the too abstract brain of modem metaphysicians".[9] Understood as the study of the human species, the idea of science of man is undoubtedly responsible for the success of the idea of perfectibility. In the eighteenth century, the idea of man's perfectibility was emphasised and was the basis, for example, of Condorcet's *Esquisse*; but this notion was soon forsaken in favour of that of progress (Binoche and Ansaldi 2004; Loterie 1998).

Science of man can still take the form of a natural history of man. As early as 1749, Buffon published a "natural history of man and animals", in which he

[9]Comte (1875a, 268); see also Comte (1875b, 353): "For Man, properly so called, when regarded in his essential reality, and not by the light of materialist or spiritualist dreams, can only be understood by means of a previous knowledge of Humanity, for on this Man necessarily depends".

developed the theory of *homo duplex* that Durkheim would take up again when he needed to distinguish the collective type of a society from the average type of the individuals that make it up.[10] Later, a natural history of man, "based on empirical observation and statistical analysis of phenomena affecting human life, was to form one of the three components of social science as conceived by Condorcet" (Baker 1975, 198; cf. 372). In 1799, the emphasis placed on observation led to the creation of a *Société des observateurs de l'homme*, considered to be the cradle of the branch of human sciences known as anthropology.

Moral and Political Sciences; Ideology The second research program is a continuation of the first one, with moral and political sciences conceived as a branch of science of man. This was a crucial step, since it is the immediate predecessor of social science.[11] It is in this context that the term *social science* was coined, in order to designate an offshoot of the moral and political sciences. The fact that one of the three classes at the *Institut National des Sciences et des Arts* was devoted to the study of these sciences testifies to their importance at that time. Yet the very close relationship between this second class of the *Institut*, abolished in 1803 by Napoleon, and the Ideologues makes the situation quite intricate, as the link between Ideology and social science is far from being obvious.

The proper subject matter of Ideology, namely the analysis of sensations and ideas, oriented it much more towards psychology than towards sociology, and in order to understand how it approached the social only tangentially, we need to take a broader view. As a matter of fact, there were three generations of Ideologues and the one we usually think of, which is limited to a few names grouped around Cabanis, was preceded by another, to which Condorcet belongs. These two generations corresponded to two different ideas of what social science should be, due to a different relationship with politics. While the first one supported actively the Revolution, the experience of the Terror led the second to mistrust politics and turn instead to physiology, which was then booming: rational Ideology was coupled with physiological Ideology, to use a distinction introduced by Tracy.

We must therefore start by going back to Condorcet, who was the mentor of the second generation of Ideologues. Many of them (Garat, Tracy, Cabanis) were members of the *Société de 1789*, which he had founded with Sieyès and where for the first time was put forward "in an institutional context, and among an influential audience, the concept of an integral social science" (Baker 1975, 285). Had Condorcet survived, it is likely that he would have kept his influence on the members of the second generation (Leterrier 1985, 15). The creation of a National Institute was, moreover, one of the projects he had devised to overhaul public education, and it is no exaggeration to say that the development of the moral and

[10] Durkheim (1914); see Desrosières (2010, 125) and Rey (2015, 198–201).

[11] It is indeed Comte's first encounter with social theory. In 1817, before meeting Saint-Simon, he planed to go to the United States to teach geometry in an engineering school that was supposed to be established on the model of the Ecole Polytechnique. He was therefore improving his English and, to the exact sciences, he added "the study of moral and political sciences" (Comte 1970, 49).

political sciences, and more particularly the constitution of a social science, very soon became the focus of all his efforts. Describing the progress of the Enlightenment, in 1782, in his *Discours de réception à l'Académie française*, he began by referring to the moral sciences, assigning to them as their subject either the human mind itself or the relations between men, and then observing that their progress would be slower than that of the physical sciences. The most comprehensive presentation of social mathematics, which was his life's work, is to be found in the *Tableau général de la science qui a pour objet l'application du calcul aux sciences morales et politiques*, published in 1793. As he conceived it, social science proper had three components (Baker 1975, 198–202 and 371–373): in addition to the natural history of man already mentioned, a theory of what social organisation should be, deduced from that analysis of sensations and ideas (rational Ideology), and finally social mathematics, a social art as much as a social science, because its ultimate goal is action, which it must guide by making social choices rational and scientific.

Comte's interpretation of Condorcet is therefore open to question. In his defence, it should simply be said that a quite similar interpretation, which ignores social mathematics in favour of the *Esquisse* and, subsidiarily, the writings on public education, is still widespread. Nevertheless, "to read the *Esquisse* in the light of what Saint-Simon and Comte made of it, rather than in the context of the whole body of Condorcet's writings is to risk a fundamental misinterpretation. [...] Condorcet's conception of social science was not historical [...his] intention was not to subordinate politics to history; indeed, it was quite the reverse" (Baker 1975, 344).[12] The contrast between social science so conceived and what it would become in Comte's hands is glaring. This was already apparent in Condorcet's assessment of Montesquieu: for the man who took the young North American democracy as his model, the admirer of the English monarchy appeared to be a conservative (Baker 1975, 261 and 353–4). On closer examination, then, it would appear that the small part played by Condorcet in the history of sociology is quite justified.

This is all the more true of the second generation of Ideologues. Condorcet was one of the last victims of the Terror: he died on 29 March 1794; Robespierre was arrested on 27 July (9 Thermidor) and guillotined the following day. The *Institut National des Sciences et des Arts* was founded a year later, on 25 October 1795, by a law that reorganised public education. The second class was divided into six sections, the titles of which give an idea of what was then covered by the expression *moral and political sciences*: analysis of sensations and ideas, morality, social science and legislation, political economy, history and geography.

[12] However Baker sometimes goes too far. (1) The function Condorcet assigns to history is the same as the one Comte attributes to it: observation of the past enables us to predict the future and act accordingly. It is an aid to decision-making. (2) When Condorcet says that politics is not based on history but on knowledge of human nature, Comte would partly agree: sociology depends on biology. Human nature sets limits, the knowledge of which enables us to control sociological hypotheses.

Mill agrees with Comte: "any general theory, or philosophy of politics supposes a previous theory of human progress, and that this is the same thing with a philosophy of history" (Mill 1981, 169).

As with Condorcet, Comte's account of the Ideologues needs to be supplemented and corrected, starting with the focus on Cabanis. Although the author of *Rapports du physique et du moral de l'homme* was, along with Tracy, the leader of this second generation, political circumstances made it a sort of nebula, where various currents intersected. Among the members of the moral and political sciences class were personalities better known for their political careers than for their links with Ideology, such as Talleyrand, Cambacérès and Roederer. If 1795 marked a turning point in the conception of social science, this was largely due to the complex relationship maintained by the members of this second class with politics (Wokler 2006). Warned by the excesses of the Terror, they were now concerned with consolidating the achievements of the Revolution. The Thermidorian reaction that followed the fall of Robespierre fulfilled their aspirations, and they took an active part in political life. Cabanis was a member of the *Conseil des Cinq-Cents* and helped to draft the Constitution of 1799. The Ideologues of this generation are usually remembered for their opposition to Napoleon. But some of its members were among the Empire's most loyal supporters. Talleyrand, of course, but also Cambacérès who, in his capacity as arch-chancellor, was for nearly 10 years the second most important person in the State. Cabanis himself was a senator and then a count of the Empire. But this involvement in political life was accompanied by a lack of interest in political theory. Whereas Sieyès conceived of social science as primarily a political science, dealing with sovereignty, assembly systems and modes of representation, for Ideologie, which had become physiological, social science would be a science of social organisation.

Mistrust of politics is not enough to explain the project of a Physiological Ideology; we must also take into account what was happening in the scientific world at that time (Jolly 2021). We are indeed witnessing the fortuitous meeting of two independent lines. On the one hand, there was the question of how to develop social science; on the other one, biology was in full expansion. Hence the idea of looking to biology for the answer to the question. This is what Cabanis proposed when he suggested making physics "the basis of the moral sciences" (Cabanis 1825, III 16).

The idea was soon forgotten, and there is no point to go back over Comte's 1822 rebuttal. According positivism, the sciences show links of dependence between them. Cabanis's project illustrates a well known phenomenon, namely the tendency of each science "to absorb the one next to it, on the ground of having reached the Positive stage earlier and more thoroughly" (Comte 1875a, 39–40). Against what he called materialism, and which we would rather call reductionism, he developed a theory of emergence, where the superior depends on the inferior, but does not result from it. Through the question of the relationship between the physical and the moral, Cabanis nonetheless had the merit of stating an inescapable problem, that of the relationship between sociology and biology. The dualism between inorganic and organic bodies is one of the cornerstones of Comtean thought, and we cannot avoid thinking about the nearness and the difference of the two sciences of organised beings.

Saint-Simon's *Mémoire sur la science de l'homme* (1813) gives a good picture of the results obtained at that time by these two research programs, highlighting in particular the weight given to physiology (Wokler 1987). As early as 1808, the author had noted that what Condorcet lacked was knowledge of physiology. Five years later, the preface to the *Mémoire* begins by summing up the state reached by the science of man: "The four most outstanding works in relation to this science seem to me to be those of Vicq-d'Azyr, Cabanis, Bichat and Condorcet". His aim was therefore to summarise these works, adding to them some necessary supplements: "The first part will deal with the individual man, the second with the human species. The first section of the first part will be a physiological summary, the second a psychological summary. The first section of the second part will contain a sketch of the history of the progress of the human mind, from its starting point to the present day. In the second, I shall present an outline of the course which the human mind will follow after the present generation" (Saint-Simon 2019a, 1070–1071).

2.4.2 Two Academic Disciplines

Although the term *social science* did indeed appear to refer to a branch of the moral and political sciences, what has just been said shows that it remained in the project stage or, at best, a mere outline. The immediate origins of sociology are rather to be found in two academic fields that were well represented in the eighteenth century and in which the Scottish Enlightenment played a leading role: political economy and conjectural history.

Political Economy I will confine myself to Comte, without considering the contributions by Scots and Ideologues.[13] His position has already been mentioned, but what has been said gives no hint of the importance that political economy had in the genesis of his thought. Usually, one remembers only his critical stance. In so doing, we forget that he began reading Smith and J.B. Say, and that sociology was born of the discovery of the limits of political economy (Comte 1970, 473, 476; 1973, 27). Contrary to what he had expected, the path followed by economists did not lead to an authentic social science, and another direction had to be taken.

Saint-Simon had already expressed some criticisms. If J.-B. Say was right when he said that political economy "is the true and only foundation of politics" (Saint-Simon 2019b, 1497), it "needs to be recast according to broader, bolder and more general views" (Saint-Simon 2019b, 1524). Saint-Simon thought he had found this broader framework in the theory of the applied sciences. He identified political economy, which he conceived as a science of application, with social science and politics, and concluded that "politics and even the whole of social science (i. e. politics and morality) is nothing but a special case of the sciences of application" (Saint-Simon 2019b, 1525). He even went on saying that the science he was envisioning

[13] On the Scots, besides the items already mentioned, see Skinner (2003) and Berry (2020). On the Ideologues, Steiner (2006) and Sonenscher (2009).

could be called "indifferently a general plan of finance, a treatise on general economics or an encyclopaedia of applied sciences" (Saint-Simon 2019b, 1451).—We are not used to see economics as a *science d'application* but, if the terminology is quite misleading, the point of view is not so unfamiliar. According to Mill, for instance, his *Principles of Political Economy* "was not a book merely of abstract science, but also of application, and treated Political Economy not as a thing by itself, but as a fragment of a greater whole; a branch of Social Philosophy, so interlinked with all the other branches, that its conclusions, even in its own peculiar province, are only true conditionally, subject to interference and counteraction from causes not directly within its scope" (Mill 1981, 243). More generally, what is at stake is the crucial question of the relationship between economics and technology.

Comte, who had been trained as an engineer, always paid attention to the links between the sciences and their applications: political art is an application of social science. However, he would not follow Saint-Simon down the path just indicated. Even recast, economics will not give us the key to social phenomena. The correct method is to take the opposite approach: we must consider economic phenomena as a subset of social phenomena and study them in their relationship to other social phenomena. In other words, positive sociology does not ignore economic life at all: a good part of the historical lessons in the *Cours* is devoted to it, and the social statics of the *Système de politique positive* contains a theory of wages and the transmission of wealth.

It is thus possible to complement the criticism already mentioned. While Saint-Simon condemned the Luddites for their inability to see that, in the long run, the introduction of machines was in their interests (Saint-Simon 2019b, 1526–27) and considered it "essential to suppress government intervention" (Saint-Simon 2019b, 1536), Comte saw in such a position the expression of the "fundamental vice that reigns in the general spirit of political economy, as it has been conceived up to now. By considering, in various cases, that social phenomena naturally tend to arrange themselves [...] economists conclude that it is superfluous and even harmful to try to influence this organisation by direct measures" (Comte 1970, 173). As for the Luddites, their behaviour shows that "from a certain point of view, all social questions are only questions of time" (*Ibid.*). "Comte's criticism of political economy for being indifferent to question of justice was leading him to deeper understanding of the essence of social science" (Pickering 1993, 110). This led him to supplement the theory of spontaneous order of human society he had developed in the 50th lesson of the *Cours* with a theory of modifiable order, "as new as [...] difficult" (Comte 1876, 422).

Conjectural History The argument put forward against Cabanis was historicity. Since what characterises the human species is the continuous action of one generation upon another, social science had to take on a historical form; but this required a new status for history. Condorcet was not the only one to have understood this and, after quoting his name, Comte added: "We should consider in the same light the efforts made in England in the last century to perfect the nature of history by giving it an explanatory or scientific character, instead of the descriptive or literary character it had had up to then" (Comte 1998, 165–6).[14] Dugald Stewart coined the term *conjectural history* to describe the originality of this new kind of history, practised by both Adam Smith and Hume.

In the eighteenth century, there was no shortage of examples of conjectural history: in England, Ferguson's *History of Civil Society* (1767), or Millar's *Origin of the distinction of Ranks* (1778); in France, Rousseau's *Discours sur l'origine et les fondements de l'inégalité* (1754) or his *Essai sur l'origine des langues* (1771); in Germany, Kant's *Conjectures on the Beginnings of Human History* (1786) or the works of Herder, such as his *Treatise on the Origin of Languages* (1771), or his *Ideas for a Philosophy of Human History* (1784–1791). This kind of history is also referred to as reasoned history or philosophical history. *Conjectural* draws attention to the fact that, as the titles quoted indicate, it often involves going back to origins for which, due to lack of direct data, we are left to conjecture. In his *Discourse on the Origins of Inequality,* Rousseau's rule of thumb was "let us begin by discarding all the facts".

The term therefore refers to a method characterised by three features. It is naturalistic, progressive and holistic. It is naturalistic in two ways. Firstly, by rejecting miracles and theological explanations (the divine origin of language was still a founding thesis for Bonald). Instead, nature, namely human nature, will be appealed to, so that conjectural history can be seen as an extension of the natural history of man. If we have no direct and certain knowledge of the object under study, the conjectural knowledge we can have of it is in this way based on observation. Observation shows us that, at every period observed, human behaviour displays permanent features that make up human nature. All we have to do is assume that they could be found also in men whose behaviour we cannot observe. The first lines of Condorcet's *Esquisse* read: "No direct observation provides us with information about what preceded this state [The first epoch: men are grouped into tribes]. It is only by examining the intellectual and moral faculties and the physical constitution of man that we can conjecture this first degree of civilisation".[15] Conjectural history is therefore based on an "incontrovertible logical maxim that the capacities of the human mind have been in all ages the same, and that the diversity of phenomena exhibited by our species is the result merely of the different circumstances in which men are placed" (Stewart 1854, I. 69).

Such a principle had been already stated a century earlier by Fontenelle, who, in order to settle the quarrel between the Ancients and the Moderns, asked us to aknowledge that "nature has in her hands a certain paste, which is always the same, which she turns over and over again, and from which she forms men, animals and

[14] Palmeri (2008, 9–14) is probably the only one to have pointed out that Comte was well acquainted with conjectural history.

[15] Condorcet (1829, 21); in the preface (13–14), he stated that the three main parts of the *Tableau* depended on the data available and on the method to be followed in dealing with them. The theory of human nature is thus largely a psychological theory; the persistence of human nature is not, however, incompatible with the fact that the human mind can make progress.

plants. The centuries make no natural difference between men", so that the differences that appear in history are due to circumstances (Lecoq 2001, 296).

The non-intentional character of some social phenomena, on which the theorists of spontaneous order have recently put so much emphasis, is a mere corollary of this first feature. It is illusory to think that we can explain a natural phenomenon by seeing it as the result of a project or intention. Nature has no aim and what we observe is the product of blind forces. There is a close link between the natural and the spontaneous: the spontaneous order of the economists was first conceived as a natural order by the physiocrats (Le Mercier de la Rivière, *L'ordre essentiel et naturel des sociétés humaines*, 1767).

The second characteristic of conjectural history, its progressive leaning, was there from the outset. If, thanks to Fontenelle, the quarrel between the Ancients and the Moderns became a genuine event in the history of the human mind, it is because it no longer focused on the question of the imitation of the Ancients, and that for the first time the idea of progress was fully brought out. The eighteenth century was to follow this path. The discovery of the New World, followed by the great voyages of exploration, taught us that there were societies very different from our own. Europe thus became aware of its uniqueness, while at the same time broadening its concept of the social world. It was thus led to adopt a universalist point of view, which Condorcet summed up in the hypothesis of mankind as a single people, and to conceive of these different kinds of society as stages in the march of civilisation.

Finally, conjectural history takes a holistic view. Society forms a complex whole, whose components (customs; legislation; economic, intellectual, religious and artistic life) are interdependent. It is therefore illusory to believe that we can study any one of them in isolation, without taking the others into account. This holistic nature is confirmed by the links between the two sciences of organisation, biology and sociology. A first step towards recognising the specificity of social facts is to affirm their organic nature. The difference between inorganic and organic bodies goes with a difference in method: In the second case, we know the whole before we know the details, which are more often than not inaccessible to us (Comte 2012, 167–170).[16]

2.5 From the Scottish Enlightenment to Comte

Until now the emphasis has been mainly laid on what was happening in France whereas, for both conjectural history and political economy, the work carried out in Scotland has been just as decisive, if not more so. The Scottish Enlightenment has recently been the subject of many studies; however, one aspect of the way in which it contributed to the genesis of social science has been almost entirely neglected, namely the decisive influence it exerted on the young Comte. To Mill, who pointed

[16] Elias (1984) emphazises the role of political economy in the sociogenesis of sociology but forgets conjectural history, in spite of its strong links with politics.

out that in Scotland "public education has a more French than English character, which explains the eminent merit of Scottish thinkers", Comte replied: this "can only increase our sympathy by the special case I have always made of this noble philosophical school, which, without being the most useful, was certainly the most advanced of all those of the last century; I am proud, thanks to you, to feel myself in continuity with it" (Comte 1975, 375 and 141).

Among the many clues to this debt, it will suffice to mention three. First of all, it is here—and not, as we are too often led to believe, with Maistre and the retrograde school—, that he found his criticism of social contract theories. Moreover, when he congratulates Smith on having noted "that at no time, nor in any country, was there a god for gravity" (Comte 2012, 303), the allusion is to the passage in which the invisible hand is first mentioned: "Heavy bodies fall, and light substances rise by the necessity of their nature, and it has never occurred to us to have recourse in these matters to the invisible hand of Jupiter" (Smith 1982, 49–50).

Comte went even so far as to present Hume "as the founder of the law of the Temporal or Active Evolution" (Comte 1876, 51). As a matter of fact, the essay *Of the Populousness of Ancient Nations* (1742) provided the opportunity for a global comparison between Antiquity and Modern Times, characterised by military activity in the former and industrial activity in the latter, leading to the conclusion that the latter were superior. It is indeed the general idea of progress that is at stake, and this is why Comte credited Hume with having put him on the path to the law of the three states. Fontenelle's position, quoted above, "was afterwards extended to Social development by Hume's indirect dissertation on the general comparison between the Industrial Civilisation of modem life and the Military Sociability of the ancient world" (Comte 1876, 504). Comte merely complemented Hume on two points. First of all, a maxim dear to Comte states that once a point of departure and a point of arrival have been set, it is easy to find the middle term; in this case, it is the metaphysical state, a mere transition. Moreover, from Condorcet, he learned that progress is first and foremost the progress of human mind, which means that we cannot stick to Hume's version. The distinction between temporal power and spiritual power, borrowed from Saint-Simon, leads to a dual characterisation of each state: theological or military, metaphysical or legalistic, positive or industrial.

In contemporary sociology, references to Comte are scarce and his status as founder has even been challenged. To find out whether this title is well deserved, we can look at the history of the discipline and draw up a list of those who have claimed his name.[17] We can also go back in time, survey the various attempts made before him to constitute social science and show how he synthesised them; in other words, take his work not as a point of departure, but as a point of arrival. This is the choice that has been made here.

Comte made our work easier, since he himself surveyed the attemps made before him to build what he then called *social physics*. But while useful, this account is not always reliable and a fresh look into the sociogenesis of sociology was needed. First

[17] See for instance Scott and Bromley (2013) and Lybeck (2013).

of all, seen as a research program, the idea of social science appeared to be one of the forms that moral and political sciences could take, which in turn was only one aspect of the science of man that was arguably one of the major concerns of the eighteenth century thinkers. But these two research programs were far from producing the expected results, and the immediate origins of sociology lie much more in conjectural history and political economy. Comte's critical appraisal of the latter made people overlook the major role it played in his initial formation. Similarly, conjectural history reminds us of the importance, that is as considerable as it is overlooked, of the Scottish Enlightenment on the genesis of Comte's thought.

From this point of view, Comte's sociology appears as the result of a progressive discovery of the specific nature of social phenomena. The distance between us and this founding moment is nonetheless so striking that it has become difficult for us to recognise ourselves in the form he gave it. The difficulty lies above all in social dynamics: in our eyes, the belief in progress on which it is based lacks positivity and is more a matter for philosophy of history. In spite of this, it cannot be denied that, with Comte, we are indeed witnessing the birth of sociology; but, as he liked to remind us, it is not on the form it had at birth that a being should be judged.

There remains the question of the relationship between social science and social philosophy. According to their sub-titles, the last three volumes of the *Cours* are dealing respectively with the dogmatic then historical parts, not of social science, but of social philosophy. The abolition of the distinction between science and philosophy is, moreover, the main general conclusion of the *Cours*; and this abolition is given as the consequence of the foundation of sociology, since with it, science finally takes possession of the human world, hitherto the domain of philosophy (Bourdeau 2023).

References

Aron, Raymond. 1965. *Main Currents in Sociological Thought*. Vol. I. Basic Books.

Baker, Keith M. 1975. *Condorcet, from Natural Philosophy to Social Mathematics*. University of Chicago Press.

Bazard, Saint-Amand, Prosper Enfantin and Émile Barrault. 1831. *Doctrine de Saint-Simon. Exposition. Première année, 1828-1829*. Au Bureau de l'organisateur.

Berry, Christopher J. 2020. *Social Theory of the Scottish Enlightenment*. Edinburgh University Press.

Binoche, Bertrand, and Saverio Ansaldi. 2004. *L'homme perfectible*. Champ vallon.

Bourdeau, Michel. 2006. *Les trois états: science, théologie et métaphysique chez Auguste Comte*. Éd. du Cerf.

Bourdeau, Michel. 2016. Du nouveau sur les rapports de Comte et de Saint-Simon. *Revue européenne des sciences sociales. European Journal of Social Sciences* 54 (2): 277–288.

Bourdeau, Michel. 2023. Auguste Comte et la philosophie comme science. *Revue Roumaine de Philosophie* 67 (1): 7–19.

Bourdeau, Michel, Mary Pickering, and Warren Schmaus, eds. 2018. *Love, Order, and Progress: The Science, Philosophy, and Politics of Auguste Comte*. University of Pittsburgh Press.

Cabanis, Pierre J. G. 1825. *Oeuvres complètes accompagnées d'une notice sur sa vie et ses ouvrages*, 5 volumes. Bossange.

Comte, Auguste. 1875a. *System of Positive Polity, Vol. 1, General View of Positivism & Introductory Principles*. Translated by John Henry Bridges. Longmans, Green.
Comte, Auguste. 1875b. *System of Positive Polity, Vol. 2, Social Statics, or the Abstract Theory of Human Order.* Translated by Frederic Harrison. Longmans, Green.
Comte, Auguste. 1876. *System of Positive Polity, Vol. 3, Social Dynamics, or the General Theory of Human Progress*. Translated by Edward S Beesly. Longmans, Green.
Comte, Auguste. 1970. In *Écrits de jeunesse 1816–1828: Suivis du Mémoire sur la 'Cosmogonie' de Laplace, 1835*, ed. Paulo E. De Berrêdo Carneiro and Pierre Arnaud. Mouton.
Comte, Auguste. 1973. *Correspondance générale et confessions. 1*. 1814-1840, ed. Paulo E. de Berrêdo Carneiro and Pierre Arnaud. Mouton.
Comte, Auguste. 1975. *Correspondance générale et confessions. 2, Avril 1841-mars 1845*, ed. Paulo E. de Berrêdo Carneiro and Pierre Arnaud. Mouton.
Comte, Auguste. 1998. *Early Political Writings*. Translated by Stuart Jones. Cambridge University Press.
Comte, Auguste. 2012. *Cours de philosophie positive: leçons 46-51*, ed. Michel Bourdeau, Laurent Clauzade, and Frédéric Dupin. Hermann.
de Condorcet, Nicolas. 1829. *Esquisse d'un tableau historique des progrès de l'esprit humain*. Au bureau de la bibliothèque choisie.
Desrosières, Alain. 2010. *La politique des grands nombres. histoire de la raison statistique*. La Découverte.
Durkheim, Émile. 1914. Le dualisme de la nature humaine et ses conditions sociales. *Scientia* 15:206–221.
Durkheim, Émile. 1928. *Le socialisme: Sa définition, ses début, la doctrine Saint-Simonienne*. Félix Alcan.
Elias, Norbert. 1984. On the Sociogenesis of Sociology. *Sociologisch tijdschrift* 11 (1): 14–52.
Evans-Pritchard, Edward Evans. 1970. *The Sociology of Comte: An Appreciation*. Manchester University Press.
Frick, Jean-Paul. 1983. Les détours de la problématique sociologique de Saint-Simon. *Revue française de sociologie* 24 (2): 183–202.
Gane, Mike. 2006. *Auguste Comte*. Routledge.
Guillin, Vincent. 2018. Comte and Social Science. In *Love, Order, and Progress: The Science, Philosophy, and Politics of Auguste Comte*, ed. Michel Bourdeau, Mary Pickering, and Warren Schmaus. University of Pittsburgh Press.
Gusdorf, Georges. 1976. *Les sciences humaines et la conscience occidentale. 6, L'avènement des sciences humaines au siècle des lumières*. Payot.
Karsenti, Bruno. 2003. Autorité, société, pouvoir: La science sociale selon Bonald. In *L'invention de la société: Nominalisme politique et science politique au XVIIIe siècle*, ed. Laurence Kaufmann and Jacques Guilhaumou. Raisons pratiques. Éditions de l'École des hautes études en sciences sociales.
Jolly, Claude. 2021. *Cabanis: l'idéologie physiologique*. Librairie philosophique J. Vrin.
Lecoq, Anne-Marie. 2001. *La querelle des Anciens et des Modernes: XVIIe-XVIIIe siècles*. Gallimard.
Leterrier, Sophie-Anne. 1985. *L'institution des sciences morales*. Editions L'Harmattan.
Lotterie, Florence. 1998. Le progrès désenchanté: la perfectibilité selon Constant, ou le malaise libéral. In *Le Groupe de Coppet et le monde moderne: Conceptions – Images – Débats*, ed. Françoise Tilkin. Presses universitaires de Liège.
Lybeck, Eric Royal. 2013. Lester Ward and Patrick Geddes in early American and British sociology. *History of the Human sciences* 26 (2): 51–69.
Macherey, Pierre. 2013. *Études de philosophie "française": de Sieyès à Barni*. Publications de la Sorbonne.
Mill, John Stuart. 1981. In *Autobiography and Literary Essays (CW of John Stuart Mill, vol. 1)*, ed. John Mercel Robson and Jack Stillinger. University of Toronto Press.

Olson, Richard George. 1993. *The Emergence of the Social Sciences: 1642-1792*. Twayne: Maxwell Macmillan international.

Palmeri, Frank. 2008. Conjectural history and the origins of sociology. *Studies in Eighteenth-Century Culture* 37:1–21.

Pickering, Mary. 1993. *Auguste Comte, an Intellectual Biography, volume 1*. Cambridge University Press.

Porter, Theodore M. 1986. *The Rise of Statistical Thinking, 1820-1900*. Princeton University Press.

Rey, Olivier. 2015. *Quand le monde s'est fait nombre*. Stock.

Saint-Simon, Claude-Henri. 1965. In *La physiologie sociale*, ed. Georges Gurvitch. Presses universitaires de France.

Saint-Simon, Claude Henri. 2019a. In *Oeuvres complètes*, ed. Juliette Grange, Pierre Musso, et al., vol. 1. Presses Universitaires de France.

Saint-Simon, Claude Henri. 2019b. In *Oeuvres complètes*, ed. Juliette Grange, Pierre Musso, et al., vol. 2. Presses Universitaires de France.

Scott, John, and R. J. Bromley. 2013. *Envisioning Sociology: Victor Branford, Patrick Geddes, and the Quest for Social Reconstruction*. SUNY Press.

Skinner, Andrew S. 2003. Economic Theory. In *The Cambridge Companion to the Scottish Enlightenment*, ed. Alexander Broadie. Cambridge University Press.

Smith, Adam. 1982. *Essays on Philosophical Subjects*. Liberty Fund.

Sonenscher, Michael. 2009. "The moment of social science": The *décade philosophique* and late eighteenth-century french thought. *Modern Intellectual History* 6 (1): 121–146.

Spaemann, Robert. 1959. *Der Ursprung der Soziologie aus dem Geist der Restauration: Studien über L.G.A. de Bonald*. Kösel-Verlag.

Steiner, Philippe. 2006. La science de l'économie politique et les sciences sociales en France (1750-1830). *Revue d'histoire des sciences* 15:15–42.

Stewart, Dugald. 1854. *The Collected Works of Dugald Stewart*, ed. Sir William Hamilton. Constable.

Winch, Donald. 1996. Malthus versus Condorcet revisited. *The European Journal of the History of Economic Thought* 3 (1): 44–60.

Wokler, Robert. 1987. Saint-Simon and the passage from political to social science. In *The Languages of Political Theory in Early-Modern Europe*, ed. Anthony Pagden. Cambridge University Press.

Wokler, Robert. 2006. Ideology and the origins of social science. In *The Cambridge History of Eighteenth-Century Political Thought*, ed. Mark Goldie and Robert Wokler. Cambridge University Press.

Chapter 3
Wilhelm Dilthey's Philosophy of the Human Sciences

Eric S. Nelson

Abstract Wilhelm Dilthey's historical and systematic research has been a crucial yet controversial source in the history of the social sciences and social philosophy. This contribution reconstructs Dilthey's critique of historical reason and the controversies surrounding it. Dilthey's early shift from theology and metaphysics towards historical interpretation and structural-functional social explanation has been seen as a precursor to subsequent developments in philosophy, history, and the social sciences. However, critics described his emphasis on the structural formation of the individual and relational individualism, the pre-reflective self-reflexivity, autobiographically situated self-reflection of lived experience in first- and second-person perspective, and the elementary role of interpretive understanding in practical life and the human sciences as remnants of idealism, romanticism, and liberalism. After outlining Dilthey's philosophical development and his models of relational individualism, self-reflexivity, and the interpretive stance, this chapter examines the "primacy of the social" assumed by Dilthey's critics and consider to what extent Dilthey's strategies might still address practical existence and its systematic study in the social sciences.

Keywords Epistemology · History · Individuality · Science · Society · Understanding

E. S. Nelson (✉)
The Hong Kong University of Science and Technology, Hong Kong, Hong Kong
e-mail: eric.nelson@ust.hk

Y. Shan (ed.), *History and Philosophy of the Social Sciences*,
https://doi.org/10.1007/978-3-032-08167-4_3

3.1 Introduction

Wilhelm Dilthey (1833–1911) is undoubtedly a crucial figure in the development of the philosophy of the social sciences in the late nineteenth- and early twentieth-centuries.[1] His pioneering historical and systematic research has been a pivotal albeit contested source in the history of the social sciences and social philosophy. His thought anticipated many of the insights of interpretive and phenomenological psychology and social science, and directly and indirectly (through his students and the appropriation of his thought) contributed to dialogical philosophy, hermeneutics, phenomenology, critical social theory, and debates in the social sciences. The present contribution offers a reconstruction of his project of a "critique of historical reason" (GS 1: ix, 116) in view of his philosophy of the social sciences and contentious issues in its reception such as individualism and holism, nature vis-à-vis spirit, rational objectivity and affective subjectivity, and scientism and relativism.

Dilthey experienced a skeptical crisis of faith in 1852 that would lead him from the study of theology to philosophy and history (BW I: 1). His skeptical turn away from theology and against metaphysics informed his trajectory that eventually led to his accounts of historical and intersubjective understanding, social-historical interpretation, and structural-functional explanation in the human sciences, which prefigure subsequent developments in history and the social sciences. This unique constellation began to emerge in his early drafts, lectures, and publications of the 1860s and 1870s. Dilthey articulated in this context three epistemic principles that would inform his later works: (1) phenomenality, as the way of givenness of the experientially real in the structure of the "there-for-me" and self-environment nexus;[2] (2) a pre-reflective and self-reflexive mode of awareness, an expression he concurrently used—due to its interconnection with the feeling of life and self (GW 18: 141)—to describe the intimacy of poetic, religious, and other types of self-awareness; and (3) self-reflection from its everyday occurrences to its heightened theoretical incarnations. Self-reflexive awareness (*Innewerden*) becomes explicit through self-reflection (*Selbstbesinnung*) and the mediation of conceptualization and production of knowledge (Makkreel 1992: 432).

Philosophy, in the wake of Hume's skepticism and Kant's critical philosophy and notwithstanding the efforts of German idealism, could no longer justifiably posit a unitary substantial self, soul, or subject (GS 1: 7–8; GS 20: 54–55). Further, the

---

[1] English language overviews of Dilthey's thought include Bulhof (1980), De Mul (2004), Makkreel (1992), Nelson (2019).

[2] Dilthey distinguished his principle of phenomenality, i.e., the first-person perspective of consciousness and inner experience within their relational interconnectedness, from phenomenalism, which maintains that phenomenality is epiphenomenal (GS 5: 91).

Kantian transcendental conditions of subjectivity proved inadequate. Through the phenomenal and structural analysis of the self-world nexus, in which the environing external world is co-given concurrently and with as much certainty as the self (GS 1: xix; GS 19: 9), Dilthey introduced the constitutive role of the first- and second-person participant perspective enacted in everyday understanding and praxis and structurally presupposed in the practice of scientific inquiry. This strategy oriented Dilthey's early twofold analysis of (1) the biological, environmental, and social formation of individuals in view of (2) the autobiographically situated self-reflexivity of lived-experiences in the first- and second-person participant perspective, and the constitutive role of intersubjective and interpretive understanding (*Verstehen*) in everyday existence and the human sciences.[3]

Dilthey's innovative approach challenged existing epistemological and scientific discourses and indicated alternative pathways in which the philosopher and scientist become objects of scientific inquiry instead of—like Hegel and Auguste Comte for the early Dilthey—standing above them. At the same time, his thought has been inadequately depicted by his critics as shaped by nineteenth-century residues of a foundationalist conception of science, idealist philosophy of consciousness, romantic feeling and individualism, and liberalism.[4] This chapter will develop a more complex and nuanced approach to Dilthey's philosophy by tracing the formation of Dilthey's epistemic project of a critical or structural experientialism and his models of relational autonomy and individualism (inspired by the German liberal reformist tradition of Kant and Wilhelm von Humboldt), the phenomenality and self-reflexivity of the self-interpreting first-person perspective, and intersubjective understanding and social scientific interpretation in relation to associated issues. It also engages questions of the primacy of social explanation over individual understanding, assumed by Dilthey's critics, and the extent to which his strategies can still address the philosophy of the practical everyday effective life-nexus (*Lebenszusammenhang*) and its systematic investigation in the social and human sciences.

[3] Karl-Otto Apel and Jürgen Habermas described Dilthey's structural strategies as "quasi-transcendental." They both highlighted the intersubjective character of understanding in Dilthey, its affinities with philosophy of social science inspired by the later Wittgenstein, and the continuing significance of understanding in the social sciences and critical social theory (Apel 1955, 1979; Habermas 1973; Habermas 1982). Dilthey was, however, primarily concerned with norms and rules embedded within concretely lived forms of life and less committed to the conceptualism of a linguistic a priori than Peter Winch, G.H. von Wright, Apel, or Habermas.

[4] Compare Gadamer (1985), Kornberg (1981), Lukács (1955).

3.2 The Young Dilthey's Contexts and New Theory of Science

Dilthey was simultaneously a historian, practitioner, and theorist of what he called the *Geisteswissenschaften*. This expression literally denotes the "sciences of spirit" and is most appropriately translated as the "human sciences." It was inspired by the British expression "moral sciences" and mediated through the German notion of "spirit" then closely associated with Hegel's idealism. Dilthey began to consistently use this expression only after experimenting with it and others. In his 1875 essay, for instance, he used "moral-political sciences" and the "sciences of humans, society, and the state" to name those sciences that connect factual (experiential), causal (theoretical), and rule-oriented (normative) relations (GS 5: 31–32).

The expression "moral sciences" signified the sciences of practical life. They did not only include normative ethics but inquiry into human nature, psychology, and practical activities in the wake of David Hume and Adam Smith. Particularly relevant was the sixth book, *The Logic of the Moral Sciences*, of John Stuart Mill's 1843 *A System of Logic, Ratiocinative and Inductive*. The title of the sixth book is translated as the "Logik der Geisteswissenschaften" in Johannes Schiel's 1849 German edition, such that Mill and British empiricism more broadly would operate as significant sources for the early development of Dilthey's philosophy of the human sciences. Mill construed the moral sciences as constructive causal sciences—with their own explanatory grounds—that followed the same method of all science.[5] Secondly, however, "spirit" had unique connotations in the nineteenth-century German context that linked Dilthey with German idealism, romanticism, and historicism. After Herder, Hegel, and the historical school, *Geist* had come to signify not merely consciousness or the mind but its social-historical formation and systematic scope. Thus, to give a brief example, Hegel maintained in his speculative philosophy of spirit that there was one unitary science in which self-knowing spirit comprehended the earlier developmental phases of nature and spirit. Dilthey maintained the sociality of spirit, retaining Hegelian categories such as "ethical life" and "objective spirit" in liberalized and pluralized form, while rejecting spirit's speculative and systematic metaphysical character.[6]

Nineteenth-century "German empiricists" and the young Dilthey advocated, in contrast to Hegel, recourse to the particularity of each empirical science. Dilthey had a complex relationship with German empiricism that stemmed from Friedrich Schleiermacher, whose works he extensively studied and whose multi-volumed

[5] GS 5: 56. On the significance of Mill's *Logic* and positivism for the early Dilthey, see Nelson (2024: 103–123).

[6] On Dilthey's reception of Hegel, compare Nelson (2023b: 90–107) and Ortega and Gasset (1958). The latter traces Dilthey's radicality vis-à-vis Hegel's philosophy of history.

biography he wrote, and his Berlin teacher Friedrich Adolf Trendelenburg.[7] Dilthey's early fusion of experiential science and Kantian critical philosophy (GS 5: 13) was in multiple ways closer to the empiricism of Mill, whose logic he extensively studied and critically appropriated, than the anti-Kantian Aristotelianism of Trendelenburg or Friedrich Ueberweg.[8] While Dilthey rejected Trendelenburg's Aristotelian realism and restricted his immanent teleology to the human formation and projection of purposes, he adopted his pluralistic Aristotelian position that each science was distinguished by its content and subject matter instead of abstractly by method.[9] Dilthey concluded that if each science is conditioned by its own contents, then form and content, method and matter, cannot be effectively separated (GS 20: 15). Theory must follow usage and praxis (GS 18: 1, 47), as specific tasks form their appropriate methods (GS 5: 44), and each individual science has its own formative epistemic presuppositions and interests (GS 18: 23).

The idea that science has one unitary form or method, shared by idealism and positivism, is a remnant of metaphysics contradicted by the fact and value of the plurality of models of scientific inquiry. Whereas idealism reified the historical projection of normative orders onto nature, naturalism ignored it. In addition to the abstract and ultimately circular concern with the idea and method of science detached from specific contexts and contents, the young Dilthey criticized the idealist Hegel and positivist Comte for determining and ranking particular sciences according to an abstract schema of unitary science that ignored their specific histories and practices. Dilthey noted in 1859 and 1875 how the theory of science required the history of the sciences (JD: 81–82; GSv5: 49). In his 1864–1868 Berlin lectures on logic, he defended the natural sciences from the exaggerated claims of spirit, which conflate subject and object, and the individual experiential sciences as the only way of approaching the things or matters themselves (GS 20: 4–5).

Dilthey shared the empiricist insistence on the primacy of experience while critiquing doctrinal truncations of experience for the sake of what he described as an "unrestricted empiria" (*unbefangene Empirie*) and "empiria without empiricism" (*Empirie, nicht Empirismus*) (GS 1: 81; GS 19: 17). Stemming from the Greek word for experience (*εμπειρία, empeiría*), empiria signified for Dilthey experience in its manifold modes of givenness. Instead of appealing to bare experience, whether atomistic or holistic, a genuine philosophy of experience proceeded through—as he wrote in 1859—a "new critique" that could elucidate its natural and historical crystallizations (JD: 80). This critique, a preliminary version of his critique of historical

---

[7] Concerning Trendelenburg's influences on Dilthey, see Kühne-Bertram (2015: 99–118).

On Dilthey and German empiricism, see Damböck (2016). This work construes Dilthey as an empiricist and naturalistic philosopher, despite Dilthey's delineation of the limits of both.

[8] As seen in GS 15: 150–160. On Dilthey's modified Kantianism, see Makkreel (1992).

[9] GS 20: 15; compare Trendelenburg (1840: iv). Aristotle differentiated the sciences by their genus and object in, for instance, *Posterior Analytics* 1.7. Dilthey differentiates sciences according to how their objects are phenomenally given, since there is no underlying metaphysical two-worlds distinction between nature and spirit. He accepts an object-oriented distinction in the first sense, while rejecting it in the second sense (GS 5: 248, 254, 264).

reason and the categories of life, diverged from Kant in two ways: it was developmental and structural rather than transcendental and it could only be adequately analyzed through human scientific inquiry.[10] Despite his public 1867 call to return to Kant (GS 5: 13) and lifelong proximity to Kant's critical project, he rejected Kant's distinction between empiria and concepts, content and form, and sense and thought as dualistic (GS 5: 77–79). Instead of making assertions based primarily on rational form (Kant) or a presumptuous conceptual identity (Hegel), the theory of science is concerned with logic, as a science of normative rules of inference, and logical categories operative within the interwoven networks of shifting concepts and things (GS 20: 4–14, 33–34).

As Dilthey stated in 1883, the human sciences presuppose the natural sciences (GS 1: 14), and the domain of history differentiates itself from the domain of nature (GS 1: 6). Dilthey's epistemic strategy of structural and asymmetrical differentiation has been misconstrued as maintaining a dualism between nature and spirit when it is markedly emergentist in the sense of structural transformation and differentiation.[11] Nature-spirit dualism is untenable as historical-ethical life presupposes natural conditions and the human sciences the natural sciences, from physics to biology and geology, even as the relations between the sciences remain mutually heterogenous and irreducible. Dilthey's earliest epistemological project of a "new theory of science" was to analyze the structural differentiation of the natural and human sciences in relation to their contents, and their consequent distinctiveness, to legitimize their respective individual autonomy (JD: 124, 287). Each science has its appropriate scope and limits that philosophers have one-sidedly and illegitimately extended to the entirety of reality. The autonomy of the individual sciences was threatened on multiple fronts by the pursuit of one unitary science in idealism as well as in materialism, naturalism, and positivism. While the former construed nature as a lower expression of spirit, the latter reduced ethical and intersubjective life—with its own specific conditions of understanding and recognition—to a historically formed conception of physical nature in which those conditions disappeared.

Dilthey engaged in an immanent critique of the naturalistic and positivistic understanding of science by turning to the historical and normative formation of the sciences. The scientist is not only an impartial spectator but an agent, who belongs to a social-historical form of life, and the scientific experiment is a historically established social activity (GS 20: 130). Each researcher, as he later described, has a distinctive habitus and art of seeing, collecting, and analyzing (GS 5: 260). Natural scientific experimentation and inquiry presupposes while justifiably—given its

[10] On Dilthey's structuralism and use of *Zusammenhang* (interactive structured nexus), see Rodi (2003) and Rodi (2018: 33–69); on the concept of structure, also compare Hamid (2021).

[11] Damböck (2016), De Mul (2004), and Hamid (2016) stress the continuity and unity operative in Dilthey's conception of the sciences against dualistic interpretations of the sciences of nature and spirit. However, Dilthey's theory of science does not only posit continuity between the sciences, but (as evident in his criticisms of Comte and Mill) structural transformation and asymmetry (Nelson 2024: 103–123).

aims—abstracting away from its own subjective and intersubjective social-historical conditions and practices (GS 1: 51). Such strategies are inadequate to the human sciences of the history and theory of science. The history of the natural sciences exposes the power and complexity of those unquestioned all too human presuppositions and conditions without which science would not exist. Dilthey's historicizing of the sciences as a socially objectivated system (GS 1: 51) influenced Karl Mannheim and the early sociology of knowledge (Mannheim 2015: 40) and provoked accusations of relativism (e.g., Rickert and the early Husserl). This assessment is premature as it ignores the second phase of Dilthey's analysis: confronting this first moment of radical relativity and skepticism is a prerequisite for the second moment of the contemporary reconstruction of science as the historically situated interconnectedness of propositions and rules normatively and purposively directed at universal validity and truth.

The sciences normatively aim at, within their historical conditions and limits, universally valid knowledge (*Erkenntnis*). The theory of science concerns its foundational, objective, and universal epistemic orientation and aspirations, under fallibilistic and finite conditions, and how scientific inquiry inevitably presupposes even as it interrogates and transforms its social-historically configured meanings, norms, purposes, and values. Dilthey's justification of the sciences spurred allegations of lingering positivistic scientism (e.g., Heidegger and Gadamer).[12] Such one-sided interpretations fail to adequately recognize the twofold character of Dilthey's early philosophy of science, and its later reformation in his mature philosophy of science and worldviews.

3.3 Empiria and Structure: Dilthey's Early Theory of the Human Sciences

Dilthey differentiated the empirical human sciences into three groups (GS 1: 40–42; GS 5: 58). First, anthropology and psychology investigate human nature and the mind. Second, the "social sciences" or the "sciences of the external organization of society" concern morals, law, economy, and the state. They are not merely cultural, as they involve material, economic, and power relations. Third, systems and sciences of culture refer specifically to myth, art, religion, and philosophy, and not the entirety of the human sciences (excepting psychology) as in the Neo-Kantianism of Windelband and Rickert. The human sciences have systematic natural and material dimensions. They are distinguished by their inexorable twofold immersion in meanings, norms, purposes, rules, and values operative within their objects of research (self-reflexive individuals, their relational patterns, and sedimented social systems)

[12] See Gadamer (1985). Gadamer juxtaposes various elements (romanticism and positivism, objectivism and relativism) without arriving at how they are interconnected and modified in Dilthey's thought.

as well as within the researchers as self-reflexive co-participants within a structured horizon or form of life.

Dilthey's early conception of psychology and anthropology developed in confrontation with John Stuart Mill and Moritz Lazarus, his early friend, mentor, and proponent of the psychology of peoples.[13] In contrast to the new experimental psychologists and Neo-Kantian philosophers, Dilthey controversially defined psychology as a human science. While Hume and Mill had classified psychology as a naturalistic moral science, Dilthey connected it with the interpretive stance of the human sciences. All higher psychological forms are, he remarked, historically mediated products (GS 5: 35). The mind is primarily a formed, acquired, and individuated nexus interwoven with rules of action and the symbolic and normative orders of culture and society (GS 18: 55). The epistemic analysis of phenomenality and self-consciousness reveals that consciousness, interiority, and selfhood experientially transpire within the self-world relational nexus and its reflexive differentiation into self and reality.[14]

There is accordingly only one structure of experience in which two directions arise: internal self-directed and external other and thing directed (GS 18: 194). Thus, to consider an historical example first, his early analyses of Christianity as a form of life (*Lebensform*, as in GS 13.2: 589) traced how different social-historical configurations of interiority-exteriority were formed, including his influential account of the emergence of a new configuration of interiority in Augustine (GS 1: 255–267). Amidst the play of phenomena, the individual is reflexively aware of itself and has a feeling of its life (GS 1: 131). Secondly, to turn to Dilthey's emerging structural psychology, naturalistic explanatory psychology divorced physiological and mental processes from the cultural and social matrices in which they are embodied and enacted, thereby truncating the experiential empiria and misconstruing the formation of the psychophysical life-nexus.

Dilthey's fusion of modified empirical and structural strategies led him to redefine and reposition psychology against dominant idealist conceptualist, teleological organicist, and causal naturalistic models. Psychology has a decisive position in epistemology as the theory of science, since all sciences are experiential, and psychology is the science of experience and its conditions (GS 1: xvii). Dilthey emphasized in his lectures from 1865–1868 the primary roles of empirical inductive inference and understanding in psychology as a human science concerned with psychophysical beings that partake in and—at times radically—alter their physical and cultural environments. First, induction does not necessitate here an underlying lawfulness and uniformity to establish its conclusions, as in Mill's account; it engages in a logic of part and whole as it confronts a heterogenous and asymmetrically structured nexus through which connections between particulars are formed, described, and analyzed (GS 18: 5–9). Psychology is not confined to Mill's choice

[13] On Mill and Dilthey, see Nelson (2024: 103–123). On the role of anthropology, see Kühne-Bertram (2015: 135–147).

[14] Compare different versions of this argument in GS 5: 90–91; GS 19: 17; and GS 20: 54–55.

between an intuitive or inductive-hypothetical science. It relies neither on the ostensive self-evidence of intuition (with its independently real truths) nor external hypothetical causal explanations. Psychology critically analyzes the general structures of the mind's interconnectedness through the methodical description of experience (GS 21: 23, 25).

Secondly, interpretive understanding of oneself and others engages in an inference—modifying Mill's account of induction—that shifts from particular to particular, or event to event, through which patterns and habits of adaptation, imitation, analogy, and comparison emerge (GS 20: 100–101). Understanding constitutes the intersubjective logic of everyday existence that the human sciences methodologically describe and analyze. Understanding, as employed by Dilthey or later Georg Simmel and Max Weber, does not need to assume a constitutive idealist subject, as his later positivist and collectivist critics presume. It indicates the often only tacit inevitability of taking the self-interpretations of co-active participants (the first-person point of view in its plural multiplicity) into account in social theory in contrast to applying social categories that tacitly assume them and are more hypothetical than the concrete living individuals they deem epiphenomenal. Typical models of the primacy of the category of the social include capital, power, nationality, the state, or tradition. They can only be normative, purposive, and reflexive because individuals are. This relational individualist thesis clashes with structural-functional sociology and systems-theories that assume social systems are autonomous autopoietic entities. Social and cultural systems are—in Dilthey's analysis—functionally and normatively effective, and scientifically explicable, by being reproduced and sedimented in relations between self-interpreting rule-following individuals, as their breakdowns in periods of catastrophe and revolution can reveal.

Two different senses of structure emerge in Dilthey's reflections: (1) external structural-functional explanations, as employed in biological and social sciences, in which particulars are taken as manifestations of uniformities and epiphenomenal to an underlying explanatory framework; (2) structural-functional description and analysis, as deployed in history and other human sciences, in which particulars are interpreted not merely as parts but as co-creative contributors and potentially reflective participants within a shifting physically and normatively (intersubjectively) reproduced relational nexus. The human sciences encompass varieties of external explanations, which numerous social theories posit as the whole, even as they aim at knowledge of a form of life that can help inform and possibly reform existing practices and institutions.[15]

From these early points of departure, Dilthey elucidated how tacit self-reflexive awareness and consequent autobiographical sense, rule-following and breaking, and self-reflection of the immanent first-person participant perspective form a reflexive asymmetry in the structural differentiation of human worlds within the encompassing natural world. As there is only one world differentiated through how objects are

[15] Dilthey identified with the German reformist lineage of Kant and Humboldt and a moderate constitutional liberalism under German monarchic conditions (compare Bulhof 1980: 12–16; Kornberg 1981; Nelson 2023: 136, 140–141).

given and experienced, there is no justification for appealing to the duality of nature and spirit or a two-worlds ontology of two realms or kinds of objects (GS 5: 248, 254). Their emergent differentiation introduces an asymmetrical heterogeneity between the content and structure of the human and natural sciences: whereas individuals intersubjectively understand or fail to understand each other and their shared and competing worlds, natural and some social phenomena are explained from a constructed hypothetical external third person perspective abstracted from that intersubjective life-nexus.

In the human sciences, humans are natural organisms and complexly mediated and self-interpreting agents formed in interactions (*Wechselwirkungen*) between a multiplicity of natural and social forces and structural systems who have become self-reflexively (without explicit awarenesses) and reflectively (with explicit awarenesses) individuated within their own form of life or effective relational nexus (*Wirkungszusammenhang*).[16] This has significant consequences.

First, Dilthey contrasts these effective interactive models, in which individuals are self-interpreting co-participants, with metaphysical substantive as well as efficient causal models. The concepts of substance and causality are metaphysically abstracted from and fail to comprehend the effective interactional life-nexus analyzed in the human sciences (GS 1: 401; GS 18: 164). Following in the wake of the skeptical and critical dismantling of such metaphysical concepts by Hume and Kant, Dilthey maintained that they only have a conditional heuristic value in the sciences (GS 1: 399). Secondly, each individual experiences itself as real even as it is a crossing point of various sedimented social and cultural systems (GS 1: 50–51; GS 5: 58; GS 20: 132). Individuals are simultaneously externally determined by their world and by how they self-reflexively and self-reflectively participate in, question and wonder, and potentially transform their world.[17]

As the first-person perspective indicates *in concreto* a multiplicity of acquired and conflicting positionalities, it is problematic to directly reduce it, as mere ideology, to either a conservative communal sense of identity and tradition or a liberal idea that individuals best pursue their own self-interests based on their own self-interpretations.[18] At the same time, to briefly summarize a complex issue, the critique of ideology presupposes the self-interpretive stance (visible in self-reflective and autobiographical statements) through its distinction between false and real

[16] On individuals as diverging intersections of roles and structures, see Marom (2014); on selves as socially and reflexively formed and acquired, see Nelson (2023: 135–152).

[17] Heidegger's 1925 Kassel lectures on Dilthey took the reflexivity of human existence to signify that it existentially questions itself in asking after reality and its sense (Heidegger 2016: 103, 110). Heidegger emphasizes its role in the formation of historical worldviews rather than the sciences, which are deemphasized as ontic.

[18] Lukács (1955) and others suspected that the first-person perspective and interpretive understanding in Dilthey and Weber are systematically ideologically distorted and reflect conservative communitarianism or liberal individualism.

consciousness and interests.[19] Dilthey likewise clearly linked the first-person perspective of how individuals experience themselves with a sense of freedom that began with bodily and environmental motility and eventually arrived at aspirations for the modern constitutional state to—echoing Humboldt—preserve individual rights and promote the richest diversity of individual ways of existing.[20]

3.4 Individuality and Science as Guiding Themes of Dilthey's *Introduction to the Human Sciences*

The individual is a prerequisite of the human sciences in two ways. First, psychophysical individuals are the basic elements of history, society, and objects of their systematic study (GS 1: 28). These sciences aim at comprehending individuals in their own relational reality, as the abstract isolated self is an ahistorical fiction (GS 1: 31). Secondly, this requires recourse to their experiences of themselves that are formed through perception, communication, and self-interpretation. The human sciences presuppose and aim at clarifying the first-person point of view of the phenomenal "there-for-me" in which an individual experiences itself, in its own immanent agency and situation, as having a relative degree of autonomy and sovereignty within its natural and social world (GS 1: 6).

Although Dilthey has been erroneously accused of being a bare romantic individualist and his work has been used to justify methodological individualism, his own approach is more nuanced as he articulated an inherently socially mediated individual in the ordinary life-nexus and the human sciences.[21] The complexity of his thought is evident at the very beginning of his first extended systematic book, the 1883 *Introduction to the Human Sciences*. This work can be read as a skeptical critique of previous metaphysical and current scientific theories of the human sciences. First, against—so to speak—the spirit of his times, Dilthey rejected taking social forms, such as the soul or spirit of the people, as real substantial entities and critiqued reifying society or the state as a machine or organism.[22] The people are not a

[19] We can only infer such a response from Dilthey's writings. Implicitly normative and self-interpretive ideas of proletariat class-consciousness, democracy, and solidarity as well as the realization of free association without coercion in a future society presuppose—as Habermas argued—the significance of intersubjective interaction and interpretation, as a knowledge constitutive interest, within society and in its analysis (Habermas 1973).

[20] On Dilthey's Humboldtian and social individualist inspirations, see Nelson (2023: 135–152) and Nelson (2023b: 90–107).

[21] On Dilthey's socially mediated and formative conception of individuality, see Nelson (2023: 135–152).

[22] Compare GS 1: 31, 41, 122; GS 6: 43; GS 16: 110; GS 22: 3. Dilthey contested for ethical and scientific reasons right-Hegelian and organicist nationalist spirit as well as left-Hegelian socialist naturalism as substantializing reifications of relational realities (GS 10: 15, 124). At the same time, Dilthey appropriated, desubstantialized, and pluralized Hegelian categories such as ethical life and objective spirit (GS 4: 258; GS 7: 150; compare Nelson 2023b: 90–107).

substance and have no body, soul, or spirit as a collective entity (GS 1: 31, 41). However, he was not an eliminativist regarding collective expressions. He retained a sense of weak relational holism through his conception of an interactive context and allowed speaking of relational wholes (from local groups to humanity) as individuals. He interpreted collective functional systems such as beaurocracy or money as sedimented objectivated relations between individuals. They function through intersubjective relations even as they appear to independently determine them. Secondly, Dilthey argued for a thoroughly relational and structural individualism as an essential element of practical existence and human scientific inquiry: socially-historical individual psychophysical life-courses are the matters themselves to be studied as such in these sciences (GS 1: 8). As there is no science of the individual as a monadic singularity without a relational nexus, these sciences investigated individuals and their motivations in view of their relations and enduring sedimented systems (GS 1: 50; GS 5: 35).

Franz Brentano and other contemporaries could perceive Dilthey both as a defender of the German "historical school" and as its critic.[23] Dilthey embraced its historiographical practices while criticizing its epistemological basis, as he critically worked through both the claims of positivism (represented by Comte and, more complexly, Mill) for a universally valid theory of historical and social laws and those of German historicism (represented by the historian Leopold von Ranke and the methodological reflections of Johann Droysen) for the concrete contextuality and particularity of each epoch that are all equal before God (Ranke 1954: 7). It should be noted, to avoid potential confusion, that Karl Popper's noteworthy critique of historicism almost a century later applies to Comte rather than Ranke (Popper 1957). Popper critiqued historicism as the use of historical laws in Hegel and Marx while Ranke's historicism rejected such laws for contextualist reasons.

Dilthey did not adopt the primarily anti-universalist contextualist or particularist position of pure empirical description that he attributed to the historical school. He sought to reconcile its historiographical practices and its demand to understand each era contextually from itself (from its own immanent self-interpretations) with the universalizing aspirations of scientific theory, particularly epistemology and psychology (GS 1: xvi). Understanding, in Ranke's practice and Droysen's reflections on historical methodology, transitioned from particular to particular and perspective to perspective, avoiding externally imposing the perspective of the present onto the past or projecting ostensibly universal and necessary developmental laws onto other eras and peoples. The historian's objectivity was to be immersed in the object and its contexts and conditions. Dilthey's contribution was to expand the concept of understanding by analyzing its elementary roles within everyday existence, epistemology, and the human sciences. Despite interpretations that seek to subsume Dilthey's project into the philosophy of consciousness, Dilthey would not formalize the first-person perspective qua Kantian transcendental subjectivity or the

[23] See Brentano (1893: 3). Beiser argues for Dilthey's status as an advocate of historicism, notwithstanding his other philosophical motives (Beiser 2019: 103–119).

phenomenological transcendental ego of Husserl. Modifying the modern epistemic paradigm, epistemological reflection begins with the contents and facts of consciousness as disclosed in the phenomenality of the there-for-me (the self-experience of the first-person perspective). These contents and facts, with their structural conditions, belong to empirical inquiry into the totality of human existence instead of the bloodless epistemic subject or reified a priori of empiricism and rationalism (GS 1: xvii). As the self is an individuated relational crossing point, and not a discrete unitary subject in need of becoming embedded, embodied, enacted, or extended, it is already and thoroughly environmentally, intersubjectively, and socially-historically structured in an effective relational web from the outset and is known through the empirical human sciences.

A primary antagonist of Dilthey's *Introduction* was Comte's positivism against which he formulated an alternative epistemology (book one) and the liberation of the sciences from metaphysics (book two). He rejected Comtean developmental laws of history and unitary sociology not only as ahistorical but also as unscientific. The anti-metaphysician Comte retained the worst qualities of previous metaphysics and the speculative philosophy of history (GS 5: 50–53). The sciences must be liberated from metaphysics and positivism as its totalizing heir, since the very idea of a unitary science and method is untrue to the matters themselves and violates the Aristotelian principle that each science is differentiated according to its distinct empirical contents. Modernity signified for Dilthey the emancipation of the individual sciences from physics to the contemporary liberation of historical consciousness and historical science (GS 1: xv). Gadamer branded Dilthey a lingering positivist due to his commitment to the idea of science without adequately distinguishing it from actual positivism (Gadamer 1985). It is important to recognize, however, that Dilthey did not criticize positivism because of its commitments to science but due to its conception of unitary science and method and its grounding in an ultimately speculative teleology of history that unfolded from theology through metaphysics to science. Nor did he deny the ideas of progress and the Enlightenment. He historically relativized and disenchanted their external teleological necessity while delineating their immanent orientational purposiveness and normative value.

Dilthey's analysis entails recognizing how scientific inquiry is a meaningful culturally and normatively guided praxis situated within the experiential life-nexus and, furthermore, how the human sciences fall completely within that human world (GS 1: xviii). The genealogical study of the formation of the sciences within the historical and intersubjective life-nexus (or, in the later works of Husserl, the life-world) is not only of historical interest. Nor need it undermine science by erroneously inferring that the knife of historicity—to adopt his later image—can only result in a destructive skeptical relativism (GS 8: 121, 234–235). It can lead to healing, since historical contextualization discloses the normative structures and purposive orientations of scientific inquiry, and the heightened doubled reflexivity of the human sciences, which are sine qua non for the theory and practice of the sciences as well as a flourishing culture of science.

Dilthey recognized the legitimate scientific aspirations expressed, to a lesser extent, in the classificatory encyclopedic systematizing of Comte and, to a greater extent, in Mill's logic of scientific inquiry (GS 1: 23). Thus, in the opening of the *Introduction*, Dilthey announced that his work would critique, balance, and formulate an alternative to both Western European positivism and the German historical school. Both approaches were incomplete: whereas the ahistorical abstractions and general laws of the former overlooked the concrete contextual particularities of historical forms of life, German historicism lacked the methodological and theoretical self-reflection and universalizing aspirations necessary for history to be scientific.

3.5 Normativity, Reflexivity, and Understanding

Another prerequisite of human scientific inquiry is normativity. The human sciences encompass three varieties of propositions: (1) perceived historical realities; (2) abstracted and theoretically constructed uniformities and regularities; and (3) normative rules and value judgments (GS 1: 26). These three varieties reflect a more fundamental differentiation, or a doubled nexus, between propositions about reality (including rules as part of reality) and propositions prescribing rules and values. The former are true or false; the latter right or not right (GS 1: 27). Whilst Mill placed norms and prescriptions outside of science under art and practical application, Dilthey concluded that the sciences require them. Dilthey also distinguished between causal explanation, which can be formulated in the natural historical and physiological conditions of the species, and the deployment of rules and purposes, which are understood in relation to historically situated agents and which even the most radical naturalistic thinkers presuppose. Such distinctions differentiate two tendencies that operate concurrently in reciprocal effective interaction or an effective nexus.

Dilthey's model of normativity within forms of life is more historically dynamic and entangled than Wittgesteinian philosophy of social science that tends toward a priori conceptualism and a static conception of society (Winch 2008). In Dilthey's analysis of rules and values, as evident with linguistic and positive legal rules, they are not simply habitually replicated. They are also misapplied and modified in their habitual repetition by different individuals. They are broken, criticized, and transformed with or without explicit awareness. Rules require judgment to be applied, as there can be no rule of applying a rule without an infinite regress. Further, they are implicitly reflexively enacted and can become explicitly reflectively questioned and changed. Human existence and accordingly the human sciences are characterized for Dilthey by a reflexive and interpretive doubling that he describes in a variety of ways. The individual is simultaneously a structural element of society, which can in part be externally explained from the third-person stance of the disinterested spectator, and a reflexive coparticipant in the intersubjective reproduction of the ethical life of society (GS 1: 37–38; GS 5: 69).

Dilthey described the immanent sociality or intersubjectivity of understanding in his early writings (GS 5: 61). Despite its importance in early works, Dilthey barely explains understanding in the *Introduction*. As clarified more adequately in later works, ethical life occurs through ordinary practices of understanding that are: (1) the precondition of social inquiry, even when they are ignored, and (2) methodologically refined in the human sciences in view of the aspirations toward theoretical objectivity and practical validity. Understanding has a structural role in human scientific inquiry. Social groups, religious believers, and readers of poetry engage in various practices of understanding that the human sciences draw on in description and narrative, structurally and comparatively analyze in conjunction with explanatory models, and arrive at theoretically informed conclusions that in turn can be taken up in practical affairs. This differs from its more constrained role in Weber's sociology who avoided direct discussions of Dilthey (a friend of his father). Whereas understanding is in Weber's account a point of departure and heuristic for social scientific explanation, it informs the entire structural interconnectedness of the human sciences for Dilthey.

Dilthey focused on understanding through lived-experience and its limits in the *Introduction*, still describing it as shifting from particular to particular (GS 1: 138, 204, 383). Understanding finds affinities through particulars, such that individuals and communities do not immediately—and potentially at all—understand what they experience as fundamentally alien. In a radical "first contact," an initial understanding leads to an action and a pattern of understanding and misunderstanding is created. Dilthey subsequently elucidated understanding more clearly as a relation between particulars (the self's awareness of others and itself) that occurs within an effective relational nexus (GS 5: 172). It includes the perspectives of: (1) the perceiver or interpreter, (2) the perceived or interpreted other, and (3) the intersubjective medium in which both are reciprocally interconnected and differentiated.

Overly emphasizing the subjective and affective state of the first perspective, which is pivotal yet partial, distorts the concept of understanding as well as Dilthey's philosophical aims. First, understanding must be coordinated with science. Historians, he remarked, only really understand history insofar as they apply the science of the subject matter itself (BW 1: 850). Secondly, understanding is presupposed and expressed in actions (GS 5: 317). It does not leap from the interiority of one subject to that of another through mere intuition or feeling. It enacts and belongs to a pattern of ethical and practical relations that distinguishes forms of life. Dilthey increasingly utilizes the Hegelian expression ethical life (*Sittlichkeit*) while analyzing its reproduction through processes of understanding, which are more individualistic and pluralistic, in contrast with Hegel's dialectic of progressing stages of the struggle for recognition.

Dilthey shared with Hegel and Marx a form of Giambattista Vico's *verum-factum* principle. If the "true is the made," humans can know the history that they themselves create. Yet, Dilthey denied the possibility of identifying knowing and making through gray concepts such as identity as well as knowing humanity through a speculative developmental philosophy of history whether in its Comtean positivist or Hegelian idealist incarnation (GS 1: 100–101, 104–105). One hierarchical order of history and society cannot be established, given human fallibility and finitude, but only recuring affinities, partial patterns, and structural similarities. Knowing the made requires proceeding through a different circle of mediations: a critical and self-reflective epistemological grounding, the middle link of logic and scientific method, and the individual empirical sciences (GS 1: 119, 125). Dilthey occasionally writes of humans understanding animals and nature that are not "made" in Vico's sense. This might leave the impression that life intuits life as such. Such ways of understanding, like all understanding, occur through the mediation of society and culture.

Dilthey's lifelong deployment of expressions such as grounding and foundation should be interpreted conditionally, as he rejected the possibility of any strong or metaphysical version of foundationalism. He maintained that there is no presuppositionless science, nor an absolute self-evident starting point in the self or things, reason or experience (GS 19: 13, 16). Science can expose and overturn a presupposition while relying on other ones (GS 19: 11). Dilthey contrasts his modified Kantian epistemic grounding, which proceeds from the immanent analysis of the experiential matrix and its structural conditions, with the metaphysical grounding that posits and reifies the subject or reality as a substantial ground, overstepping the bounds of fallibility and finitude. In 1859, he called for a new critique (JD: 80). His 1883 task is a "critique of historical reason" that concerns the empirical and structural conditions and formations of history, society, and culture through which human life is comprehended (GS 1: 116).

3.6 Individuality, Science, and the Hermeneutics of Historical-Normative Life

As Dilthey scholars have argued, it is overly simplistic to radically distinguish an earlier psychological phase from a later hermeneutical one, as Dilthey's works in the shift from an epistemic-psychological position to a more thoroughly interpretive and social-historical stance.[24] Dilthey's later philosophy was undoubtedly transformed in response to his psychological (e.g., Hermann Ebbinghaus) and Neo-Kantian opponents and new philosophical developments (e.g., Husserl). Yet it is difficult to identify a fundamental break between a psychological and hermeneutical Dilthey in contrast to shifts in emphasis and sources. Dilthey himself stressed the continuity of his project, and his later interpretive or hermeneutical position still prioritized individual lived-experience, expression, and reciprocal (i.e., intersubjective) understanding within the life-nexus and systemized forms of interpretation in human science. Dilthey's adaptation of the Hegelian language of objectification, objective spirit, and ethical life gave a greater sense of solidity and unity to society than his earlier use of the external organization of society.[25] However, contrary to prevalent varieties of Hegelianism and organicism, he liberalized these concepts by

[24] For instance, Bulhof (1980), De Mul (2004), Makkreel (1992), compare Gadamer (1961: 241–276).

[25] Note, for example, GS 4: 258; GS 7: 150; Nelson (2023: 138).

stressing that each such expression names an effective relational nexus, with a multiplicity of intersecting elements, rather than an underlying identity or unity (GS 7: 135, 185).

Dilthey's mature magnum opus, *The Formation of the Historical World in the Human Sciences*, was published posthumously by his editors based on the 1910 article with this title and other materials. It commences, much like his earlier works, with the individual, as a mediated reality, in an investigation (first published in 1905) of the acquired psychological structural nexus and the thesis that the human sciences must be accompanied by logical-epistemic self-reflection to grasp knowledge in its enactment and medium (GS 7: 3–4). His critical epistemic aims and the pivotal role of the individual's first-person interpretive stance continue to distinguish his discourse from standard forms of psychology, sociology, and social theory. The elderly Dilthey persisted in contesting forms of one-sided reductive monism and two-world theories. They are inadequate to grasp how self and reality emerge from one nexus, how perception and conception are a differentiated continuum, and how all science is a consequence of the pursuit of self-knowledge. First, givenness and reality appearing in the reflexive medium of lived-experience establishes the possibility and continuity of individual self-reflection and objective, universal, valid knowledge (GS 7: 7–8). Second, perception is already selective and interpretive, as it relies on implicit awareness and preliminary forms of understanding, and even the most abstract conceptual forms need to be referred to experience. Third, reflexive self-questioning concerning oneself and one's world motivated inquiry in the past and the present.

Self-reflexivity is a structural condition of purposiveness, as a category of human activity and the formed or acquired psychic nexus (GS 6: 207). Individuals remember, experience, and anticipate with an immanent subjective sense that develops into a sense of goal-oriented purposiveness.[26] Individual purposiveness, which Dilthey designated an internal or immanent teleology, is sedimented in the purposiveness of social and cultural systems and projected, as if it were an externally real teleology, onto nature. Individuals form rules and purposes that are sedimented and reproduced in social systems, which in turn powerfully shape them, and attributed to nature to comprehend and order it. Leaving aside the issue of nature, Dilthey clarified the links between the first-person interpretive perspective (of life reflexively understanding itself) and structural-functional social explanations, in which systems (such as the state, law, or the market) can be explained as operating with their own purposiveness independently of individual agents and their self-understandings. Yet, systems are interdependently heteropoietic and open rather than strictly autopoietic and closed in relation to their environments and the individuals that enact them. The social sciences require, consequently, both interpretive understanding

[26] See GS 7: 8–9. Although Dilthey rejected external and metaphysical uses of teleology (GS 6: 207), as teleological purposiveness relates to individual willing (GS 18: 165), he suggested in other contexts that immanent purposiveness is prefigured in the biological reflex and mobility (GS 9: 188) and allowed for its mediated use in the life sciences and the history of the species (compare De Mul 2019: 41–60; Makkreel 2019: 21–40).

and functional-structural explanation, since society operates through the interaction of purposive systems that presuppose and are realized through norms and individual behaviors (GS 7: 5, 7).

What then of science? It has an interpretive dimension in being born of self-reflection and oriented toward knowledge. Scientific knowledge is more than the systematic connection of factual, intermediate, and theoretical propositions. Scientific inquiry is one conditional element of a historically generated social nexus. As a purposive system aiming at knowledge, it is not only socially enacted but also normatively structured. Normative frameworks operate as a priori fore-structures in each scientific era or culture; yet they are fallible and revisable over time depending on their relation to experience through processes of experimentation and testing as well as other historical changes. The sciences presuppose, utilize, and transform the norms of logic, mathematics, and scientific method through their purposive application to the experientially given. Its highest expression is experiential verification in accordance with the norms of thought (GS 7: 13).

Dilthey's late philosophy, notwithstanding its differences from his earlier thought, continues to offer an immanent justification of the sciences and a liberal scientific culture through their own history and normativity as disclosed by human scientific inquiry. First, Dilthey historicized and normatively legitimated the two-fold task of the Enlightenment: scientific progress vis-à-vis its social transmission and progress in the freedom of subjectivity and the individual formation of life (GS 7: 346–347). Such progress is interpreted as a free self-determination, oriented toward realizing its task, and neither deterministically nor teleologically guaranteed. Second, Dilthey was a proponent of one scientific culture with a diversity of individuated forms rather than—as his thought was taken in the Anglophone world—two opposing cultures of the natural sciences and the humanities.

It is from this perspective that Dilthey contested as partial and one-sided both the reduction of nature to spirit and spirit to nature (GS 5: 3). Naturalism and positivism, paradoxically despite their intentions, undermine the theory and practice of science. To briefly summarize, eliminating the reflexive and reflective individual who co-participates in social life disregards a pivotal prerequisite of society, culture, and science, while tacitly assuming it. The naturalistic and social theoretical elimination of the first-person self-interpreting individual as an epiphenomenal flickering circuit is consequently conceptually incoherent and practically destructive. Dilthey's strategy, which emphasizes the connection between self-knowledge and scientific inquiry, remains pertinent to our own situation, even if it would need to be radically modified in response to current experiences and conditions.

3.7 Understanding and Meaning as Hermeneutical Concepts

The human sciences implicitly and explicitly use what could be designated "hermeneutic concepts" (Kühne-Bertram 2015: 55–79). Dilthey's analysis of experience leads to interpretive concepts such as lived-experience, expression, and understanding. We experience and understand others and groups primarily through their expressions and sedimented systematic relations. Dilthey had explicitly maintained since the mid-1890s, and arguably tacitly in his earlier works, that hermeneutical and critical methods form the basis of human scientific inquiry (GS 5: 262).

According to Dilthey's modification of the Aristotelian principle of the individuality of the sciences, differences in method follow from differences in ways of experiencing and givenness (GS 5: 264). What is experienced through individual and intersubjective understanding requires sciences thereof. What then is understanding? It is the rediscovery of the I in the you (GS 7: 191), as individuals comport themselves towards themselves and others through understanding (GS 7: 196). Each life has, in the temporal relational reality of its own interpretive positionality, its own sense (GS 7: 199), and interpretive understanding opens each world (GS 7: 205). Understanding concerns the concrete relation of self and other within a context of signification. Human actions, events, and systems of relations are understood in objectivated contexts of meaning, value, and purpose (GS 7: 23–24).

Dilthey differentiated common elementary and higher complex forms of understanding. Elementary understanding encompasses its bodily-perceptual forms. Dilthey further advanced his earlier strategy of critically establishing normative claims through their very historicity and temporality, which other contemporary philosophers could only construe as either external to history and time or skeptically refuted. One exception was the early Heidegger. He enthusiastically embraced Dilthey's hermeneutical conception of historicity and temporality without, however, adopting its normative dimensions (Heidegger 2016: 103–158). More complex forms of understanding occur in the context of their temporal formation and relations sedimented in objective spirit. To simplify, (1) meanings are generated from memory and reflection on the past; (2) values are engendered through the feeling of self or life interpretively situated in the present; and (3) purposes are formulated from the projection and anticipation of future goals (GS 7: 222).

Individuals understand themselves, others, and their social realities in their context of significance and objectified networks of meaning, value, and purpose. Traces of the latter are ubiquitous in the human transformation of material objects and environments as well as in symbolically formed entities. Persisting effective meaningful constructions such as society, state, or nation are born and abstracted from intersubjective interconnections, such that they function as relatively self-directed purposive systems. But they are mystified when grasped as real substances or subjects that determine and explain individuals as their mechanical circuits or biological organs. The individual empirical human sciences require a weaker version of social holism—with a greater sense of the contingency and diversity of forces—and, in comparison with atomistic or methodological individualism, a more robustly social interpretation of individuals. These sciences must not presume but examine how social relations and networks practically constitute and are constituted by individual participation in practices, associations, and institutions.

3.8 Conclusion

José Ortega y Gasset suggested that the primary question in interpreting Dilthey's philosophy is how to comprehend the complex relationship between "pure reason" and "vital reason," or rationalism and vitalism (Ortega y Gasset 1958: 175). Dilthey's critique of historical reason indicates the formative continuity and asymmetrical differentiation of forms of life and rational inquiry such that there is no radical duality between life and reason. This chapter has outlined several pivotal and controversial moments in the interpretation of Dilthey's philosophy of the human sciences. His project of a critique of historical reason is, on the one hand, critical, epistemic, and structural; and, on the other hand, it is bound to interpretively proceed through the social and historical empiria that disclose its effective normative and purposive forces and structures. The empirical, historical, and embedded normative orientation of the human sciences entails that they cannot consist primarily of a priori axiologies of values or conceptual analysis of games and rules, as they are in jeopardy of becoming in Neo-Kantian (Rickert 1926) and Neo-Wittgensteinian (Winch 2008) discourses. Dilthey's philosophical strategies were meaningful sources for Apel and Habermas that can also serve as a correction to their formalism.[27]

Dilthey's philosophy of the interconnectedness and differentiation of the sciences avoids the one-sidedness of reductive monistic theories (whether on the side of nature or spirit) and the bifurcated reality posited by dualistic theories. He is one of the few philosophers of social science to develop strategies for describing and analyzing the immanent interconnections between the empirical, social systemic, and normative dimensions of the human world and its scientific investigation.

Dilthey's thought has been called hermeneutical in a wide sense, even though he used this term in an expanded yet still narrower sense than subsequent philosophical hermeneutics. Later hermeneutical philosophy and critical social theory have lost pivotal dimensions of interpretive understanding and hermeneutics. They lack Dilthey's sense for interpretive participants within concrete intersecting relations in asserting the universality of the ontological question of being (Heidegger), language and tradition (Gadamer), or quasi-transcendental formal structures of communication (Apel and Habermas).[28] Numerous social theories continue to assume the primacy of the social and prioritize constructed collective entities, such as power or society, over the intersubjective and interpersonal dynamics of social life and reflexive first-person perspective. Contemporary thinking might still learn from Dilthey's moderate skepticism and his unique strategies, not to stay with them, but to self-reflectively engage current questions concerning the lifeworld and the sciences.

[27] On Dilthey and transcendental-pragmatic philosophy, see Apel (1979), Habermas (1973).

[28] See GS 5: 262; Apel (1979), Gadamer (1985), Habermas (1973).

References

1. Dilthey's Works

Dilthey, Wilhelm. 1914–2005. *Gesammelte Schriften*, 26 vols (Stuttgart: B. G. Teubner and Gottingen: Vandenhoeck and Ruprecht) [cited as GS plus volume number].

Dilthey, Wilhelm. 1960. *Der Junge Dilthey: Ein Lebensbild in Briefen Und Tagebüchern 1852–1870*, 2nd edition. Stuttgart: Teubner and Göttingen: Vandenhoeck and Ruprecht [cited as JD].

Dilthey, Wilhelm. 2011–2022. *Briefwechsel*, 4 vols. Gottingen: Vandenhoeck and Ruprecht [cited as BW].

2. Other Works

Apel, Karl-Otto. 1955. Das Verstehen: Eine Problemgeschichte als Begriffsgeschichte. *Archiv für Begriffsgeschichte* 1:142–199.

Apel, Karl-Otto. 1979. *Die Erklärung-Verstehen-Kontroverse in Tranzendentalpragmatischer Sicht*. Frankfurt: Suhrkamp.

Beiser, Frederick. 2019. Dilthey's Defense of Historicism. In *Interpreting Dilthey: Critical Essays*, ed. Eric S. Nelson, 103–119. Cambridge: Cambridge University Press.

Brentano, Franz. 1893. *Ueber die Zukunft der Philosophie*. Wien: Holder.

Bulhof, Ilse N. 1980. *Wilhelm Dilthey: A Hermeneutic Approach to the Study of History and. Culture*. The Hague: Martinus Nijhoff.

Damböck, Christian. 2016. *'Deutscher Empirismus': Studien zur Philosophie im deutschsprachigen Raum 1830-1930*. Cham: Springer Verlag.

De Mul, Jos. 2004. *The Tragedy of Finitude: Dilthey's Hermeneutics of Life*. Yale University Press.

De Mul, Jos. 2019. Leben erfaßt hier Leben: Dilthey as a Philosopher of (the) Life (Sciences). In *Interpreting Dilthey: Critical Essays*, ed. Eric S. Nelson, 41–60. Cambridge: Cambridge University Press.

Gadamer, Hans-Georg. 1961. Hermeneutik und Historismus. *Philosophische Rundschau* 9 (4): 241–276.

Gadamer, Hans-Georg. 1985. Wilhelm Dilthey nach 150 Jahren: Zwischen Romantik und Positivismus. In *Dilthey und Philosophie der Gegenwart*, ed. Ernst W. Orth. Freiburg: Karl Alber.

Habermas, Jürgen. 1973. *Erkenntnis und Interesse*. 2nd ed. Frankfurt: Suhrkamp.

Habermas, Jürgen. 1982. *Zur Logik der Sozialwissenschaften*. Frankfurt: Suhrkamp.

Hamid, Nabeel. 2016. Dilthey on the Unity of Science. *British Journal for the History of Philosophy* 24 (4): 635–656.

Hamid, Nabeel. 2021. Law and Structure in Dilthey's Philosophy of History. *British Journal for the History of Philosophy* 29 (4): 633–651.

Heidegger, Martin. 2016. *Vorträge, Teil 1: 1915-1932*. Frankfurt: Klostermann.

Kornberg, Jacques. 1981. Historicism and Liberalism: Wilhelm Dilthey's Philosophy of History. *Canadian Journal of Political and Social Theory* 5 (3): 16–30.

Kühne-Bertram, Gudrun. 2015. *Konzeptionen einer Lebenshermeneutischen Theorie des Wissens: Interpretationen zu Wilhelm Dilthey, Georg Misch und Graf Paul Yorck Von Wartenburg*. Würzburg: Königshausen & Neumann.

Lukács, Georg. 1955. *Die Zerstörung der Vernunft: Der Weg des Irrationalismus von Schelling zu Hitler*. Berlin: Aufbau Verlag.

Makkreel, Rudolf A. 1992. *Dilthey: Philosopher of the Human Studies*. Princeton: Princeton University Press.

Makkreel, Rudolf A. 2019. Dilthey's Conception of Purposiveness: Its Kantian Basis and Hermeneutical Function. In *Interpreting Dilthey: Critical Essays*, ed. Eric S. Nelson, 21–40. Cambridge: Cambridge University Press.

Mannheim, Karl. 2015. *Ideologie und Utopie*. Frankfurt: Klostermann.

Marom, Amnon. 2014. Universality, Particularity, and Potentiality: The Sources of Human Divergence as arise from Wilhelm Dilthey's Writings. *Human Studies* 37 (1): 1–13.

Nelson, Eric S. 2019. Wilhelm Dilthey in Context. In *Interpreting Dilthey: Critical Essays*, ed. Eric S. Nelson, 1–18. Cambridge: Cambridge University Press.

Nelson, Eric S. 2023. Individuation and Self-Awareness in Wilhelm Dilthey. In *Varieties of Self-Awareness*, ed. Saulius Geniusas, 135–152. Cham: Springer.

Nelson, Eric S. 2023b. Liberalizing Second Nature: McDowell, Dilthey, and the Sociality of Reason. In *McDowell and the Hermeneutic Tradition*, ed. Daniel Feige and Thomas Spiegel, 90–107. London: Routledge.

Nelson, Eric S. 2024. Wilhelm Dilthey, John Stuart Mill, and the Logic of the Human Sciences. *Revue Roumaine De Philosophie* 68 (1): 103–123.

Ortega y Gasset, José. 1958. *Kant, Hegel, Dilthey*. Madrid: Revista de Occidente.

Popper, Karl R. 1957. *The Poverty of Historicism*. Boston: Beacon Press.

Rickert, Heinrich. 1926. *Kulturwissenschaft Und Naturwissenchaft*. 7th ed. Tübingen: Mohr.

Rodi, Frithjof. 2003. *Das Strukturierte Ganze: Studien zum Werk von Wilhelm Dilthey*. Weilerswist: Velbrück Wissenschaft.

Rodi, Frithjof. 2018. *Diltheys Philosophie des Lebenszusammenhangs: Strukturtheorie-Hermeneutik-Anthropologie*. Munich: Herder.

Trendelenburg, Friedrich Adolf. 1840. *Logische Untersuchungen*. Vol. 1. Leipzig: Hirzel.

von Ranke, Leopold. 1954. *Über die Epochen der neueren Geschichte*. Darmstadt: Wissenschaftliche Buchgesellschaft.

Winch, Peter. 2008. *The Idea of a Social Science and Its Relation to Philosophy*. London: Routledge.

Chapter 4
Georg Simmel on Philosophical Sociology as a Life-Science

Olli Pyyhtinen and Thomas Kemple

Abstract The chapter examines the philosophical underpinnings of Georg Simmel's sociology, arguing that the various fragments of his thinking of social reality cannot be properly understood if they are disconnected from his philosophical concerns and the whole of his work. The emphasis Simmel's sociology places on process resonates with his life-philosophy generally, and with contemporary problems of modern existence more particularly. While also situating Simmel's ideas in their cultural and societal context, above all in relation to the revolution in modes of thinking and everyday perception connected to the mechanized culture of modernity, the primary goal of the chapter is to explore how his work might help us understand our present condition. Here, especially Simmel's perspectives on the entanglement of humans with the broader, more-than-human world of natural beings, inorganic matter, and technological networks are foregrounded.

Keywords Life-philosophy · Philosophical sociology · Process · Relation · Sociological relationalism · Vitalism

It has been common in the secondary literature to view Georg Simmel (1858–1918) primarily as a sociologist, who, in the last ten years of his life supposedly made a dramatic and sudden 'turn' away from sociology towards philosophy (see e.g. Landmann 1968: 8; 1976: 4–5). For instance, sociologist Friedrich Tenbruck states that Simmel published his magnum opus *Sociology* at a time, in 1908, when 'sociological study was already left far behind him and he had definitely turned his attention to philosophical and aesthetic questions' (Tenbruck 1958; cited by Frisby 1981: 24). However, it may very well be that the turn is not quite as radical as Tenbruck

O. Pyyhtinen (✉)
Tampere University, Tampere, Finland
e-mail: olli.pyyhtinen@tuni.fi

T. Kemple
University of British Columbia, Vancouver, BC, Canada
e-mail: kemple@mail.ubc.ca

Y. Shan (ed.), *History and Philosophy of the Social Sciences*,
https://doi.org/10.1007/978-3-032-08167-4_4

assumes. David Frisby would seem to hit the mark better by suggesting that while the publication of *Sociology* signified, perhaps, 'the end of Simmel's *major* occupation with sociology' (Frisby 1981: 4; italics added), Simmel never abandoned sociology completely. After all, he was trained in and lectured in philosophy throughout his career, even when addressing recognizably sociological topics.

In this chapter, we argue that reading Simmel's sociological writings detached from his philosophical concerns produces an unnecessarily narrow, one-sided, and ahistorical understanding of his contributions, since the sociological and philosophical aspects of his thought were closely intertwined. As Simmel's student Albert Salomon ([1963] 1997: 93–94) suggested, his 'sociological studies [...] were never separated from his philosophical concerns. He became a philosopher sociologist, one and indivisible'. We might add that his life-long studies in culture and aesthetics, while distinctive, are in various ways both philosophical and sociological. Therefore, it is not possible to fully comprehend Simmel's sociology and the development of his sociological ideas adequately without paying attention to his philosophy. In the chapter, we examine the philosophical underpinnings of Simmel's sociology and his move to a philosophy of life, indeed toward a sociology understood as a 'life-science'. We investigate how one of the key principles of his work, the concept of *Wechselwirkung* ('reciprocal effect'), stems from his philosophical concerns and broadens into a general metaphysical principle that extends to the whole of reality. We also examine how the emphasis his sociology places on process resonates with his life-philosophy generally, and with contemporary problems of modern existence in particular. Throughout, we explore the potential relevance of Simmel's work for understanding our present condition, especially the entanglement of humans with the broader, more-than-human world of natural beings, inorganic matter, and technological networks.

Even as sociologists have selectively claimed Simmel as one of their own, his work has typically been excluded from the discipline's holy trinity formed by Émile Durkheim, Max Weber, and Karl Marx, whose theories and ideas are conventionally understood to form the classical cornerstone of sociological theory. While enjoying a remarkable reputation and large readership during his lifetime in North America, France, and Italy through translations, anthologies, and text-books, for many decades of the twentieth century Simmel remained in the margins of the discipline (see Levine et al. 1976a, b on the American case). Both the unusually wide range of topics and his unorthodox style, (unjustly) described as 'essayistic', 'fragmentary', as well 'impressionistic', have contributed to the relatively less attention his work has received compared to the oeuvres of Durkheim, Weber, and Marx.

However, it is not always fruitful to read Simmel's work alongside Durkheim, Weber, and Marx. Indeed, Simmel's sociology does not share many of the same key concepts and problems with them, or when it does, he starts with different premises and ends with different results (Kemple 2018a: 39–42, 56–58, 71–74, 2018b). Where Durkheim, Weber, and Marx tend to settle on ideas about social facts, actions, and conflicts, respectively, Simmel's work transforms the entire landscape of sociological problems by developing a relational mode of thought from the ground up (Pyyhtinen 2018). This, we suggest, is the richest legacy of his work for contemporary sociology and social theory more broadly. Contrary to the largely substantialist and objectifying tradition of sociology (see Elias 1978), Simmel places the *social relation* in the centre of sociological thought as the ultimate unit of analysis. Of course, already in Simmel's time countless others also paid attention to social relations, including Durkheim on symbolic rituals, Weber on intersubjective actions, and Marx on metabolic exchanges, among others. However, relations have typically been seen either as the product of the actions of individuals or as the effects of society and its structures. Simmel is unique in the tradition of sociology to insist on *understanding the relationship itself in relational terms*, and to adopt this relational view as a perspective on the whole of reality and in the scientific study of social life: in his sociology as in his philosophy, he consistently views phenomena as relations and as being based on relations (Pyyhtinen 2010, 2016, 2018, 2021). Before taking up these issues and considering their implications, however, we begin with a biographical overview of his work and its legacy.

4.1 'The Brightest Man in Europe'

Almost without exception, accounts of Simmel's personal history mention Berlin, and his biography explicitly informs his work in distinctive ways. Simmel lived in the city for most of his life ever since his birth in 1858, as the youngest child to a family of seven children. Berlin's rapid economic and demographic growth since the end of the Franco-Prussian War in 1871 into a world metropolis coincided with Simmel's coming of age and early adulthood. Adapting to the hustle and bustle of the modern metropolis was undoubtedly a daily challenge as much for him as it was for the 'typical metropolitan person' in what is perhaps his most famous essay, 'Die Großsätdte und das Geistesleben' (Engl. 'Metropolises and Mental Life'), published in 1903. The rapidly shifting stimulations demand heightened intellectuality and also make the nerves flicker between hyperactivity and numbness, between overstimulation and a blasé attitude. Berlin was therefore the primary source and enduring object of Simmel's observations on metropolitan modern life, though often in allusive and implicit ways. For him, the metropolis was the stage for the dramas of money, consumption, encounters with strangers, and fashion, providing a framework for the experience of modernity. Simmel conceived the metropolis as both cause and effect, even as a kind of 'third' factor reconciling opposites or connecting extremes, such as in the way money weaves a network of metropolitan interactions. City life, for example, is simultaneously composed of impersonal contents and supports intellectual individuation. Moreover, it makes individuals both independent and dependent on an increasing number of others while generating a universal human freedom common to all individuals along with the qualities of individual uniqueness.

It was with sadness and regret that Simmel left Berlin in 1914 to take up a professorship at the University of Strassburg, having to leave behind thereby also the intellectual and art circles of the city he was such a crucial part of. Students from various walks of life and academic disciplines, including women and men from North America and other European countries, had crowded his lectures to witness the virtuoso at work. He was influential especially among the new generation of students in the first years after 1900, before he even had managed to secure himself an academic post. To this generation, he was the living embodiment of what it is to be a philosopher in the broadest sense of the term (Matthias, [1928] 1993: 193), significantly contributing to the rehabilitation of philosophy (Fechter [1918] 1993, p. 157). Simmel also had a following among artists and within wider cultural circles. Emil Ludwig ([1918] 1993: 156) suggested that out of all German academics it was Simmel who was the foremost 'teacher of artists'. He exerted a far-reaching influence on major cultural and political movements, such as German modernism and the political avant-garde (Leck 2000). Simmel also had regularly hosted an intellectual salon in his home with his wife, the painter, philosopher, and poet Gertrud Simmel (née Kinel), which was visited by poets such as Rainer Maria Rilke and Stefan George, the philosopher Edmund Husserl, and social scientists such as Max and Marianne Weber. After leaving Berlin to take up the chair in pedagogy at the Kaiser Wilhelm University in Strasbourg Simmel wrote some of his most important and original works, dying there 4 years later, in 1918, of liver cancer.

In European intellectual culture, Simmel was one of the most renowned thinkers of his time. George Santayana famously described Simmel as 'the brightest man in Europe' (cited in Levine and Silver 2010: x). Simmel was also an author of bestselling books. For example, the first edition of *Hauptprobleme der Philosophie* ('Main Problems of Philosophy') sold out its 10,000 copies within 2 weeks of its publication, and his published lectures on Kant went through four editions in his lifetime. Nevertheless, Simmel's personal academic career was anything but a success story relative to his popularity among contemporaries, the scale of his output, or his subsequent importance. Setbacks followed one after the other: both his first dissertation and his habilitation thesis were rejected; he spent 17 years as Privatdozent (Instructor) without regularly paid employment and 13 years as Ausserordentlich (non-tenured) Professor; he was repeatedly passed over for professorships; and he was only appointed professor at the age of 56, just before the outbreak of the First World War and 4 years before his death. The reasons for the uneven progress of his career have been sought in Simmel's Jewish background, the unorthodox nature of his thinking, and his ambiguous position between the roles of sociologist and philosopher, scientist and intellectual, but none of these offer an adequate explanation.

Although Simmel saw himself first and foremost as a philosopher, he has gone down in the history of ideas as a classic of sociology. Already in his letters, he complained of the fact that he was known, especially abroad, exclusively as a sociologist (Goodstein 2017). As a sociologist, he was much more famous in Germany than Durkheim, for instance, who remained practically unknown in the German sociology community for a long time. In German sociology, Simmel's influence was perhaps the greatest on Theodor Litt (1880–1962), Alfred Vierkandt (1867–1953), and Leopold von Wiese (1876–1969), all of whom took up above all Simmel's focus on relations, but he was an important interlocutor also for Karl Mannheim (1893–1947). Something of Simmel's popularity and influence is evident in the fact that he was a prime model and source of inspiration for intellectuals of both the radical left and radical right, such as Marxist philosopher Georg Lukács (1885–1971) and National Socialist sociologist Hans Freyer (1887–1969), who were among his students.

Simmel's influence was not limited to Germany alone. His sociology was particularly well received in the United States, where it was adopted by the first generation of the Chicago School. His fame was based especially on the distribution of the translations of his texts. By the 1910s no other European sociologist had more pieces translated into English than Simmel, allowing his ideas to reach a broad audience. Their wide dissemination was guaranteed by the fact that most of them were published in *The American Journal of Sociology*, as translations by Albion Small, the founder and first editor-in-chief of the journal, and provided basis for a sociology textbook by Robert Park, another Chicago professor who had studied with Simmel in Berlin. While the earliest North American reception in the early decades of the twentieth century was strongly associated with the University of Chicago, later, by the 1950s, there were many more centres, with, for example, Talcott Parsons teaching Simmel to his students at Harvard, Robert Merton at Columbia University, Alfred Schutz and Albert Salomon at the New School for Social Research in New York City, and Kaspar D. Naegele at the University of British Columbia (Levine et al. 1976a, 1976b). Curiously enough, after having fallen into nearly complete oblivion by the 1940s in German sociology, it was partly from across the Atlantic that Simmel was re-imported to Germany as a classic of urban studies, role theory, conflict theory, and analyses of small groups (Rammstedt 1994: 112; Köhnke 1996: 14n).

In France, Simmel was introduced as a sociologist to the community of social scientists above all by Durkheim's student Célestin Bouglé, who translated Simmel's essay 'Das Problem der Sociologie' into French as 'Le problème de la sociologie' in 1894. Later, the publication of *Mélanges de philosophie relativiste* (1912), a collection of Simmel's essays in French translation initiated by Henri Bergson, presented Simmel above all as a relativist philosopher. Interestingly, in Italy, Simmel was initially translated and appreciated as a philosopher (Portioli 2012). Beyond Europe and the United States, Simmel also exerted early influence for example on Japanese sociology especially in the 1920s and 1930s. Masamicha Shimmei (1959: 202) reports that among Japanese scholars Simmel's sociology 'became a sort of cult'. Simmel's work has also had a long reception history in South America, for example in Argentina (Maioli 2010) and Brazil (Waizbort 2008).

However, at its height Simmel's domestic and international fame was not lasting. He went from an acknowledged and influential author to near oblivion only two or three decades after his death. Otthein Rammstedt reports that in the years 1933–45 no new editions of Simmel's works or monographs on his work were published (Rammstedt 2012: 305). Simmel's influence waned in North American sociology, too. Donald Levine and his co-authors note that almost no new translations of Simmel's work were published in the United States for nearly three decades after the early 1920s (Levine et al. 1976a, b). Some interest in his work survived throughout the 1930s and 1940s but remained limited. It was especially his exclusion from the canon of classical sociologists that Talcott Parsons built with his *Structure of*

Social Action ([1937] 1968) that doomed Simmel to insignificance in American sociology for about two decades. However, by the end of the 1950s and immediately thereafter the interest in Simmel's work experienced a renaissance and a new bloom. For example, in 1958 *The American Journal of Sociology* published a special issue titled 'Émile Durkheim—'Georg Simmel, 1858–1958' to honour the centennial of Durkheim and Simmel. As the editor of the issue Peter H. Rossi (1958: 579) suggests, 'Each in his own way has contributed so much to the development of social science that it would be ungrateful not to mark the occasion of the one-hundredth anniversary of their births without some special tribute'. For Rossi, the 'greatness' of Simmel and Durkheim lies, first, in the fact that 'each played an important part in his own time in giving life and body to our discipline in its earliest days'; and, second, in their continuing relevance: 'their works today continue to stand as inspiring models of scientific endeavor, providing sharp insights into scientific problems still under consideration' (ibid.: 579). A year later, in a paper published in the *American Sociological Review*, Theodore Abel (1959, p. 474) celebrated Simmel 'as the founder of modern sociology'.

Around that time, in the 1950s, interest in Simmel began to increase in Germany as well, partly due to American influence, though on the whole the German reception has been very different from the American one. The North American Simmel reception has, from the outset, characteristically centred on the operationalizability of Simmel's theses, and on various fragments of his theory of social reality irrespective of their connection to the whole of his work. In updating Levine's review of the influence of Simmel on American sociology, Broćić and Silver note the increase in citations above all in the disciplinary subfields of symbolic interactionism and social network analysis, but also in studies of urban culture and in economic and organizational sociology (Broćić and Silver 2021). The German reception has been characterized by a more philological and historicist approach, with a focus that situates Simmel's work within the context of the time it was written rather than on the usability of his theses (Schnabel 1974).

Today, Simmel's status as a sociological classic is no longer questioned, even when it is placed in a secondary position behind the canonical founders Marx, Durkheim, and Weber. The selective and sporadic reception of his work has resulted in distortions and biases, however. For a long time, Simmel was read mainly as an author of fascinating sociological sketches and cultural-philosophical essays, often fetishized as stand-alone pieces separate from his oeuvre as a whole. In the 1980s and 1990s, Simmel was internationally revived especially as a cultural theorist and a theorist of modernity, if not also as a fore-runner of postmodern thinking, until a new wave of Simmel scholarship emphasized how his sociology is intertwined with its philosophical premises and thus his contribution to both the social sciences and the humanities (Helle 2019). While one should not disregard the brilliance of such individual pieces as the sociology of the senses, conflict, or sociability, let alone Simmel's perceptive analyses of money, style, the stranger, the metropolis, or fashion, these texts do not present the whole picture, and they need to be considered in relation to one another and to his overall project. Likewise, the attention his remarks on dyads and triads have generated among network researchers in the U.S., or his ideas on urban change and conflict in cultural studies, should not detract from the epistemological concerns and systematic treatises they have been taken from (Broćić and Silver 2021). Simmel's sociology and cultural philosophy cover only a part of a remarkably large oeuvre of some 20 books and over 200 essays. Throughout his career, he worked on questions of epistemology, ethics, psychology, and the philosophy of art, and increasingly towards the end of his life on metaphysics and aesthetics, as well as publishing scholarly treatises on Kant, Goethe, Bergson, Rembrandt, Schopenhauer, and Nietzsche, in addition to a work that fundamentally rethinks the 'problem areas' of sociology itself. In recent years, the completion of the 24-volume collected works, *Georg Simmel Gesamtausgabe*, published by Suhrkamp, has created important conditions for correcting mistaken and misleading assumptions in the reception. The series, edited by Otthein Rammstedt, has made available, among other things, Simmel's previously hard-to-find writings and presents even familiar works in a new light. The *Gesamtausgabe* edition also includes two thick volumes of Simmel's letters (although they do not cover the full extent of his correspondence, since a considerable number of letters have either been lost or destroyed). Still today a lot of work remains to be done uncovering the distribution and circulation of the legacy of Simmel's work, and tapping into its potential for advancing contemporary thought in new directions.

4.2 The Philosophical Underpinnings of Simmel's Sociological Relationalism

Several commentators have asserted the essential role of the notion of *Wechselwirkung* ('reciprocal effect') not only in Simmel's sociology but also in his thought in general. The word *Wechselwirkung* does not translate into English without difficulties. For example, the term 'interaction', which is an often-seen translation, problematically implies that *Wechselwirkung* would be based primarily on (social) 'action' or (human) interaction. However, relations of *Wechselwirkung* cannot be explained by reference solely to the action of individual agents and their motivations, ambitions, or ends, but presents rather a 'third' term, as it were, in relation to the actions of each entity. What is more, for Simmel, *Wechselwirkung* consists not only of what the individuals and other agents involved actively 'do' (*tun*), but also of how they and other entities are influenced and influence one another (*wirken*). It refers to the dynamics and interplay of affecting and being affected. In this regard, Simmel also insists, for example, that individuals affect each other not only actively, but passively, too, and that human as well as non-human or more-than-human agents are in constant relationship with one another.

In *The Philosophy of Money*, Simmel insists that '[t]he starting point of all social formations can only be the reciprocity of effects [*Wechselwirkungen*] between person to person' (1989b, GSG 6: 208; 2004b: 173; translation altered). In this regard, money exchanges (*Tausch*) can be understood as a special instance of 'reciprocal effects', not all of which are strictly economic. As a science of society, sociology should take interpersonal and even impersonal forms of *Wechselwirkung* as its object of inquiry. While 'everything which takes place in society' (1992, GSG 11: 23) may pass for a subject matter in various other social sciences, this cannot be the case in sociology, Simmel insists, but sociology must investigate 'what in society is really "society"' (p. 25). That is, it needs to study the constitutive processes of society, the forms of association (*Vergesellschftungsformen*), and the principles on which this unity rests. And, according to Simmel, the production and reassembling of society can be grasped only by examining the *Wechselwirkungen* of its elements, not all of which are human or even social, strictly speaking (1989a, GSG 1: 370). As he puts it in *Soziologie*: 'Should there be a science, whose object of study is society and nothing else, it can examine only [...] reciprocal effects [*Wechselwirkungen*], [...] ways and forms of association [*Vergesellschaftsformen*]' (1992, GSG 11: 19). The notion that reciprocal effects constitute and give rise to a variety of forms of association, broadly understood, makes up the relational basis of Simmel's sociology, and highlights what is distinctive and original in his work.

Simmel regards the independent and autonomous forms (social classes, family, politics, the economy, the church, etc.) that make up the traditional objects of sociology as merely secondary phenomena compared to the *Wechselwirkungen* between individuals and other entities (1989a, GSG 2: 130; see also 1989b, GSG 6: 208–209; Frisby 1992: 10). Accordingly, society, too, is nothing autonomous or absolute for him. He does not regard society as the defining feature of the social, as a fundamental substance or general framework for social phenomena that the notion of the social would necessarily presuppose. On the contrary, society is for him only a result of the relations among individuals and groups that has to be constantly produced and connected rather than being always and already assumed to be there. The relational turn that Simmel's work initiates and radicalizes in the emerging science of sociology resonates, for example, with the work of his contemporary Gabriel Tarde and later Norbert Elias. At the same time, his relationalism also figures with varying degrees in recent calls for a relational programme for sociology (see e.g. Emirbayer 1997; Crossley 2011; Donati 2011; Dépelteau and Powell 2013; Powell and Dépelteau 2013; Dépelteau 2018; Broćić and Silver 2021: 99–101). Rather than constituting a new subfield or paradigm influenced by Simmel, this relational turn should be understood as a return to the fundamental premises that guided his sociology from its beginning.

To be sure, Simmel did not invent the concept of *Wechselwirkung*. Petra Christian (1978: 115–120) has traced the evolution of the concept back to Kant and examined how it gradually takes on new meanings from Kant to Hegel, Goethe, Schleiermacher, and Dilthey and finally to Simmel. In making its way into Simmel's sociology, it had already appeared in his epistemology, where it is connected to the notion of truth. Simmel insists on the relativity of truth. Whereas Simmel himself calls his view 'relativism' (*Relativismus*), we hold that it is more accurate to call it *epistemological and ontological relationalism*: while relativism in its conventional form sees that truth is weakened by its relativity, Simmel, on the contrary, aspires to disclose how truth is founded upon relations and how reality itself is composed of relationships. In a letter to Heinrich Rickert from the year 1916, Simmel notes that his version of 'relativism' should not be identified with the subjectivism and scepticism commonly associated with the term: 'What I understand by relativism is an entirely positive metaphysical worldview, and [it has] so little scepticism [in it] as does Einstein's or [Max von] Laue's physical relativism' (2008, GSG 23: 638). A few lines later, he elaborates on this a bit further, explaining that for him the relativity of truth 'does not mean at all that truth [*Wahrheit*] and falsehood [*Unwahrheit*] are relative to one another; but truth means *a relation between contents* of which none possesses truth in and by themselves' (p. 638; italics added).

Simmel explains also in *The Philosophy of Money* how '[r]elativity does not mean—as in common usage—a diminution of truth'. Rather, he insists, relativity 'is the positive fulfillment and validation of the concept of truth. Truth is valid, not in spite of its relativity but precisely on account of it' (1989b, GSG 6: 116; 2004a: 116). Interestingly, Simmel connects his discussion of the relativity of truth to the sociocultural changes of his time as well as to transformations in the scientific worldview. He asks how it is possible to think in the modern world of flux without falling into subjectivism and scepticism. He provides an answer in a rare short autobiographical text, drafted around 1910 or a few years before, later named 'Anfang einer unvollendeten Selbstdarstellung' ('Beginning of an Incomplete Self-portrait'):

> The recent dissolution of everything substantial, absolute, eternal into the flux of things, into historical mutability, into merely psychological reality is, as it appears to me, secured against an unstable subjectivism and scepticism only if one sets in the place of the substantial fixed values *the living reciprocity of elements*, which themselves are subject to the same dissolution ad infinitum. The central concepts of truth, value, objectivity, etc. expressed themselves to me as reciprocities, as contents of a relativism that now no longer signified the sceptical dissipation of all that is solid but precisely protection against it via a new concept of solidity. (2004a, GSG 20: 304–305; emphasis added)

Here Simmel emphasizes his view that reality is the product of reciprocal effects (*Wechselwirkungen*), the result of 'the living reciprocity of elements', and thus his 'relativism' should be understood as a cosmic 'absolute' (2004a, GSG 20: 305). He thinks that 'relativism' or relationalism provides a new ground for solidity: the dissolution of absolute objectivity can be carried in such a way that no 'absolute is [...] required as a conceptual counterpart to the relativity of things' (1989b, GSG 6: 97; 2004a: 104). The strength of his 'relativism'/relationalism, compared to other epistemological principles, lies precisely here. While other principles suffer from the difficulty of subjecting their own content to the judgment they pronounce upon knowledge and existence in general, this does not apply to 'relativism'/relational ism (GSG 6: 116; 2004a: 116). Rather, Simmel claims, 'the relativity of things is the only absolute' (GSG 6: 307; 2004a: 238).

Simmel regards 'relativism'/relationalism not just as a metaphysical principle, but also as 'the most appropriate expression of the contemporary contents of science and emotional currents', adding that it 'decisively exclude[s] the opposing world picture' (1989b, GSG 6: 13; 2004a: 56). Interestingly, his epistemological and ontological relationalism also suggests a link to process: the dissolution of absolutes into relations is necessarily 'an infinite process' or 'a never-ending process' (1989b, GSG 6: 118; 2004a: 117). It is established on a fleeting basis: truth, value, and objectivity appear as dynamic reciprocities, not as fixed entities or stable substances. Simmel's sociological relationalism presents itself as a kind of 'third' way with respect to established sociological approaches to social knowledge and reality. For example, he he takes a middle position between methodological individualism, on the one hand, which assumes that individuals are the only agents active in social reality, and therefore the starting point for sociological analysis, and methodological holism or collectivism on the other, which argues that the behaviour of individuals can only be understood and explained by relating it to the larger social whole and its structures. He argues that the starting point of sociology should not be individuals and their activities, nor even the society containing individuals, but relations and the forms that emerge from relations. Contrary to the way Simmel is often depicted in textbooks, he does not focus mainly on studying microsocial interactions of individuals in private settings to the exclusion of macro-social processes and large-scale institutions, but shows rather how individuals shape and are shaped by processes on multiple scales. Finally, Simmel's epistemological and ontological relationism avoids the extremes of subjectivism, in which mental processes are the only certain facts and the sole foundation of all knowledge, and objectivism, where knowledge is ultimately grounded in realities which are independent and external to the knowing mind. In all of these ways, he can be said to have followed the example of the science of his time, since the composition of beings in relations is a particular insight of modern science. Already in the second volume of *Einleitung in die Moralwissenschaft* ('Introduction to Moral Science'), published in 1893, Simmel argues that modern forms of thought are determined by the shift from substance to relation and from being and identity to becoming and event.

4.3 Extending *Wechselwirkung* Beyond the Human Sphere to Vitalism and Processuality

The standard view in the secondary literature, voiced for example in the early 1920s by Simmel's student Siegfried Kracauer (1995: 225), is that Simmel's ideas bear 'almost no relationship to the natural sciences'. While it is true that Simmel cannot be characterized as a natural philosopher, his concept of *Wechselwirkung* need not be restricted to apply to human relationships alone, but can be stretched to cover relations between entities of all kinds as well, be they human or nonhuman. Not only is Simmel's conception of society analogous to his notion of matter, but the first can be interpreted as an extension of the latter. Of course, at no point in his career did Simmel pursue the study of natural science or physics in any systematic way, nor can Simmel be labeled as a natural philosopher. On the contrary, 'It is always man [...] who stands at the center of Simmel's field of vision', as Kracauer argues (ibid.: 226). Nevertheless, by consistently combining his metaphysical speculations with his sociological observations, what he himself called his 'philosophical sociology' or 'sociological metaphysics', Simmel is always concerned with how human social figurations are enmeshed without more-than-human processes and other-than-human networks (Harrington and Kemple 2012).

To consider society as a process, indeed, as 'a living reciprocity of elements', rather than as an entity in a state of rest, is to pay attention to social life as a process of becoming and *in statu nascendi*, in its nascent state. This means studying not so much the historical origin of society, but its logical origin, that is, the underlying dynamics, the relations that account for its coming into being at 'every day and on every hour' (Simmel [1907] 1993, GSG 8: 277). As Simmel writes in his last major statement about sociology, *Grundfragen der Soziologie* (Fundamental Questions of Sociology): 'The association between people folds, unfolds, and refolds itself without rest; it is an eternal flux and pulsation that links individuals to one another also where it does not give rise to actual organizations' ([1917] 1999b, GSG 16: 69; Simmel 1950: 10; translation modified). Simmel shows a preference for the processual term 'association' (*Vergesellschaftung*) over the substantive noun 'society' (*Gesellschaft*), as the former expresses much better the dynamic nature of social life. We are not dealing with a 'substance' when we talk about society, he suggests, but with a process or an 'event' (*Geschehen*) through which society comes to existence and occurs (1999b, GSG 16: 70; Simmel 1950: 11).

Interestingly, Simmel already conceived matter along these lines in his doctoral thesis *Das Wesen der Materie nach Kants Physischer Monadologie* ('The Nature of Matter According to Kant's Physical Monadology'). In it, Simmel criticizes Kant for hypostasizing matter and maintains that matter and the forces that produce it are not separate but fold into each other. They may only separate momentarily: 'Process and result, event and the thing [that] occurred, [are] kept apart only to flow into one another again' (Simmel, [1881] 1999a, GSG 1: 35). Related to this notion, some pages earlier he notes that matter is not a finished product, but a 'molecular process'. It is not a mode of being but of becoming:

> If matter emerges out of forces, then one should no longer treat it as purely passive stuff upon which other forces can exercise their undisturbed interplay; for the product of these energies is no finished product, but a continuous process, not being [...], but becoming [...]. (ibid.: 26)

It is easy to see the remarkable resemblance between Simmel's later sociological conception of society and his early philosophical understanding of matter. Just as Simmel does not consider matter in molar terms, he understands society not as a whole that is, but as a molecular process that emerges, becomes, and changes. In the essay 'The Sociology of Sociability', a written version of the opening lecture Simmel gave at the first conference of the German Sociological Association held in October 1910 in Frankfurt, he himself explicitly draws the analogy between human beings and other kinds of matter:

> The energy effects of atoms upon each other bring matter into the innumerable forms which we see as 'things.' Just so the impulses and interests which a person experiences in oneself and which push him or her out toward others bring about all the forms of association by

> which the mere sum of separate individuals are made into a 'society'. ([1911] 2001, GSG 12: 177; 1971: 127–8)

Instead of starting from the finished product, Simmel tries to explain society by the processes that have produced it. Accordingly, the term 'realistic-dynamic' (*realistischdynamisch*) which Simmel used to describe his position in his doctoral thesis is also an apt label for how he sees society. For Simmel, the ancient controversy between realism and nominalism turns on a badly stated problem. He regards society neither as a self-subsistent entity with properties of its own, nor as merely a name for an aggregate of individuals. There is without doubt a reality to which the term 'society' refers, but it is not that of an independent entity, but designates rather relations of interaction, reciprocal effects (*Wechselwirkungen*) (Simmel, [1908] 1992, GSG 11: 17–19): society comes into being through the synthetic realities of mutual influences between individuals (Levine 1971: xxxiii–iv). Social life is produced and sustained in and by the mutual relations between components which are always in the process of becoming and never finished, of constantly involving new connections while dropping others.

On the face of it, the difficulty of pursuing such a radically relational approach lies, of course, in the fact that interactions do not make the world each time anew, as if each encounter would have to begin again each time from scratch; rather, interactions take place in at least partly pre-given settings. Nevertheless, despite his emphasis on relations and processes, Simmel does not simply dissolve society into constantly changing relations without consistency or into fluxes and flows without form. When suggesting that society owes its existence to relations of interaction he does not dismiss the stability of society or the durability of forms of association. On the contrary, what is wonderful about Simmel's relational take on society is that it simultaneously meets two opposite aims: it both attends to how society is produced in and by relations, and accounts for its endurance and stability. What is more, this view is not restricted to human social groupings alone, but can be generalized to apply to all entities. It allows us to consider entities as bundles of relations without disregarding their possible endurance. And, importantly, it gives us a means to consider their durability without resorting to notions of a static substance or to the assumption that an immutable essence sustains them. While some of the features of any human or non-human entity may endure, even when many of its constitutive relations are eliminated, something can persist and endure only because and insofar as there is a sufficiently large number of other relationships that remain (relatively) intact. In a sense, Simmel takes substance out of substantialism: there is indeed something in things that cannot be subtracted from them without their ceasing to exist, but this something is not reducible to their essence or substance, but consists, rather, of their relations. Another way to put this difficult point would be to say that there is no substance to things other than their relations or, more exactly, other than their emergence and actualization in relations. For Simmel, relations are one with the essence or the substance of a thing.

The nexus between Simmel's sociology and his realistic-dynamic conception of matter can be stretched a bit further. To the extent that matter can be conceived as relational, one can argue that the relations between material bodies are social. Although human sociality is the conventional focus of sociology, the social need not be assigned exclusively to the privileged sphere of human relationships. Social relations occur not only between humans but between nonhuman things as well. Without saying so explicitly himself, Simmel replaces the model of autonomous entities and independent solid objects with a model that is inherently and expansively social: matter is based on, composed of, and produced by forces interacting with one another. In his doctoral thesis, Simmel maintains that 'force[...] must only be a relation [*Relation*] between monads', a relation that, he emphasizes, should not be 'hypostasized into an extended substance' (1999a, GSG 1: 19). He goes on to claim that every piece of matter should be explained with the forces that have produced it (*Wenn Kräfte die Materie bilden, so muss sich eben jegliche Erscheinung der Materie aus diesen, den angenommene Urkräften, erklären lassen*) (1999a, GSG 1: 26). If force is relational, and relation is no thing, then force cannot be a thing either. It is more like the reciprocal effect that counts for the social. Instead of hypostasizing force into a thing, Simmel claims that it 'seems like an invisible fluid [*Fluidum*]' (1999a, GSG 1, p. 20). His later contention in *The Philosophy of Money* and in *Sociology* that all social formations have to be considered on the basis of processual relations of reciprocal effect can therefore be regarded as an extension of his realistic-dynamic account of matter, which he later elaborates on in terms of the flux and flow of life itself.

Simmel's sociological vitalism, and his life-philosophy or life-science of social relations more broadly, has its roots in his early studies of Kant but above all in his encounter with the ideas of Henri Bergson, as he makes explicit in his 1912 essay 'On Some Contemporary Problems in Philosophy' (Simmel 2019). Simmel's son Hans reports that his father once admitted: 'The fact that Bergson knows more than me makes me happy, but the fact that I know less than him is, however, agonizing' (Simmel 1976, p. 263). As if to rectify this deficiency, in 'The Fragmentary Character of Life' published in 1916 and reworked as the first chapter of his last book, *View of Life*, Simmel maintains that the spatiotemporal existence of all nature must be understood in terms of a vitality, that is, as 'a constant stream of energies interacting with everything, in an unending unity of elements in ever different combinations' (Simmel 1999b: 203; 2012). This fluidity is not the ground of beings but process, motion, and becoming, incessant intermingling or intermixing of all that exists.

In one of his last essays, 'On the Nature of Historical Understanding' published the year he died in 1918, he extends his earlier ideas concerning the perception of other individuals in social space and historical time—from the strange to the familiar—to encompass a vitalist conception of non-human beings and even inorganic existence as well:

> [W]e experience the other person, the Thou, both as the most remote and impenetrable creature imaginable, and also as the closest and most familiar. On the one hand, the ensouled Thou is our only peer or counterpart in the universe. It is the only being with whom we can mutually understand and feel as 'one'. If, for example, we feel ourselves in harmony with the natural world, this is because it is comprehended under the category of the Thou. Consider Saint Francis, who spoke to both animals and inanimate nature as his brothers. On

> the other hand, the Thou also has an incomprehensible autonomy and sovereignty. It resists any decomposition or analysis into the subjective representation of the ego. It has that absoluteness of reality which the ego ascribes to itself (Simmel 1980: 106–107; translation modified).

Here the Thou or 'other self' is explicitly conceived as a primal and irreducible entity, an *Urphänomen* that may not always be human or necessarily have human characteristics (Simmel 1980: 105). The non-human Thou need not be endowed with religious or mystical qualities in order to be understood as a companion, peer, or counterpart to the self. Another individual's experiences cannot be understood purely in subjective terms, just as an other's way of being is not always comprehensible only objectively but also as 'ensouled' in some sense. As Ephraim Podoksik expresses this aspect of Simmel's later thought, 'each particular self then could be understood as containing the totality of the world in itself, yet being unique in its reflection of this totality, and thus could be compared to a kind of monad' (Podoksik 2010: 144). A life subsists through a perpetual movement that never reaches its destination, and exists within the unending tension between more-life and more-than-life.

Simmel's later writings demonstrate an awareness not just of the recent insights of contemporary philosophy and science but also of the increasingly mechanized culture of modernity. 'On the Nature of Historical Understanding' argues that the new life-philosophy requires shifting perspective from the conventional historical-psychological notion of interpretation, which attempts to recover an actor's or author's intentionality or to identify some causal factor, to an objective-immanent interpretation, where combinations of elements may be understood in multiple ways depending on which criteria are applied (coherence, simplicity, completeness, and so on). Simmel uses the analogy of a rebus or puzzle to explain this shift in interpretive standpoint:

> A creation of the mind that is intended to be understood can be compared to a puzzle [*Rätsel*] that the creator has constructed with a code word [*Lösungswort*]. If someone attempting to solve the puzzle should find a second word, one that, taken objectively, solves the puzzle with equal logical and poetic success, the second word is as perfectly 'correct' a solution as the one intended by the poet [or creator of the puzzle]; indeed the latter has not the slightest priority over that one or other, in principle, unlimited, code words that might still be found (Simmel 1980: 113; translation modified).

As Friedrich Kittler (1990: 270) has pointed out in discussing this passage, Simmel imagines a model of cultural hermeneutics which does not search for the intentions of authors, attribute causality to events, or translate meanings between utterances, but which serves rather as a kind of cryptographic technique for transposing signs between different media. In this model, which according to Kittler is facilitated by new technologies of communication and information storage such as the telegraph and the gramophone, the decoding-interpreter does not treat the encoding author either as an idol whose intentions must be deferred to or as a peer in the search for a common understanding, but rather as a combatant in the struggle for control over the techniques of coding itself.

Simmel inaugurates these reflections on the technological acceleration and instrumental character of the modern world in the last chapter of *The Philosophy of Money*, 'The Style of Life', where he argues that 'objective culture' is increasingly overwhelming and diminishing the resources of 'subjective culture' (Kemple 2018b). In a passage that had a significant influence on a generation of critical theorists (including his student, the Hungarian philosopher and literary theorist Georg Lukàcs), Simmel elaborates on the contradictory character of money as the reification of the general form of all life in the capitalist world. Money has become the means and the model by which things derive significance from relationships between individuals while also being incorporated into each individual's innermost identity: 'These counter-tendencies, once started, may press forward to an ideal of completely pure separation in which all the material contents of life become increasingly objective and impersonal, so that the remainder that cannot be reified [*der nicht zu verdinglichende Rest*] becomes all the more personal, all the more the indisputable property of the self' (Simmel 2004b: 474, translation modified; quoted in Lukács 1971: 156–57). The impersonal and generic style of life promoted by the capitalist money economy pushes the individual to resist being reduced to a thing among other things, and thus to strive to preserve what is most intimate or personal to one's inner sense of self. As if to reflect on his own scene of writing, Simmel goes on to consider how the typewriter produces mechanically uniform letters that seem to convey the pure, ideal, and objective contents of writing more efficiently than the contingencies of handwritten manuscripts. Typewritten texts seem to allow or even to compel us to avoid all personal, idiosyncratic, original, and creative modes of expression, he suggests, which then become all the more jealously guarded (Simmel 2004b: 474–475; cf. Kittler 1999: 183–98). In other words, the more-than-human networks of money exchanges and the communication channels that facilitate these exchanges both objectify social relations and induce ever more subjectifying styles of self-reflection and personal presentation.

Whether his focus is on historical and sociological epistemology, the philosophy of art, or metaphysics and ethics, Simmel's final writings are devoted to advancing the *vitalist* turn in modern thought, including science. In the final pages of the essay 'On the Nature of Historical Understanding', for example, he embraces the principles of life-philosophy which acknowledge the changing nature of existence against the mechanistic assumptions of the industrial age which embrace a crude realism: 'Historical interpretation is a distinctive mental construct, the truth of which with respect to its object is something vital, functional, and developed, not a mechanical reproduction of a photographic plate' (Simmel 1980: 125–126; translation modified). This argument is not restricted to history but extends to artistic forms as well, such as painting. Like the question Simmel (2009: 40–52) asks in the famous excursus from the first chapter of his monumental *Sociology*, 'How is Society Possible?', in his 1916 monograph *Rembrandt* he provides an answer to the question 'What Do We See in a Work of Art?' (Simmel 2005: 142–151; Kemple 2020). Here too his ideas hinge on a Kantian understanding of how knowledge, perception, and representation cannot provide an exact copy, a perfect replica, or a mechanical reproduction of reality. A painting or drawing does not just convey the corporeal presence of the original, as in a photograph, a movie, or some other mechanical record. Instead, the material reality of the image articulates the vision of the artist while reflecting back on, and appealing to, the observing eye of the viewer. The drawing of a fur collar, for example, as in Rembrandt's masterful sketch of his mother, cannot replicate a real fur collar in the way that a photograph reproduces light and shadow in realistic proportions through chemical and mechanical means (as in a photograph of Simmel's wife Gertrud wearing a mink stole; see Kemple 2018a: 12–17). In constantly refining his technique over time, Rembrandt learned to see that every object and each feature has its own splendour—the collar as much as the veil, the eyes of his mother as much as her smile, all lovingly portrayed through the artist's masterful manipulation of brush, paint, and canvas. In Simmel's (2005: 6) formulation of this effect: 'the entire human being … is inherent in each separate experience'. His aim in this study is above all to discover the secret of Rembrandt's genius in representing movement, at one point asking: 'what is it then that moves within the picture?' (Simmel 2005: 38) In the cinema, he notes, the images themselves move while the spectators remain an immobile mass public settled into their seats with their attention fixed on the screen (Fritsch 2009: 131–142 points out the historical context of these remarks). By contrast, while the painted figure itself does not move, something within 'the viewer Simmel' moves through the act of the painting and by completing the gesture immanent in the work, both before and after the moment represented (ibid.). The achievement of Rembrandt's artistry lies not in how it realizes a static form imposed from without, as in earlier artistic styles, and even less in his technical accuracy of representing reality, but rather in how it appeals to a dynamic originating from inside 'the totality of life in its perceptive function' (Simmel 2005: 42). Although Simmel himself does not acknowledge commonalities in the constructed character of painting and drawing, on the one hand, versus photography and cinema, on the other, in each visual medium certain details are arranged and displayed so as to pierce through the general style or cultural atmosphere of the image (Barthes 1981: 94–96).

Each of Simmel's later writings is concerned with the threshold of the modern revolution in intellectual habits and everyday perception, a revolution that is accompanied or reinforced by a fundamental shift in the ethical view of human beings and the scientific study of life. This larger cultural and historical context is explicitly invoked in the opening pages of his last book, *View of Life* (2010), where he suggests in broadly philosophical terms that it is in the very nature of human relationships to transcend their limits, and that a defining feature of the modern age lies in the advanced technical capacity to break through boundaries. For example, he notes that the expansion of the sensible world through the invention of the telescope and the microscope in the early modern period has given rise to a new awareness of our boundedness as human beings, at the same time propelling our drive to burst through all limitations. Where the telescope brings far-away worlds into close proximity and renders the invisible visible, the microscope discloses the fine details of minute structures that disrupt our natural perception of space and disorient our ordinary sense of time. To illustrate this point, he cites a long passage from 'a most thoughtful biologist' of the day who imagines how a fictional creature with 'eyes like the structure of a giant telescope' might view the world (Simmel 2010: 4). Such a being could be expected to have a longer lifespan, produce very different kinds of objects, or conceive of time and space in ways that unsettle or surpass human perception and capacities. Simmel's unacknowledged source here is the baron Felix von Uexküll's 1913 essay collection *Bausteine zu einer biologischen Weltanschauung* (Elements of a Biological Worldview), which considers the world of animal perception in a way that abandons the conventional anthropocentric perspective of the life-sciences (Agamben 2004: 39). In the essay Simmel quotes from, Uexküll (1913: 141) describes an 'emerging worldview' (*kommende Weltanschauung*) already evident in modern physics and the avant-garde artistic movements of his day. Just as Uexküll considers how animal perspectives on the world as well as innovations in optical technologies displace ordinary human-centred worldviews, Simmel reflects on how artists, philosophers, scientists, and other modern 'adventurers of the spirit' open up other perspectives on life and create new ways of experiencing the world.

4.4 Conclusion

In this chapter, we have discussed the philosophical underpinnings of Simmel's sociological work, insisting that the various fragments of his theory of social reality cannot be properly understood if they are disconnected from his philosophical concerns and the whole of his work. We also situated those ideas in their cultural and societal context, especially in relation to the revolution in modes of thinking and everyday perception connected to the mechanized culture of modernity. However, our aspiration was not only to contextualize Simmel's work historically, but throughout we also stressed and addressed how Simmel's ideas might help us in understanding the complexities of the present and what kind of new perspectives they might offer. Even though Simmel was primarily preoccupied in his work with the world-relation of humans, there are moments in his work which exceed the all-too-human logos and nomos and acknowledge the entanglement of humans within non-human relationships and the more-than-human world of natural entities, inorganic matter, and technology. One of them is his sociological vitalism or more broadly life-philosophy and its attention to the nonliving in life, and thus his move toward an expansive conception of sociology as a kind of life-science. As a finite living organism, the human being is placed on a multinatural continuum across all species and among non-human beings or other-than-human networks. It holds for all living beings that death is immanent in life, and thus the human individual too is dying as a biological organism, as some-body that is alive. Such an embedded, material understanding of human beings as deeply seated in the processes of life avoids any abstract, universalist, scientistic, or crudely eurocentric notions of 'the human'. Taking our lead from Simmel's later writings, we consider how reading his most radical ideas apart from his own explicitly stated intentions provokes questions which most concern us today. Thinking our world with and beyond Simmel may help us understand the affective, multispecies, technological, and planetary processes we humans are entangled with and better see our own share in producing them.

References

Abel, Theodore. 1959. The Contribution of Georg Simmel: A Reappraisal. *American Sociological Review* 24 (4): 473–479.
Agamben, Giorgio. 2004. *The Open: Man and Animal*, Translated by Kevin Attell. Redwood: Stanford University Press.
Barthes, Roland. 1981. *Camera Lucida: Reflections on Photography*, Translated by Richard Howard. New York: Hill & Wang.
Broćić, Miloš, and Daniel Silver. 2021. The Influence of Simmel on American Sociology. *American Review of Sociology* 47:87–108.
Christian, Petra. 1978. *Einheit und Zwiespalt. Zum hegelianisierenden Denken in der Philosophie und Soziologie Georg Simmels*. Berlin: Duncker & Humblot.
Crossley, Nick. 2011. *Towards Relational Sociology*. London and New York: Routledge.
Dépelteau, François, ed. 2018. *The Palgrave Handbook of Relational Sociology*. New York: Palgrave Macmillan.
Dépelteau, François, and Christopher Powell, eds. 2013. *Applying Relational Sociology: Relations, Networks, & Society*. New York: Palgrave Macmillan.
Donati, Pierpaolo. 2011. *Relational Sociology: A New Paradigm for the Social Sciences*. London & New York: Routledge.
Elias, Norbert. 1978. *What Is Sociology?* Translated by Stephen Mennell & Grace Morrissey. London: Hutchinson & Company.
Emirbayer, Mustafa. 1997. Manifesto for a Relational Sociology. *The American Journal of Sociology* 103 (2): 281–317.
Fechter, Paul. [1918] 1993. Erinnerungen an Simmel. In *Buch des Dankes an Georg Simmel. Briefe, Erinnerungen, Bibliographie. Zu seinem 100. Geburtstag am 1. März 1958*, ed. Kurt Gassen & Michael Landmann, 2. Aufl, 157–162. Berlin: Duncker & Humblot.
Frisby, David (1981). Sociological Impressionism. A Reassessment of Georg Simmel's Social Theory. London: Heinemann.
Frisby, David. 1992. *Simmel and Since. Essays on Georg Simmel's Social Theory*. London and New York: Routledge.
Fritsch, Daniel. 2009. *Georg Simmel im Kino: Die Soziologie des frühen Films und das Abenteuer der Moderne*. Bielefeld: Transcript.
Goodstein, Elizabeth. 2017. *Georg Simmel and the Disciplinary Imaginary*. Stanford: Stanford University Press.
Harrington, Austin, and Thomas Kemple. 2012. Georg Simmel's 'Sociological Metaphysics': Money, Sociality, and Precarious Life. *Theory, Culture & Society* 29 (4): 6–25.
Helle, Horst. 2019. Zehn neue Bücher zur Aktualität Georg Simmels. *Soziologische Review* 42 (4): 593–614.
Kemple, Thomas. 2018a. *Simmel*. London: Polity Press.
Kemple, Thomas. 2018b. Infinite Ends and the Tempo of Life: Notes on the Marx/Simmel Divergence/Convergence. *Dissonância: Critical Theory Journal* 2 (2): 137–166.
Kemple, Thomas. 2020. Framing, Painting, Seeing: Simmel's *Rembrandt* and the Sense of Modernity. In *The Routledge International Handbook of Simmel Studies*, ed. Gregor Fitzi, 196–209. London: Routledge.
Kittler, Friedrich. 1990. *Discourse Networks 1800/1900*. Translated by Michael Metteer, Chris Cullens. Stanford University Press.
Kittler, Friedrich. 1999. *Gramophone, Film, Typewriter*. Translated by, intr. G. Winthrop-Young, M. Wutz. Stanford: Stanford University Press.
Köhnke, Klaus Christian. 1996. *Der junge Simmel in Theoriebeziehungen und sozialen Bewegungen*. Frankfurt am Main: Suhrkamp.
Kracauer, Siegfried. 1995. *The Mass Ornament. Weimar Essays*. Translated by and ed. Thomas Y. Levin. Cambridge Mass: Harvard University Press.
Landmann, Michael. 1968. Einleitung des Herausgebers. In *Das Individuelle Gesetz. Philosophische Exkurse*, ed. Michael Landmann, Georg Simmel, 7–29. Frankfurt am Main: Suhrkamp.

Landmann, Michael. 1976. Georg Simmel: Konturen seines Denkens. In *Ästhetik und Soziologie um die Jahrhundertwende: Georg Simmel*, ed. Hannes Böhringer and Karlfried Gründer, 3–17. Frankfurt am Main: Vittorio Klostermann.
Leck, Ralph M. 2000. *Georg Simmel and Avant-Garde Sociology. The Birth of Modernity, 1880–1920*. New York: Humanity Books.
Levine, Donald N. 1971. Introduction. In *On Individuality and Social Forms. Selected Writings*, ed. Georg Simmel, ix–lxv. Chicago and London: The University of Chicago Press.
Levine, Donald N., F. Carter, and E. M. Gorman. 1976a. Simmel's Influence on American Sociology, I. *American Journal of Sociology* 81 (4): 813–845.
Levine, Donald N., F. Carter, and E. M. Gorman. 1976b. Simmel's Influence on American Sociology, II. *American Journal of Sociology* 81 (5): 1112–1132.
Levine, Donald N., Silver, Daniel. 2010. Introduction. In Georg Simmel Georg: The View of Life: Four Metaphysical Essays, with journal aphorisms. Trans. Jon A. Y. Andrews & Donald N. Levine, ix–xxxii. Chicago and London: University of Chicago Press.
Ludwig, Emil. [1914, 1918, and 1931] 1993. Erinnerungen an Simmel. In *Buch des Dankes an Georg Simmel. Briefe, Erinnerungen, Bibliographie. Zu seinem 100. Geburtstag am 1. März 1958*, ed. Kurt Gassen and Michael Landmann, 2. Aufl, 152–157. Berlin: Duncker & Humblot.
Lukács, Georg. 1971. *History and Class Consciousness*. Translated by R. Livingstone. London: Merlin Press.
Maioli, Esteban. 2010. La recepción de la obra de Georg Simmel en la Argentina. VI Jornadas de Sociología de la UNLP. Universidad Nacional de La Plata. Facultad de Humanidades y Ciencias de la Educaci.n. Departamento de Sociolog.a, La Plata.
Matthias, Leo. [1928] 1993. Erinnerungen an Simmel. In *Buch des Dankes an Georg Simmel. Briefe, Erinnerungen, Bibliographie. Zu seinem 100. Geburtstag am 1. März 1958*, ed. Kurt Gassen and Michael Landmann, 2. Aufl, 192–194. Berlin: Duncker & Humblot.
Parsons, Talcott. [1937] 1968. *The Structure of Social Action*. New York: Free Press.
Podoksik, Efraim. 2010. Georg Simmel: Three Forms of Individualism and Historical Understanding. *New German Critique* 109, 37(1): 119–145.
Portioli, Claudia. 2012. Les chemins de la pensée de G. Simmel en Italie. *Sociologie & Sociétés* 44 (2): 263–287.
Powell, Christopher, and François D.pelteau. 2013. *Conceptualizing Relational Sociology: Ontological and Theoretical Issues*. New York: Palgrave Macmillan.
Pyyhtinen, Olli (2010). Simmel and 'the Social'. Basingstoke and New York: Palgrave Macmillan.
Pyyhtinen, Olli. 2016. The Real as Relation: Simmel as a Pioneer of Relational Sociology. In *The Anthem Companion to Georg Simmel*, ed. Thomas Kemple and Olli Pyyhtinen, 101–120. London and New York: Anthem Press.
Pyyhtinen, Olli. 2018. *The Simmelian Legacy: A Science of Relations*. Basingstoke and New York: Palgrave Macmillan.
Pyyhtinen, Olli. 2021. Relations Along, not Between: Incorporating Becoming into Relational Sociology (via Life-Philosophy). *Simmel Studies* 25 (1): 37–72.
Rammstedt, Otthein. 1994. Georg Simmel. *Soziologie: Mitteilungsblatt der Deutschne Gesellschaft für Soziologie*. Sonderheft, 3/1994: 103–113.
Rammstedt, Otthein. 2012. On the Genesis of a Collected Edition of Simmel's Works, 1918– 2012. Trans. Austin Harrington. *Theory, Culture & Society (Annual Review)* 24 (7–8): 302–316.
Rossi, Peter H. 1958. Emile Durkheim-Georg Simmel, 1858–1958: An Editorial Foreword. *The American Journal of Sociology* 63 (6): 579.
Salomon, Albert. [1963] 1997. Georg Simmel Reconsidered. Edited, with an Introduction and Notes by Gary D. Jaworski, 91–108 in Gary D. Jaworski, *Georg Simmel and the American Prospect*. New York: State University of New York Press.
Schnabel, Peter-Ernst. 1974. *Die Soziologische Gesamtkonzeption Georg Simmels. Ein wissenschaftshistorische und wissenschaftstheoretische Untersuchung*. Stuttgart: Gustav Fischer.
Shimmei, Masamicha. 1959. Georg Simmel's Influence on Japanese Thought. In *Georg Simmel, 1858 –1918. A Collection of Essays, with Translations and a Bibliography*, ed. Kurt Wolff, 201–215. Columbus: The Ohio State University Press.
Simmel, Georg. 1950. *The Sociology of Georg Simmel*. Translated by, ed. Kurt H. Wolff. New York: The Free Press of Glencoe.

Simmel, Georg. 1971. *On Individuality and Social Forms. Selected Writings*. Chicago and London: The University of Chicago Press.

Simmel, Georg. 1980. On the Nature of Historical Understanding (1919). *Essays on Interpretation in Social Science*. Translated by, intro. Guy Oakes, 97–126. Manchester University Press.

Simmel, Georg. 1989a. *Aufsätze 1887–1890; Über sociale Differenzierung; Die Probleme der Geschichtsphilosophie*. Georg Simmel Gesamtausgabe Band 2, ed. Heinz-Jürgen Dahme. Frankfurt am Main: Suhrkamp. (GSG 2)

Simmel, Georg. 1989b. *Philosophie des Geldes*. Georg Simmel Gesamtausgabe Band 6, ed. David Frisby & Klaus Christian Köhnke. Frankfurt am Main: Suhrkamp. (GSG 6)

Simmel, Georg. 1992. *Soziologie: Untersuchungen über die Formen der Vergesellschaftung. Georg Simmel Gesamtausgabe Band 11*. Ed. Otthein Rammstedt. Frankfurt am Main: Suhrkamp. (GSG 11)

Simmel, Georg. 1993. *Aufsätze und Abhandlungen 1901–1908 Band II. Georg Simmel Gesamtausgabe Band 8*, ed. Alessandro Cavalli & Volkhard Krech. Frankfurt am Main: Suhrkamp. (GSG 8)

Simmel, Georg. 1999a. *Das Wesen der Materie nach Kants Physischer Monadologie; Abhandlungen 1882–1884; Rezensionen 1883–1901*. Georg Simmel Gesamtausgabe Band 1, ed. Klaus Christian Köhnke. Frankfurt am Main: Suhrkamp. (GSG 1)

Simmel, Georg. 1999b. *Der Krieg und die geistigen Entscheidungen; Grundfragen der Soziologie; Vom Wesen des historischen Verstehens; Der Konflikt der modernen Kultur; Lebensanschauung. Georg Simmel Gesamtausgabe Band 16*. Eds Gregor Fitzi & Otthein Rammstedt. Frankfurt am Main: Suhrkamp. (GSG 16)

Simmel, Georg. 2001. *Aufsätze und Abhandlungen 1909–1918 Band I. Georg Simmel Gesamtausgabe Band 12*. Eds Rüdiger Kramme & Angela Rammstedt. Frankfurt am Main: Suhrkamp. (GSG 12)

Simmel, Georg. 2004a. *Postume Veröffentlichungen; Ungedrucktes; Schuldpädagogik. Georg Simmel Gesamtausgabe Band 20*. Eds Torge Karlsruhen & Otthein Rammstedt. Frankfurt am Main: Suhrkamp. (GSG 20)

Simmel, Georg. 2004b. *The Philosophy of Money*. Translated by T. Bottomore, D. Frisby. London: Routledge.

Simmel, Georg. 2005. *Rembrandt. An Essay in the Philosophy of Art*. Translated by Alan Scott & Helmut Staubmann. New York and London: Routledge.

Simmel Georg. 2008. *Briefe 1912–1918; Jugendbriefe. Georg Simmel Gesamtausgabe Band 23*, ed. Otthein Rammstedt & Angela Rammstedt. Frankfurt am Main: Suhrkamp. (GSG 23)

Simmel Georg. 2009. *Sociology: Inquiries into the Construction of Social Forms* (2 Volumes). Translated by and edited by Anthony J. Blasi, Anton K. Jacobs & Matthew Kanjiranthinkal. Brill.

Simmel Georg. 2010. *The View of Life: Four Metaphysical Essays, with journal aphorisms*. Translated by J. A. Y. Andrews, D. N. Levine. Chicago: University of Chicago Press.

Simmel, Georg. 2019. On Some Contemporary Problems of Philosophy (1912). Trans. Olli Pyyhtinen, Thomas Kemple. *Simmel Studies* 23 (2): 111–119.

Simmel, Hans. 1976. Auszüge aus der Lebenserinnerungen. In *Ästhetik und Soziologie um die Jahrhundertwende: Georg Simmel*, ed. H. Böhringer and K. Gründer. Frankfurt am Main: Vittorio Klostermann.

Tenbruck, Friedrich. 1958. Formal Sociology. In *Georg Simmel, 1858–1918. A Collection of Essays, with Translations and a Bibliography*, ed. Kurt H. Wolff, 61–99. Columbus: The Ohio State University Press.

Uexküll, Jakob Baron von. 1913. *Bausteine zu einer biologischen Weltanschauung*. München: F. Bruckmann.

Waizbort, Leopold. 2008. Simmel in Brazil. Trans. Anthony Doyle. (Originally published in *Portuguese in Dados – Revista de Ciências Sociais,* 50(1): 11–48. 2007). http://socialsciences.scielo.org/pdf/s_dados/v4nse/scs_a05.pdf. Last Checked

Chapter 5
Émile Durkheim as Negative Foil: The Creation of a Contested Legacy Through Collective (Mis)representation

Paul Carls

Abstract Émile Durkheim, Max Weber, and Karl Marx are generally considered the founding fathers of sociology and have had great influence in related fields such as anthropology, political science, and philosophy. Durkheim's work has been influential in certain areas of the social sciences and humanities, yet beyond this recognition his reception is more checkered. In many fields Durkheim remains a marginal figure, if not one that is actively derided. This contribution explores this legacy and what it reveals about the social sciences and humanities. It first discusses the most prominent tropes, or recurring themes, related to Durkheim's work that have negatively influenced its reception. It then focuses on examples of particularly harsh treatment of Durkheim by prominent philosophers, sociologists, and political scientists, including John Searle, Charles Tilly, Theodor Adorno, Zygmunt Bauman, and Jacques Derrida. Each author's assessment does two things. First, it plays into the various aforementioned tropes. Second, it uses Durkheim as a negative foil to promote a preferred research paradigm. These responses reveal less about Durkheim than they do about the social sciences and humanities themselves, including which research paradigms are dominant and what their biases are. The essay thus seeks to explore the interpretative and conceptual obstacles that have hindered Durkheim reception. The essay will defend Durkheim's legacy, demonstrate the ways he remains relevant, and argue for a reevaluation by scholars in relevant fields.

Keywords Social fact · Post-modernism · John Searle · Charles Tilly · Marxism · Critical theory

P. Carls (✉)
Université de Montréal, QC, Canada
e-mail: paul.carls@umontreal.ca

Y. Shan (ed.), *History and Philosophy of the Social Sciences*,
https://doi.org/10.1007/978-3-032-08167-4_5

5.1 Introduction

Émile Durkheim, Max Weber, and Karl Marx are generally considered the founding fathers of sociology and have had great influence in related fields such as anthropology, political science, and philosophy. Durkheim, known principally for his four major works, *The Division of Labor* (1893), *The Rules of Sociological Method* (1895), *Suicide* (1897), and *The Elementary Forms of Religious Life* (1912), was the only to hold an actual chair of sociology. His influence has ebbed and flowed over time, reaching a nadir in the mid-twentieth century before regaining traction in the 1970s and 1980s that carried over into the twenty-first century (Fournier and Carls 2024). Durkheim is one of the most cited authors in the social sciences and humanities (Korom 2020; Ollion and Abbott 2016; Thomson Reuters 2009), giving the impression that Durkheimian inspired research is abundant. Durkheim's work has been influential in certain areas of the social sciences and humanities, and his status as a founding father and classical author in sociology guarantees that he will be known. Unlike Marx and Weber, however, beyond this limited recognition, Durkheim's reception is more checkered. Rare indeed are the researchers who proclaim themselves to be "Durkheimian," while Durkheim himself often remains a marginal figure, if not one that is actively derided and used as a negative foil for supposedly superior theoretical and analytical frameworks. Given the breadth, profundity, and originality of his work, as well as his continued relevance to discussions in multiple fields (see Carls 2022a), this state of affairs is indeed puzzling.

Research on Durkheim's reception offers some explanations. There are institutional reasons: despite being well-placed within the university system following Durkheim's death in 1917, members of Durkheim's team in France (with the exception of Marcel Mauss and Maurice Halbwachs, and to some extent Célestin Bouglé) failed to adequately defend and expand upon the Durkheimian research agenda (Fournier and Carls 2024; Mucchielli 1995; Smith 2020; Turner 2007); following Wilhelm Dilthey's distinction between the natural and social sciences, the social sciences in Germany were founded on anti-positivist grounds and reacted poorly to Durkheim's methodological positions (Keim 2021; Fitzi et al. 2017); in the first half of the twentieth century in the United States, the Chicago school of sociology, which was oriented towards pragmatism and symbolic interactionism, was dominant and had no interest in Durkheim (Platt 1995). The prominence Marxist thought and its derivatives such as Critical Theory enjoyed internationally in numerous disciplines is likewise relevant, and Durkheim's antipathy towards Marx and Marxism (see Durkheim 1958a [1928]; 1982c [1897]) was of no help. More recently, criticisms from feminists and post-colonial authors have highlighted Durkheim's stance on women and his general avoidance of the colonial project and issues relating to race, and challenged the idea of a "classical" cannon altogether, limiting his appeal in these camps (Susen 2020). Lastly, Durkheim's work has not resonated also due to a general preference for analyses that preserve the idea of self-aware, autonomous individual agency over and against structuralist ones (Bourdieu 1968; Morrison 2001; Pizarro Noël 2010).

Many authors have also explored issues related to the (mis)interpretation of Durkheim's work (Bourdieu and Passeron 1967; Carls 2022a; Collins 1988; Fitzi et al. 2017; Fournier 2013; Gane 2011 [1988]; Heilbronn 1985; Keim 2021; Lukes 1973, 1982, 2026; Platt 1995; Richman 1997; Rawls 1996, 2004; Schmaus 1994; Turner 1995). Perhaps the most important of these interpretations is the idea that Durkheim hypostatized a metaphysical group mind that wholly subsumes the individual, denying their agency. His supposed naïve positivism, perceived conservatism, and the claim that he pays no attention to power and politics are equally common. Relatedly, in the mid-twentieth century Durkheim's work became tethered to the structural-functionalism of Talcott Parsons. While the latter was a giant in his time, by the 1970s he had fallen strongly out of favor. Together, these interpretations cast a shadow that Durkheim has had a difficult time escaping. Beginning in the 1970s and 1980s, authors in different fields attempted to de-parsonize Durkheim in a way that had been successful for Max Weber, but such efforts remained somewhat limited (Emirbayer 1996; Mallory and Cormack 2018; Ramp 2008a; Smith 2020). In fact, numerous authors in different fields have made similar rehabilitation efforts, seemingly to no avail (see Alexander 1989; Belvedere 2023; Carls 2023, 2024, 2026a, 2026b; Collins 1988, 2004; Cladis 1992, 2005; Datta 2008; Emirbayer 1996; Fournier and Carls 2026; Fish 2005; Gane 2011 [1988]; Kertzer 1988; Lukes 2007; Malczewski 2019; Marcucci 2021; Miller and Lukes 2020; Morrison 2001; Müller 2019; Paoletti 2012; Pearce 1989; Peter 2013; Pickering 2000; Rawls 2004; Ramp 2008b; Riley 2002; Rosati 2009; Salmon 2016; Sawyer 2002; Schmaus 1994, 2004; Stedman-Jones 2001; Tiryakian 1962; Weiss 2012).

While the institutional and interpretive explanations are in many ways complimentary, this contribution explores the latter and looks at the historical reception of Durkheim's ideas, which has largely been dominated by a misinterpretation or misrepresentation of key sociological and philosophical concepts relating to Durkheim's social ontology and sociological method. These issues of reception involve primarily his social realism, notion of social causation, and the relation of the individual to society. These misunderstandings have turned Durkheim himself into a collective representation of bad thinking and led to his rejection in multiple fields, even when he is directly relevant and indirectly influential. This text will address these misinterpretations, correct them, and show how they continue to influence debate over Durkheim. It first discusses the most prominent tropes, or recurring themes, related to Durkheim's work that have negatively influenced its reception. It then focuses on the particularly harsh treatment of Durkheim by well-known philosophers, sociologists, and political scientists, including John Searle, Charles Tilly, Theodor Adorno, Zygmunt Bauman, and Jacques Derrida, and shows how each author's assessment does two things. First, it plays into the various aforementioned tropes. Second, it uses Durkheim as a negative foil to promote a preferred research paradigm. Ultimately, the text will defend Durkheim's legacy, demonstrate the ways he remains relevant, and argue for a reevaluation in fields where he makes substantive contributions.

5.2 Tropes and Fallacies in Durkheim Reception

5.2.1 The Group Mind Fallacy

Perhaps the most prominent misinterpretation of Durkheim's thought deals with his social ontology and takes the form of the "group mind" fallacy. This fallacy relates to Durkheim's definition of the social fact as ways of acting, thinking, or feeling that exist outside of the individual's consciousness and that exert external constraint over the individual (1982a [1895], 50–59). Many critics take the notion of externality to mean that social facts exist independent of all individuals in a group, leading to the charge that he "reified" or "hypostasized" society (Lukes 1982, 3–4), or that he argues for the existence of a "metaphysical 'group mind'" (Jones 1986, 78). Durkheim's social realism (i.e. social facts are real and exert a real force) and his dictum to treat them as 'things' contribute to this trope and the domineering rigidity it implies. Durkheim's discussion of the coercive power the social fact has over individuals furthermore leads many commenters to conclude that it is an oppressive entity that deprives individuals of autonomy and agency. As a result, many also conclude that Durkheim sees the individual as a simple subset of society and that there is ultimately no place for the individual in his work (Lukes 1982, 16–17). Durkheim's social fact is thus derided as oppressive, untenable, and unrealistic, and this element of his work has been vigorously contested and maligned from the time Durkheim first introduced his ideas (Belvedere 2023; Fournier 2013; Gane 2011 [1988]; Lukes 2014; Pinto 1993; Schmaus 1994; Smith 2020).

The group mind fallacy is justifiable only with a superficial understanding of Durkheim's text and/or faulty logic. The idea that social facts exist outside the minds of all individuals is an absurdity that makes no sense. As Durkheim often repeats, the social fact only exists as incorporated in individual minds (1982a [1895], 62–63; 1995 [1912], 252, 351; 2010a [1898], 24–26; 2010b [1906], 59). In one sense, thus, because the social fact is internal to a number of individuals, it is necessarily external to any one individual taken in isolation. In another sense, the social fact is the result of the *sui generis* fusion of individual minds. It is a process in which individual representations abstract into each other to create new concepts or ideas for thought, what Durkheim calls collective representations and which he identifies as the unique constitutive element of society (1982b [1901], 34; see also Némedi 1995; Schmaus 1994). Although this synthesis takes place within the individual mind, it takes place strictly speaking 'outside' the individual, through social relations, but also in the special substratum of the mind where one finds the collective consciousness (Durkheim 1982b [1901], 45; see also Carls 2022b; Stedman-Jones 2001, 139–140). The social fact thus resides in the individual mind as a product of the collective; it is both internal and yet external to the individual, personal and impersonal at the same time. The awkward externality of the social fact identified by critics results from the work of the social scientist, who extracts and abstracts social facts from individuals in order to study them in an objective way (Durkheim 1982a [1895], 70).

This understanding of the social fact as a collective representation also relates to Durkheim's notion of constraint. Although they are immaterial states of consciousness, collective representations as social facts still have causal force, which stems from the social forces they contain, and which they obtain through various social processes (see Carls 2022b; Schmaus 1994, 46–48; Takla and Pope 1985). At various points Durkheim mentions the 'prestige' (1982b [1901], 44) invested in collective representations or the mental energy collective representations release (1984 [1893], 53; 1995 [1912], 210) as the source of their causal force. This aspect of collective representations points to a different understanding of constraint than the restrictive one that Durkheim is known for. Durkheim (1982b [1901], 44–45, 47) himself recognizes the limits of his initial understanding of coercion, noting individuals' "special attachment" to social institutions, with social facts as enabling as opposed to simply 'constraining.'

Lastly, while Durkheim was not a methodological individualist, he often recognizes the subjective interpretation of social facts, and even their contestation, leaving an important place for individual agency, albeit within the social structures he identifies (1982b [1901], 47; 1995 [1912], 437; 2010c [1906], 64–65, 78–79; see also Carls 2022b, 233). The group mind fallacy is thus a straw man argument used by Durkheim's opponents to hide the true complexity and depth of his thought.

5.2.2 *Durkheim's "Naïve" Positivism*

A second trope regarding Durkheim interpretation deals with Durkheim's methodology, which he describes as his "scientific rationalism" (1982b [1901], 33), but which has come to be understood somewhat problematically as his positivism.[1] Durkheim's social realism, his insistence on the scientific neutrality of the researcher, the ability for objective social scientific knowledge, and the idea that one can identify general laws of human behavior—the idea that sociology can be like a natural science—has led to much criticism. Placing sociology on par with the natural sciences puts Durkheim at odds with Dilthey, who separates the social sciences from the natural sciences due to the former's hermeneutic and interpretive dimension. This move establishes the distinction between the *Erklären*, or explanatory, tradition, which models the social sciences off of the natural sciences, and the *Verstehen*, or interpretative tradition, which focuses on the subjective, historically shaped, and

[1] It is important to note that positivism has different meanings in philosophy and the social sciences. If in the social sciences positivism connotes a general adherence to scientific neutrality and the use of objective empirical data (typically of a quantitative nature) to draw general conclusions, in philosophy the term indicates a radical extension of empiricism, with a strict reliance on directly observable knowledge only. Durkheim was critical of this type of positivism, advocating a rationalist view of the world including the ability to determine cause and effect and general laws in the social sphere, and philosophical positivists would likely reject many elements of Durkheim's social realism (e.g. unobservable "collective forces") as unfounded metaphysics (Schmaus 1994, 13, 62–63; see also Stedman-Jones 1996).

meaningful nature of human experience. Dilthey's distinction forces the researcher to reflect on the interpretive limits of their results, which are tied to a particular perspective and preclude the possibility of an objective or correspondence-like truth. Accordingly, many critics argue that Durkheim does not engage in such critical reflection; namely, Durkheim does not reflect on the "hermeneutic circle," or "the fact that particular interpretations or readings only make sense within a total interpretative framework, which is in turn made up of particular interpretations" (Lukes 1982, 14; see also Abend 2008, 112–116; Taylor 1971, 6, 39–40). On what criteria then can one judge the objectivity or truthfulness of social scientific knowledge? The critique is that Durkheim's attempt at objective, bias-free knowledge is unattainable and naïve (Lukes 1982, 11–13). Jules Monnerot, an essayist associated with the Georges Bataille-led *Collège de sociologie*, was notably among the first to make these points in his 1946 book *Les faits sociaux ne sont pas des choses* (Coenen 1981). More recently, feminists and post-colonialists similarly contend that Durkheim's methodology is marked by an unreflexive and abstracting or universalizing "imperial gaze" (Connell 1997), or that it is bounded by the limits of its own historical formation. These features prevent addressing issues related to race, gender, and colonialism, thereby excluding and effacing marginalized voices (Bhambra et al. 2014; Pederson 2001).

Durkheim's methodological approach is, however, nowhere near as naïve as critics contend. For starters, Durkheim did reflect on the interpretative nature of reality; on Durkheim's own account, reality exists only as apprehended through different perspectives (i.e. historically formed, socially constructed collective representations). In fact, Durkheim (1977 [1938]) provides an extensive sociological account for the historical development of scientific rationalism as a particular way of understanding reality. He, thus, does provide a sustained reflection on the hermeneutic circle; simply he maintains that the social and historical situatedness of the scientific perspective does not prevent it from disclosing truth about the social world (see also Carls 2026a). Building on this fact, Susan Stedman-Jones (1996) notes that Durkheim's injunction to conduct social science in a neutral, objective way implies a self-aware intentionality, or a particular interpretative perspective, in this case a scientific perspective (see also Carls 2022b, fn 22, 242). Furthermore, Durkheim takes human society to be a part of nature, and as such, products of human behavior are also part of nature. This move implies that social facts are a part of the natural world; they thus have an objective existence and can be studied in ways similar to other natural phenomena. In an important sense then, Durkheim's sociology engages in kind-realism and seeks to identify natural kinds located in the social world, or the essential structures of social life that shape social behavior in similar ways across groups (Meadwell 2023; Schmaus 1994). In another sense, social facts are the result of, and derive their causal force from social interaction, and are thus grounded in the natural world; hence Durkheim's instruction to treat social facts like 'things' imbued with causal force (Durkheim 2010d [1911]; see also Carls 2022b; Karsenti 2012; Sawyer 2002). Such a move to naturalize society undoubtedly leaves many still claiming that as an interpretative endeavor, objective knowledge like what one has in physics is not possible in the social sciences. Nevertheless, there are those who have pushed back against such claims of indeterminacy (Abend 2008; Carls 2022b, 2023; Elder-Vass 2010; Meadwell 2023; Searle 2006a). This point is rather important, insofar as Durkheim makes collective representations, which inherently contain meaning, the focus of his sociology (1982b [1901], 34). Far from being a problem, however, this element of Durkheim's work points to its dynamic nature. Ultimately, one can say that Durkheim's (1995 [1912], fn 41, 230) claim "that ideas are realities—forces—and that collective representations are forces even more dynamic and powerful than individual representations," in fact, offers an original and evocative synthesis of the *Erklären* and *Verstehen* traditions in the social sciences (Carls 2026a; Schmaus 2010).

Finally, despite critics seeking historical contextualization and a certain decanonization, it is clear that Durkheim made significant contributions to the philosophy of science and social theory that extend beyond his specific philosophical and scientific context. Such critics furthermore, while pointing out blind spots, often have difficulty offering alternative epistemologies that go beyond what is articulated by classical (and other) sociologists (Susen 2020, 285–286). Durkheim's work, thus, remains relevant to unsettled epistemological and methodological debates (see also Lukes 2014).

5.2.3 *Durkheim, Power, Conflict*

Another common critique of Durkheim that touches partly on his social ontology is that his work has little place for conflict, politics, or power (Cotterrell 1999, 204; Giddens 1971; Lukes 1982, 22–23; Milbrandt and Pearce 2011, 273–274). It is true that Durkheim, unlike Marx for example, does not make social conflict a primary focus of his work. When Durkheim does study conflict, he tends to look at it as "socially functional" or "pathological" (Lukes 1982, 23). In looking at power, Durkheim almost always assumes he is talking about a singular, legitimate power to which everyone consents and which needs no justification; contestations to this power are illegitimate and thus pathological (Cotterrell 1999, 205). While he was aware of these issues (see Durkheim 1958b [1950], 63–64; 1995 [1912], 425; 2008 [1899]), and condemned imperialism and racism in his various writings (Steinmetz 2026), Durkheim does not give much sustained attention to instances where two or more groups compete for hegemony within a shared social space (see however Mukherjee 2010), nor the conquest and subjugation of foreign social groups in, for example, the project of colonization (Bhambra and Holmwood 2021; see, however, Durkheim 1915 [1915]; Dawson 2022; Steinmetz 2026).

While these critiques contain validity, Durkheim's focus on the consensual aspects of social life allows for deep insights that would likely not be possible otherwise. Such insights are, furthermore, easily applied to situations of social and political conflict and asymmetrical power relations, granted that one follow Randall Collins (1988, 109) and liberate Durkheim's monodimentional concept of "society" by defining it in a generic sense, "as any instance of prolonged sociation, whatever its boundaries in space or in time." This move makes Durkheim's social ontology more dynamic and allows his work to be applied in situations of conflict (for examples, see Carls 2019, 2023; Collins 2004; Kertzer 1988).

5.2.4 *Durkheim the Conservative?*

A final trope I will discuss, which indeed is a fallacy, is the idea that Durkheim is a conservative thinker, a view that stems essentially from elements of his methodology, his analytical focus and conceptual repertoire, or the empirical conclusions he drew. While Durkheim's political leanings and the implications of his work have been interpreted in manifold ways (Lukes 1973, 2–3; see also Fournier 2013; Smith 2020), the one that seems to have stuck is that his thinking is conservative in nature, that it is preoccupied with order, authority, and conformity, and serves the interests of the status quo. The Marxist Paul Nizan famously makes such an argument in his 1932 book *Les chiens du garde*, calling Durkheim (among others) an agent of the French Third Republic who sought to craft and disseminate the state's moral ideology "with great authoritarian rigor" and whose sociological method seeks to shroud these attempts at defending bourgeois, capitalist democracy in "the venerable appearance of science" (quoted in Lukes 1973, 357). In an essay published in 1939 in the prominent American philosophical journal *Ethics*, the Danish sociologist Svend Ranulf (1939) went so far as to claim that Durkheim's work, with its focus on "solidarity," laid the groundwork for the emergence of fascism in Europe. Robert Nisbet (1952) also famously read into Durkheim's sociology strands of nineteenth century French conservative thought that reacted against revolutionary, individualistic tendencies. Durkheim's association with Parsonian structural-functionalism has alimented this trope, and many other notable commenters have come to similar conclusions, including Lewis Coser (1960), Jonathan Haidt (2012), and authors discussed later in this essay (see also Milbrandt and Pearce 2011, 272–275; Rosati 2004). Alternatively, post-colonial criticisms have problematized Durkheim's avoidance of the issues of imperialism and colonialism, and argued that his use of social evolutionary logic, which categorizes societies as 'primitive' or 'advanced,' is normatively biased and worked towards the legitimation of the colonial project (Bhambra and Holmwood 2021; Connell 1997; Susen 2020, 32). A final critique comes from feminists, who point to Durkheim's views on women and argue that Durkheim's sociological project is in many respects fundamentally gendered to the exclusion of women (Christi 2012; Gane 1993; Lehmann 1994; Marshall 2002; Pederson 2001; Smith 2020).

Whatever Durkheim's actual normative positions were, it would be difficult to describe them as conservative. Excepting his views on women,[2] Durkheim

[2] For a nuanced view and defense of Durkheim's work on this point see Milbrandt and Pearce (2011, 274–275), Milbrandt (2026), Riley (2026).

normatively is likely closer to John Rawls in most respects than to any other prominent political thinker (Carls 2019; Cladis 1992, 2005; Herzog 2018). Regarding his sociology, Durkheim is, in fact, arguably best understood as representative of the French liberal tradition rather than the conservative tradition. Unlike the English liberal tradition, which opposes the isolated individual to society, the French liberal tradition integrates individualism into a sociological or communitarian framework, a feature that explains many elements of Durkheim's work (Paoletti 2026). It is likewise worth noting the many ways Durkheim's work challenged the conservative tendencies of his time. For example, not only did his analysis of religion question the legitimacy of established religious doctrines, but, contrary to post-colonial portrayals, Durkheim (1995 [1912], 2–3) also made the radical move of equating 'primitive' and 'complex' religions as fundamentally equal and deserving of the same respect. It should also be remembered that it is possible to separate normative from scientific engagements; when, for example, Durkheim argues that moral authority is a structure of every society, he is making an empirical observation, not a political statement about the need to respect this authority.

5.2.5 *Durkheim as a Collective Representation*

Combined, these four tropes have turned Durkheim into a collective representation of bad social scientific research: sloppy metaphysical thinking, naïve positivism, ignorance of the most gripping aspects of social life (power and conflict), and a conservative and authoritarian worldview that demands conformity and legitimates established forms of power. Durkheim's poor reputation has led to a lack of recognition (Alexander 1988; Bourdieu and Passerson 1961) as well as to what Philip Smith (2020, 151) calls "under recognition and sins of omission" among those building on Durkheim's thought. This is perhaps why many authors have found it easier to associate themselves with more respectable members of the Durkheimian legacy, such as Marcel Mauss or Maurice Halbwachs, than with Durkheim himself (Fournier and Carls 2024). Elements of this straw man Durkheim continue to inform large swaths of readers, and Durkheim's detractors often mobilize it as a means to not just dismiss his work, but also to promote their own supposedly superior theoretical and analytical frameworks.

5.3 Criticisms from Philosophers and Social Scientists

5.3.1 Searle

The American philosopher John Searle is one of the most interesting (and regrettable) critics of Durkheim, as both conceptually and thematically his work stands in close proximity to Durkheim's. In a 2006 special issue of *Anthropological Theory* Searle responded to commentary on his 1995 book *The Construction of Social Reality* as well as his overall theory of society. Many commentators, particularly Neil Gross (2006), much to Searle's chagrin, had pointed to the great similarities between his theory of institutional facts, collective representations, and collective intentionality, and Durkheim's theory of social facts, collective representations, and institutions. In his response, Searle (2006c, 57) notes that upon first reading Durkheim he felt that Durkheim's conception of social facts was "inadequate," and "so impoverished [he] did not read any further in his work." Upon a second look, Searle (2006c, 58) concluded that Durkheim's work was "much worse than [he] had originally thought."

Searle lists twelve complaints against Durkheim, but the most important ones relate to (1) Durkheim's conceptualization of the social fact and (2) the supposed absence of deontology in Durkheim's work. Concerning the first point, Searle takes the view that for Durkheim social facts are marked by a negative coercion and constraint; social facts are subsequently not enabling or empowering, whereas Searle's social institutions are. Searle also claims that Durkheim is inconsistent on the relation between social facts and the individual; Durkheim notes the role of the individual in social facts, but ultimately argues that they surpass the individual to exist in some metaphysical level of reality. Such a view causes problems for Searle. First, collective intentionality can only work if it is located in individual minds. Second, Durkheim does not grasp the fact that social reality is mostly perceived subjectively by individuals, but that this fact does not preclude the possibility of having objective knowledge of the social world since it is composed of empirically objective (although socially constructed) facts. Concerning the second point, Searle claims that Durkheim has no account of deontic power or deontology. What Searle means here is that humans assign special statuses to objects or people that they do not inherently possess, and thereby these objects or people gain new social functions. Because we accept these social functions, "we accept a series of obligations, rights, responsibilities, duties, entitlements, authorizations, permissions, requirements and so on," what Searle calls "deontic powers" (Searle 2006a, 18). Searle also claims that Durkheim's notion of social fact is too vague, stating that Durkheim "lacks the conceptual resources" to distinguish between different types of coercion (2006c, 65). Based on this analysis, Searle rejects any comparison to Durkheim.

All of Searle's accusations are incorrect. As regards the ontological status of social facts, Searle's critiques are merely variants of the group mind fallacy and are corrected when one looks at a complete and honest exposition of Durkheim's thought. As concerns deontology, Searle's critiques stem from him simply refusing to read Durkheim; Durkheim's work, in fact, is rather centrally focused on how people and objects attain new statuses through symbolic transformation. To be fair, Searle's analysis of language is more detailed than Durkheim's, and he is more systematic in discussing the types of institutions he is analyzing. A common criticism of Durkheim is indeed that his notion of social fact is too broad (Jones 1986, 78). However, in later works Durkheim (2010b [1906]; 2010d [1911]; 1995 [1912]; 1983 [1955]) began to differentiate different types of social facts (e.g. moral facts) as well as identify different types of collective representations (e.g. scientific, value, moral, mythological), although he did not systematize his thoughts in a later methodological text. So, some of Searle's points are well taken. Notwithstanding, Durkheim addresses issues that Searle does not. What is, for example, the ontological status of collective intention? And how do individuals come to accept social norms and institutions? Searle does not address the latter point, while Durkheim's analysis of socialization and ritual does (Lukes 2007). As to the former point, Durkheim's argument that ideas are *sui generis* products of social interaction remains a relevant and fruitful solution to the question of collective agency (see Carls 2022b; Sawyer 2002).

As Lukes (2007, 191–192; see also Lukes 2006) points out, there is much common ground between Durkheim and Searle, including not only their thematic and conceptual overlap (also noted by Gross), but also especially their adherence to realism and their claim that objective knowledge of social facts is possible. In their attempts to identify the basic structures of social institutions, I would also add that they are both structuralists. The importance both Searle and Durkheim give to the deontic power of society is yet another major point of convergence. Why then does Searle exhibit such disdain towards Durkheim? Searle (2006b, 43) claims that comparisons to Durkheim "make [him] extremely nervous" due to the latter's supposedly flawed, metaphysical notion of social fact. So in this case, Durkheim was victim to not only the group mind fallacy, but also to the stigma he has accrued over time as a collective representation of sloppy thinking. For Searle, thus, Durkheim was a risky association to distance himself from.

5.3.2 *Tilly*

The American sociologist Charles Tilly, well-known for his analyses of conflict and domination, had a low impression of Durkheim's work, and even claims that one of his goals in conducting research into conflict, protest, and collective action was to refute Durkheim's ideas (Tilly 1981, 107). His strongest critique of Durkheim comes in his 1981 chapter "Useless Durkheim," which must be read in the context of his other writings, specifically his 1984 book *Big Structures, Large Processes, Huge Comparisons*. These texts are aimed at the Parsonian structural-functionalism that was prevalent at the time and that Durkheim was associated with (Smith 2020, 172). Tilly's problem with this school of thought is that it downplays the issues of conflict and power that he places at the center of his work. As expected, we find these criticisms in Tilly's treatment of Durkheim.

In Tilly's 1984 book he describes "eight Pernicious Postulates" that emerged from conventional nineteenth century thinking, of which Durkheim is guilty of most, if not all. These postulates encompass the idea that "'society' is a thing apart," the ability of this "society" to influence individual thought and behavior, or relate to the issues of social order, development, integration, and conflict (Tilly 1984, 11–12). In one sense, these postulates present the image of a homogeneous "society" that is distinct from other identifiable "societies" (i.e. nation-states) with the power to influence individual thought and behavior. In another sense, the postulates make illegitimate juxtapositions between order/disorder, integration/disintegration etc. At issue is that these postulates do two things: (1) they downplay the darker side of the nineteenth century's socio-economic transformations, which include domination, control, capital accumulation, and exploitation; (2) they legitimate new forms of control and repression for the established powers. These same postulates reappear in Tilly's "Useless Durkheim" chapter. There, the homogeneous collective conscience is threatened by increasing differentiation, which is a cause of social disorder and a number of social ills such as criminality or class conflict. Tilly notes that despite his best efforts at finding statistical or other empirical evidence to confirm Durkheim's claims, he could find none. In the same breath, he praises other, more realistic theoretical models from Marx, Weber, and John Stuart Mill. Durkheim is thus a "useless" relic of the Victorian era that has nothing to say to contemporary societies.

In Tilly's treatment of Durkheim we again find appeals to several tropes. Insofar as Tilly portrays Durkheim's "society" as a unified, homogeneous entity that molds individual minds, there are elements of the group mind fallacy. There is also the idea that Durkheim's work has no place for questions of conflict, power, and politics, which is fair enough, although it should be noted that Tilly is here preoccupied primarily with Parsonian structural-functionalism. There are also elements of the conservative Durkheim fallacy, in which Durkheim is preoccupied by conservative themes, such as social disorder and the loss of authority and social integration. To the extent that the eight Pernicious Postulates stem from a "bourgeois reaction to nineteenth century changes" (Tilly 1984, 17), Durkheim's work represents the thinking of the status quo trying to understand and defend itself from radical change.

There are certainly merits to Tilly's criticisms, and his desire to "strip down" and "reconstruct" (1984, 17) key aspects of classical theories is indeed laudable; it is in fact what Durkheimians since the 1980s have done. Notably, Mustafa Emirbayer provides a direct counter to Tilly's assertions in his 1996 article "Useful Durkheim," painting a much more positive picture by focusing on Durkheim's later work and its relevance for the study of culture and civil society. Keith Sawyer (2002) also reexamines Durkheim's theories of social change in light of emergence theory. Randall Collins' development of Durkheim's work through microsociology and conflict sociology, especially Collins' aforementioned 'liberation' of Durkheim's notion of "society," is another important breakthrough. Perhaps all of this is "one more version of Durkheim" that Tilly (1981, 107) complains Durkheim's defenders trot out whenever he is resoundingly refuted; or perhaps, in fact, Durkheim is not as "useless" as Tilly would like us to believe.

5.3.3 *Adorno*

The German philosopher Theodor Adorno is well known for his association with the Frankfurt School and the development of Critical Theory. His most direct engagement with Durkheim comes from his 1967 introductory essay to the German translation of *Sociologie et Philosophie* (Adorno 1967), only the second such 'introduction' to Durkheim's work in the German language after René König's 1961 introduction to the German version of *The Rules* (Peter 2013, 73). This fact is quite surprising given that the text is well-known for the contempt with which Adorno treats Durkheim and also due to the fact that virtually all commenters note Adorno's profound misrepresentation of Durkheim's work (Christ 2017; Peter 2013; Smith 2020, 143).

Adorno's complaints center around two main points: (1) Durkheim's treatment of the individual, and (2) the concealed political agenda located in his work. On the first point, Adorno claims that Durkheim denies subjectivity and argues that his notion of collective consciousness is a hypostatized, metaphysical abstraction. The notion of the social fact, with its positivist emphasis on objectivity and obdurate "*chosisme*," cannot be questioned and thus contains authoritarian tendencies that constrain and deny individual agency. Durkheim furthermore sanctifies or deifies every moral norm, including bad ones, contributing to their unquestionable legitimacy. The individual is thus merely an atom to be molded by the greater society. On the second point, whereas Durkheim claims his concepts and methodology are objective and scientific, they are actually politically engaged; Durkheim's work is a reaction against Marxism and serves to justify the existing power relations of the status quo. Durkheim's notion of collective consciousness masks the issue of false consciousness and his work is itself an expression of false consciousness (see also Hagens 2006; Peter 2013). Adorno often describes Durkheim as secretly representing the interests of the bourgeoisie, whose goal is to hold together society and prevent class conflict. In this sense, Adorno (1967, 15) goes so far as to say that faced with the increasing complexity of modern society, Durkheim's obsession with obligation and duty is derived from a nostalgia for past social homogeneity that parallels fascist ideology. Durkheim (or perhaps Adorno himself?) is, in fact, so preoccupied with the idea of obligation that he exhibits a Freudian anal character, and is subsequently obsessed with order, pedantry, and cleanliness, and possesses authoritarian tendencies—in sum, Durkheim is the perfect "bourgeois" (1967, 35). According to Adorno, Durkheim's optimism and faith in the beneficial power of the collective consciousness was ultimately liquefied in the concentration camps during WWII (1967, 17). Durkheim's work thus leaves us unable to understand the Holocaust, the most extreme social fact of all, and is for that reason an irrelevant relic of the past.

Adorno's discussion of Durkheim mobilizes all four of the tropes. In many ways, Adorno restates Nizan's complaint that Durkheim is a bourgeois defender of the status quo, blissfully unaware of the normative biases that occupy his work. He also restates the group mind fallacy, with special emphasis on the oppressive nature of the collective consciousness. To the extent that the latter concept obscures any notion of false consciousness, a Marxist term designating the proletariat's adoption of the ideology of the ruling class, Durkheim's work also downplays any notion of power or conflict. Durkheim's work thus serves as a pseudo-scientific edifice to a system designed to homogenize individuals through coercion into a bourgeois, capitalist system where opposition is pathologized by scientific authority. Adorno obviously desires ideas that are more disruptive, conflictual, subversive, and liberating. His own work in the Frankfurt School, with its Marxist and Freudian underpinnings, offers such an antidote to Durkheim's supposedly flawed musings. Given Adorno's harsh and unjustified assessment of Durkheim's work, it is surprising that a new edition of the German language translation for *Sociologie et Philosophie*, one with an introductory essay that shows Durkheim in an honest and positive light, has not been published. The fact that one has not says much about Durkheim's stature in Germany.[3]

5.3.4 Bauman

The Polish sociologist and philosopher Zygmunt Bauman is known for his associations with Marxism, post-modernism, and anti-positivism. It seems hardly surprising that he would be unfavorable to Durkheim, and his engagement with the latter in his 2005 text, "Durkheim's Society Revisited," does not disappoint. Bauman's harsh critique of Durkheim echoes many of Adorno's arguments and involves both Durkheim's vision of society as well as the supposed unacknowledged political and normative goals that motivate his work.

According to Bauman (2005, 366), Durkheim's sociology is characterized by a "denigration of the individual" as well as a "deification of society." His methodology, which embraces social realism, positivism, and objectivity, rejects methodological individualism and grants the social scientist a privileged view of the social world over and against individual determinations. The individual's relation to society is furthermore characterized by a seemingly unquestioning submission to social rules, regulations, and norms. In order for this submission to be justified, Durkheim deifies society, or makes society into God. Bauman sees these moves by Durkheim as part of a hidden normative agenda. As Bauman contends, the nineteenth century saw the collapse of the doxa and routines that characterized the *ancien régime*. The project of modernity, of which Durkheim's thought is exemplary, was an attempt to

[3] Jürgen Habermas' discussion and use of Durkheim in volume II of *The Theory of Communicative Action* would be an exception.

recapture the certainties and regularities that had preceded it and to stave off the disenchantment of social processes and institutions that continued well into the twentieth century. As Bauman (2005, 366) argues, "[m]odernity took off as a bold effort to manage the world of humans and to make it manageable," by which he means transparent, unambiguous, calculable, and predictable. Durkheim's sociology, which (according to Bauman) posits the existence of a central authority that knows the reality of society as an uncontestable empirically defined and objective truth and to which individuals were (right or wrong) obliged or otherwise coerced to submit, corresponds to these modern ambitions. Bauman thus contends that for Durkheim, social life, including importantly morality, is reduced to nothing more than a routine submission to power and social norms, whatever those might be.

The implications of this assessment are rather severe. As Smith (2020, 216) notes, in Bauman's other works he argues that such attempts at calculation, control, and regulation led up to the Holocaust. In his assessment of Durkheim, Bauman (2005, 372) hints at the same, noting that at the time Durkheim was writing, modern society "was yet to show its other ugly, pugnacious, totalitarian face." The implication is that Durkheim's work is at best naïve and at worst "actively dangerous" and hence morally condemnable (Smith 2020, 216).

Bauman prefers other, non-Durkheimian, ways of understanding society, one in which the individual is unconstrained by bourgeois systems and conventions. He views society as imagined, wholly constructed, in constant flux, and hence pliable and moldable. In addition, he claims that since the 1960s a kind of society more in line with his vision emerged, largely eclipsing the world to which Durkheim's sociology corresponds (Bauman 2005, 373). The societal transformations, decompositions, and disenchantments that began in Durkheim's time only accelerated, leading to a liquid modernity. There are no more 'hard' social facts, categories, or institutions, and everything has become de-centralized, fluid, and even contradictory. In sum, "[s]ociety has lost the status of the ultimate ethical authority it might have had at the time when Durkheim embarked on the search for sources of morality" (2005, 378–379). Durkheim's work is thus no longer relevant, and we should move on to more adequate ways of imagining social being.

Bauman's assessment of Durkheim's work is in many ways a restatement of Adorno's critiques and exhibits a mixture of all four tropes. Like Adorno (and Tilly), Bauman restates Nizan's evaluation of Durkheim's work: Durkheim is an unwitting mouthpiece for the bourgeois establishment who seeks to provide a scientific veneer for an authoritarian system. Power is homogenous and legitimate, any conflicts or contestation are to be dismissed, and Durkheim is blissfully unaware of how his normative biases corrupt his methodology and conclusions. One difference between Adorno and Bauman comes from the post-modern dimension to Bauman's thought. Bauman's accusation that Durkheim 'denigrates' the individual is indeed a version of the group mind fallacy, although with a special twist. Bauman's argument is in fact common among post-structural authors who tend to disavow the social to promote the idea of a self-aware, autonomous individual unbound by structure (Morrison 2001). Ultimately, Bauman's analysis contains numerous errors and misrepresentations of Durkheim's ideas, which are in turn motivated largely by Bauman's own political inclinations and activism.

5.3.5 Derrida

The French philosopher Jacques Derrida is the founder of deconstruction and one of the most important post-modern or post-structural thinkers. He never engages with Durkheim's work directly, but he does engage with structuralism via Claude Lévi-Strauss, an author with close ties to Durkheim's legacy, but who like many others took pains to distance himself from his predecessor (Smith 2020, 151). In his 1966 text "Structure, Sign, and Play in the Discourse of the Human Sciences," Derrida identifies a specific form of structure he is working against, one that is a prominent feature of Durkheim's analyses. On these grounds it will be possible to construct a fruitful exchange between Derrida and Durkheim, albeit one that points to fundamental differences as opposed to similarities (e.g. Riley 2002; see also Carls 2024).

In his discussion of structuralism, Derrida (2005 [1967], 353), alluding to Friedrich Nietzsche's death of God, speaks of a "rupture" with "the entire history of the concept of structure," by which he means attempts at understanding social phenomena that refer to "a point of presence, a fixed origin […] constituted on the basis of a fundamental immobility and reassuring certitude" (2005 [1967], 352). Notably, this concept of structure is marked by onto-theology, which, following Martin Heidegger, refers to the philosophical tradition since Plato that understands objects of reality (being) through metaphysical or supersensory ideas. These ideas are considered more real than the natural world of experience and are seen as emanating from an authoritative foundational concept (such as the notion of God). Onto-theology thus combines the study of being (ontology) with an approach that grounds being in an ultimate first principle or center (theology) (Halteman 1998). Importantly, Durkheim's work on morality and religion identifies precisely this kind of onto-theological structure as a universal and enduring feature of human societies. The *sui generis* fusion of individual minds that creates social facts necessarily abstracts from subjective experience to create shared concepts, while rituals serve to shroud the most important of these ideas with the illusion of transcendence and the sacred. In this way a moral authority, or 'fixed origin,' or 'center,' is perpetually created as the necessary result of social interaction.

Derrida contends, however, that this type of thinking collapsed under its own weight; a series of substitutions, including "transcendentality, consciousness, God, [and] man," revealed "the structurality of structure" (2005 [1967], 353). What began to be thought was the arbitrary and empty nature of the center as well as the laws guiding its system of replacement. As a result, the very notion of a "center" was called into question. There was no longer any privileged position, transcendent being, fixed origin, or any other anchoring point of reference (i.e. authority) to provide a limiting structure for interpretation, meaning, or truth. Like Derrida, Durkheim, in noting the perpetual return of the sacred through ritual, thinks through the infinite substitutability of the 'center' (Durkheim 1995 [1912], 7), and also discusses the death of God as a watershed moment in Western history (1995 [1912], 429); simply, he comes to a different conclusion as to humanity's ability to live without structure.

Derrida goes on to juxtapose structuralism with post-structuralism, and to privilege the latter as superior. The structuralists, including Lévi-Strauss (and by extension Durkheim), are driven by "an ethic of archaic and natural innocence […] nostalgia, and even remorse […]. Turned towards the lost or impossible presence of the absent origin, this structuralist thematic of broken immediacy is therefore the saddened, negative, nostalgic, guilty, Rousseauistic side" of thinking (Derrida 2005 [1967], 369). Based on this assessment, one can conclude that Derrida would see Durkheim's sociology much as other authors discussed here: an authoritarian attempt, rooted in the past, to re-establish order and certainty in a world liberated by old constraints. Opposed to this wholly negative endeavor is the "Nietzschean *affirmation*, that is the joyous affirmation of the play of the world and of the innocence of becoming, the affirmation of a world of signs without fault, without truth, and without origin which is offered to an active interpretation" (2005 [1967], 369). In a nod to Michel Foucault's *The Order of Things*, in which Foucault criticizes attempts by the social sciences at discovering a universal 'human nature,' Derrida's preferred post-structuralism "affirms play and tries to pass beyond man and humanism" (2005 [1967], 370). In other words, Derrida embraces the world of play, uncertainty, and active interpretation that began in the 1960s in the wake of the death of God.

In Derrida's assessment one find elements of the group mind fallacy mixed with the conservative fallacy, in which structuralism implies an oppressive authority that seeks to reestablish an outdated social order and that demands conformity. Derrida's analysis, however, is not rooted in epistemologically grounded observations about the world, but is moral in nature: freedom is contrasted with guilt and constraint, joyous affirmation with saddened and archaic nostalgia. Part of Derrida's moral valuation lies, as with Bauman, in an ideological bias towards explanations that favor the self-aware, autonomous individual over structuralist ones (Morrison 2001). Durkheim's work is indeed far more nuanced than detractors like Derrida acknowledge, and it would be possible to turn the tables and demonstrate how deconstruction is itself a moral fact with its own structure and set of values judgments and constraints (see Carls 2024). Regardless, Derrida's rejection of structuralism is emblematic of much of the post-structuralist movement, which views structuralists like Durkheim as stodgy relics of the past.

5.4 Conclusion

These reactions to Durkheim's thought are hyperbolic, often without merit, and political in nature. The main opposition comes from those working in the Marxist tradition, methodological individualists, anti-positivists, post-structuralists, and post-colonial authors. Durkheim's adversaries have been effective in mobilizing a set of tropes and fallacies in order to construct a straw man for what they oppose, while simultaneously promoting rival authors and traditions as offering superior research paradigms. While these tropes have been refuted on numerous occasions, they endure; those maligning Durkheim have maintained a solid grip on his image, so much so that even those who do research in the Durkheimian paradigm are reticent to acknowledge their work as such.

These responses reveal less about Durkheim than they do about the biases and research paradigms that dominate the social sciences and humanities. It would appear that these disciplines have been dominated by those hostile to or apathetic to Durkheim's work. This state of affairs has unfortunately led to a number of missed opportunities as well as to the (re)discovery of ideas already found in Durkheim. While any number of examples could be presented (see Carls 2022a, 126–127), perhaps Paul Ricoeur's (1965) famous discussion of the "masters of suspicion" can sufficiently illustrate. Ricoeur bestows the title of master of suspicion upon Nietzsche, Marx, and Freud due to their critiques of the Cartesian cogito and for their application of a hermeneutics of suspicion towards religion, seeking to uncover the hidden reality behind religious symbols. Durkheim's work, which shows the self to be a socially and historically situated entity and which uncovers the concealed reality of religion in social forces, fits these criteria quite well. So why did Ricoeur not mention Durkheim? One reason is simply that Ricoeur had a low view of Durkheim's work, rejecting his social ontology and objective methodology, both of which, following Ricoeur, deny individual agency and subjective understandings of action (Deprez 2015). Durkheim, here again, seems to fall victim to elements of the group mind fallacy. More broadly, however, there are straightforward political reasons behind Ricoeur's discussion. Nietzsche, Marx, and Freud are useful to postmodernism, Marxism, and Critical Theory, movements that have specific political aims. Durkheim's brand of scientific rationalism and structuralism, especially as refracted by the likes of Nizan, Adorno, and Bauman, is not (for more, see Carls 2026a).

It is difficult to imagine that the vitriol given to Durkheim would be taken seriously for any other major sociologist or philosopher. Despite numerous rebuttals, there has been little progress in rehabilitating Durkheim for a general audience. Going forward, Durkheimians should not be shy in calling these tropes out. Durkheim discovered fundamental truths about human social behavior and if only for this reason he will resist being completely forgotten. His work contains great, unexploited potential and provides inspiration for innovative work and constructive engagement with a variety of leading authors and intellectual traditions. It remains there for the taking for those who dare to do so.

References

Abend, Gabriel. 2008. Two Main Problems in the Sociology of Morality. *Theory and Society* 37 (2): 87–125. https://doi.org/10.1007/s11186-007-9044-y.

Adorno, Theodor. 1967. Einleitung. In Émile Durkheim. *Soziologie und Philosophie*. Translated by Eva Moldenhauer, 7–44. Frankfurt am Main: Suhrkamp Verlag.

Alexander, Jeffrey. 1988. Introduction: Durkheimian Sociology and Cultural Studies Today. In *Durkheimian Sociology: Cultural Studies*, ed. Jeffrey Alexander, 1–21. Cambridge: Cambridge University Press.

Alexander, Jeffrey. 1989. *Structure and Meaning: Relinking Classical Sociology*. New York: Columbia University Press.

Bauman, Zygmunt. 2005. Durkheim's Society Revisited. In *The Cambridge Companion to Durkheim*, ed. Jeffrey Alexander and Philip Smith, 360–382. Cambridge: Cambridge University Press.

Belvedere, Carlos. 2023. *Collective Consciousness and the Phenomenology of Émile Durkheim*. London: Springer Nature.

Bhambra, Gurminder and John Holmwood. 2021. *Colonialism and Modern Social Theory*. Cambridge: Polity Press.

Bhambra, Gurminder, Robbie Shilliam, and Daniel Orrells. 2014. Contesting Imperial Epistemologies: Introduction. *Journal of Historical Sociology* 27 (3): 293–301. https://doi.org/10.1111/johs.12059.

Bourdieu, Pierre. 1968. Structuralism and Theory of Sociological Knowledge. Translated by Angela Zanotti-Karp. *Social Research* 35 (4): 681–706.

Bourdieu, Pierre, and Jean-Claude Passeron. 1967. Sociology and Philosophy in France since 1945: Death and Resurrection of a Philosophy without Subject. *Social Research* 34 (1): 162–212.

Carls, Paul. 2019. Durkheim on Religious Pluralism: Modern Democracy and Religious Conflict. *Critical Research on Religion* 7 (3): 292–311. https://doi.org/10.1177/2050303218823069.

Carls, Paul. 2022a. The Unrecognized Genius of Durkheim. *Journal of Classical Sociology* 22 (1): 123–130. https://doi.org/10.1177/1468795X211046121.

Carls, Paul. 2022b. The Social Fact in Durkheim's Late Work: Structural Hermeneutics, Positive Sociology, and Causality. *Journal of Classical Sociology* 22 (2): 222–246. https://doi.org/10.1177/1468795X20980660.

Carls, Paul. 2023. *Multiculturalism and the Nation in Germany: A Study in Moral Conflict*. London: Routledge.

Carls, Paul. 2024. Deconstruction as a Moral Fact: Durkheim versus Derrida on Structure, Authority, and Democracy. *Durkheimian Studies* 28: 3–32. https://doi.org/10.3167/ds.2024.280101.

Carls, Paul. 2026a. *Durkheim as Master of Suspicion*. In: Fournier and Carls (2026).

Carls, Paul. 2026b. *Amor Societatis*: An Essay on Durkheimian Existentialism and the Re-enchantment of Social Life. In *Phenomenology & Posthumanism(s)*, ed. Allen Porter. London: Springer.

Christ, Julia. 2017. Critique of Politics: Adorno on Durkheim. *Journal of Classical Sociology* 17 (4): 331–341. https://doi.org/10.1177/1468795X17736127.

Christi, Marcela. 2012. Durkheim on Moral Individualism, Social Justice, and Rights: A Gendered Construction of Rights. *The Canadian Journal of Sociology* 37 (4) 409–438. https://doi.org/10.29173/cjs12253.

Cladis, Mark. 1992. *A Communitarian Defense of Liberalism: Durkheim and Contemporary Social Theory*. Stanford: Stanford University Press.

Cladis, Mark. 2005. Beyond Solidarity? Durkheim and Twenty-First Century Democracy in a Global Age. In *Cambridge Companion to Durkheim*, ed. Jeffrey Alexander and Philp Smith, 383–409. Cambridge: Cambridge University Press.

Coenen, Herman. 1981. Developments in the Phenomenological Reading of Durkheim's Work. *Social Forces* 59 (4): 951–965. https://doi.org/10.2307/2577975.

Collins, Randall. 1988. The Durkheimian Tradition in Conflict Theory. In *Durkheimian Sociology: Cultural Studies*, ed. Jeffery Alexander, 107–128. Cambridge: Cambridge University Press.

Collins, Randall. 2004. *Interaction Ritual Chains*. Princeton: Princeton University Press.

Connell, Raewyn. 1997. Why Is Classical Theory Classical? *American Journal of Sociology* 102 (6): 1511–1557. https://doi.org/10.1086/231125.

Coser, Lewis. 1960. Durkheim's Conservatism and Its Implications for His Sociological Theory. In *Émile Durkheim, 1857-1917: A Collection of Essays*, ed. Kurt Wolff, 211–232. Columbus: Ohio State University Press.

Cotterrell, Roger. 1999. *Émile Durkheim: Law in a Moral Domain*. Edinburgh: Edinburgh University Press.

Datta, Ronjon Paul. 2008. Politics and Existence: Totems, *Dispositifs*, and Some Striking Parallels Between Durkheim and Foucault. *Journal of Classical Sociology* 8 (2): 283–305. https://doi.org/10.1177/1468795X08088875.

Dawson, Matt. 2022. Durkheim and the possible connections between social theory and colonialism. *Journal of Classical Sociology* 22 (4): 416–422. https://doi.org/10.1177/1468795X221105713.

Deprez, Stanislas. 2015. Ricoeur et la sociologie. *Revue Philosophique de Louvain* 113 (4): 619–643. https://doi.org/10.2143/RPL.113.4.3136931.

Derrida, Jacques. 2005 [1967]. Structure, Sign, and Play in the Discourse of the Human Sciences. In *Writing and Difference*. Translated by Alan Bass, 351–370. London and New York: Routledge.

Durkheim, Émile. 1915 [1915]. *'Germany Above All': German Mentality and the War*. Translated by Joseph Swain. Paris: Librairie Armand Colin.

Durkheim, Émile. 1958a [1928]. *Socialism and Saint-Simon*. Translated by C. Sattler. Yellow Springs, Ohio: Antioch Press.

Durkheim, Emile. 1958b [1950]. *Professional Ethics and Civic Morals*. Translated by Cornelia Brookfield. Glencoe, IL: The Free Press.

Durkheim, Émile. 1977 [1938]. *The Evolution of Educational Thought*. Translated by Peter Collins. London: Routledge.

Durkheim, Émile. 1982a [1895]. *The Rules of Sociological Method and Selected Texts on Sociology and Its Method*. Translated by W. D. Halls. New York: The Free Press.

Durkheim, Émile. 1982b [1901]. Preface to the Second Edition of *The Rules of Sociological Method*. In Durkheim 1982a [1895]: 34–47.

Durkheim, Émile. 1982c [1897]. Marxism and Sociology: The Materialist Conception of History. In *Durkheim* 1982a [1895]: 167–174.

Durkheim, Émile. 1983 [1955]. *Pragmatism and Sociology*. Translated by J. C. Whitehouse. Cambridge: Cambridge University Press.

Durkheim, Émile. 1984 [1893]. *The Division of Labor in Society*. Translated by WD Halls. New York, NY: The Free Press.

Durkheim, Émile. 1995 [1912]. *The Elementary Forms of Religious Life*. Translated by Karen Fields. New York: Free Press.

Durkheim, Émile. 2008 [1899]. Anti-semitism and Social Crisis. *Sociological Theory* 26 (4): 321–323. https://doi.org/10.1111/j.1467-9558.2008.00331_2.x

Durkheim, Émile. 2010a [1898]. Individual and Collective Representations. In *Sociology and Philosophy*. Translated by D. F. Pocock, 1–34. New York: Routledge.

Durkheim, Émile. 2010b [1906]. The Determination of Moral Facts. In *Sociology and Philosophy*. Translated by D. F. Pocock, 35–62. New York: Routledge.

Durkheim, Émile. 2010c [1906]. Replies to objections. In *Sociology and Philosophy*. Translated by D. F. Pocock, 63–79. New York: Routledge.

Durkheim, Émile. 2010d [1911]. Value Judgments and Judgments of Reality. In *Sociology and Philosophy*. Translated by D. F. Pocock, 80–97. New York: Routledge.

Elder-Vass, David. 2010. *The Causal Power of Social Structures: Emergence, Structure and Agency*. Cambridge: Cambridge University Press.

Emirbayar, Mustafa. 1996. Useful Durkheim. *Sociological Theory* 14 (2): 109–130. https://doi.org/10.2307/201902.

Fish, Jonathan. 2005. *Defending the Durkheimian Tradition: Religion, Emotion and Morality*. London: Routledge.

Fitzi, Gregor, Nicola Marcucci, and Hans-Peter Müller. 2017. Interview by Gregor Fitzi and Nicola Marcucci with Hans-Peter Müller on the reception of Émile Durkheim in Germany. *Journal of Classical Sociology* 17 (4): 399–422. https://doi.org/10.1177/1468795X17736132.

Fournier, Marcel. 2013. *Émile Durkheim: A Biography*. Translated by David Macey. Cambridge, UK: Polity Press.

Fournier, Marcel, and Paul Carls. 2024. Durkheim's Team: The *Année Sociologique*. In *The Oxford Handbook of Émile Durkheim*, ed. Hans Joas and Andreas Pettenkofer, 145–163. Oxford: Oxford University Press.

Fournier, Marcel, and Paul Carls, eds. 2026. *The Elgar Companion to Émile Durkheim*. Cheltenham: Edward Elgar Publishing.

Gane, Mike. 1993. *Harmless Lovers? Gender, Theory and Personal Relationships*. London: Routledge.

Gane, Mike. 2011 [1988]. *On Durkheim's Rules of Sociological Method*. New York: Routledge.

Giddens, Anthony. 1971. Durkheim's Political Sociology. *The Sociological Review* 19 (4): 477–519. https://doi.org/10.1111/j.1467-954X.1971.tb00643.x.

Gross, Neil. 2006. Comment on Searle. *Anthropological Theory* 6 (1): 45–56.

Hagens, Tobias Garde. 2006. Conscience Collective or false consciousness? Adorno's critique of Durkheim's sociology of morals. *Journal of Classical Sociology* 6 (2): 215–236. https://doi.org/10.1177/1463499606061734.

Haidt, Jonathan. 2012. *The Righteous Mind*. New York: Pantheon Books.

Halteman, Matthew. 1998. *Ontotheology*. Routledge Encyclopedia of Philosophy. Taylor and Francis: https://www.rep.routledge.com/articles/thematic/ontotheology/v-1

Heilbron, Johan. 1985. Les métamorphoses du durkheimisme, 1920-1940. *Revue française de sociologie* 26 (2): 203–237. https://doi.org/10.2307/3321573.

Herzog, Lisa. 2018. Durkheim on Social Justice: The Argument from 'Organic Solidarity.' *American Political Science Review* 112 (1): 112–124. https://doi.org/10.1017/S000305541700048X.

Jones, Robert Alun. 1986. *Emile Durkheim: An Introduction to Four Major Works*. Beverly Hills, CA: Sage.

Karsenti, Bruno. 2012. Durkheim and the Moral Fact. In *A Companion to Moral Anthropology*, edited by Dider Fassin, translated by Amy Jacobs, 21–36. Malden: Wiley-Blackwell.

Keim, Wiebke. 2021. Frenchman, Jew, Positivist: Reading the *Rules* and Mapping Émile Durkheim in Germany. *Durkheimian Studies* 25: 154–186. https://doi.org/10.3167/ds.2021.250107.

Kertzer, David. 1988. *Ritual, Politics, and Power*. New Haven: Yale University Press.

Korom, Philipp. 2020. The Prestige Elite in Sociology: Toward a Collective Biography of the Most Cited Scholars (1970-2010). *The Sociological Quarterly* 61 (1): 128–163. https://doi.org/10.1080/00380253.2019.1581037.

Lehmann, Jennifer. 1994. *Durkheim and Women*. Lincoln: University of Nebraska Press.

Lukes, Steven. 1973. *Émile Durkheim: His life and works*. Middlesex: Penguin Books.

Lukes, Steven. 1982. Introduction. In *The Rules of Sociological Method and Selected Texts on Sociology and Its Method*. New York: The Free Press.

Lukes, Steven. 2006. Searle and his Critics. *Anthropological Theory* 6 (1): 5–11. https://doi.org/10.1177/1463499606061729.

Lukes, Steven. 2007. Searle versus Durkheim. In *Intentional Acts and Institutional Facts: Essays on John Searle's Social Ontology*, ed. Savas L. Tsohatzidis, 191–202. Dordrecht: Springer.

Lukes, Steven. 2014. Introduction. In *The Rules of Sociological Method and Selected Texts on Sociology and Its Method*. New York, NY: The Free Press.

Lukes, Steven. 2026. Durkheim's Enemies. In: Fournier and Carls (2026).

Malczewski, Eric. 2019. Durkheim and the Nation. *İstanbul Üniversitesi Sosyoloji Dergisi* 39 (1): 41–64. https://doi.org/10.26650/SJ.2019.39.1.0013.

Mallory, Peter, and Patricia Cormack. 2018. The Two Durkheims: Founders and Classics in Canadian Introductory Sociology Textbooks. *Canadian Journal of Sociology* 43 (1): 1–24. https://doi.org/10.29173/cjs29386.

Marcucci, Nicola, ed. 2021. *Durkheim & Critique*. London: Palgrave Macmillon.

Marshall, Barbara. 2002. Snips and Snails and Theorists' Tales: Classical Sociological Theory and the Making of 'Sex.' *Journal of Classical Sociology* 2 (2): 135–155. https://doi.org/10.1177/1468795X02002002855.

Meadwell, Hudson. 2023. Durkheim and Realism. *Journal for the Theory of Social Behavior* 53 (4): 520–534. https://doi.org/10.1111/jtsb.12383.

Miller, Katherine J. L., and Steven Lukes. 2020. The Other Side of Freedom; On the Sociality of Ethics. *Anthropological Theory* 20 (4): 414–437. https://doi.org/10.1177/1463499619890762.
Milbrandt, Tara, and Frank Pearce. 2011. Émile Durkheim. In *The Wiley-Blackwell Companion to Major Social Theorists*, ed. George Ritzer and Jeffrey Stepnisky, 236–282. West Sussex: Blackwell.
Milbrandt, Tara. 2026. Revisiting Durkheim on Women, Feminism, and Equality. In: Fournier and Carls (2026).
Morrison, Ken. 2001. The Disavowal of the Social in the American Reception of Durkheim. *Journal of Classical Sociology* 1 (1): 95–125. https://doi.org/10.1177/1468795X0100100104.
Mucchielli, Laurent. 1995. Heurs et malheurs du durkheimisme. Problèmes historiographiques, enjeux épistémologiques et pédagogiques d'une mémoire. *Politix* 8 (29): 55–79. https://doi.org/10.3406/polix.1995.1901.
Mukherjee, Romi. 2010. *Durkheim and Violence*. Hoboken: Wiley-Blackwell.
Müller, Hans-Peter. 2019. *Das soziologische Genie und sein solides Handwerk*. Wiesbaden: Springer.
Némedi, Dénes. 1995. Collective Consciousness, Morphology, and Collective Representations: Durkheim's Sociology of Knowledge, 1894-1900. *Sociological Perspectives* 38 (1): 41–56. https://doi.org/10.2307/1389261.
Nisbet, Robert. 1952. Conservatism and Sociology. *American Journal of Sociology* 58 (2): 167–175. https://doi.org/10.1086/221116.
Ollion, Etienne, and Andrew Abbott. 2016. French Connections: The Reception of French Sociologists in the USA (1970-2012). *European Journal of Sociology* 57 (2): 331–372. https://doi.org/10.1017/S0003975616000126.
Paoletti, Giovanni. 2012. *Durkheim et la philosophie: Représentation, réalité et lien social*. Paris: Garnier.
Paoletti, Giovanni. 2026. Durkheim and the Liberal Tradition. In: Fournier and Carls (2026).
Pearce, Frank. 1989. *The Radical Durkheim*. London: Unwin Hyman.
Pederson, Jean Elisabeth. 2001. Sexual Politics in Comte and Durkheim: Feminism, History, and the French Sociological Tradition. *Signs* 27 (1): 229–263. https://doi.org/10.1086/495678.
Peter, Lothar. 2013. Dialektik der Gesellschaft versus 'conscience collective.' Zur Kritik Theodor W. Adornos an Durkheim. In *Émile Durkheim: Soziologie, Ethnologie, Philosophie*, ed. Tanja Bogusz and Heike Delitz, 73–94. Frankfurt: Campus.
Pickering, William F. 2000. *Durkheim and Representations*. London: Routledge.
Pinto, Louis. 1993. Le détail et la nuance: La sociologie vue par les philosophes dans la *Revue de métaphysique et de morale*, 1893-1899. *Revue de Métaphysique et de Morale* 98 (1/2): 141–174.
Pizarro Noël, François. 2010. De l'anti-réalisme de la sociologie contemporaine. In *Actualité de la théorie critique*, ed. Benoît Coutu, 203–214. Montréal: Éditions Libres du Carré Rouge.
Platt, Jennifer. 1995. The United States Reception of Durkheim's 'The Rules of Sociological Method.' *Sociological Perspectives* 38 (1): 77–105. https://doi.org/10.2307/1389263.
Ramp, William. 2008a. Durkheim Redux. *Journal of Classical Sociology* 8 (2): 147–157. https://doi.org/10.1177/1468795X08088869.
Ramp, William. 2008b. Transcendence, Liminality and Excess: Durkheim and Bataille on the Margins of 'Sociologie Religieuse.' *Journal of Classical Sociology* 8 (2): 208–232. https://doi.org/10.1177/1468795X08088872.
Ranulf, Svend. 1939. Scholarly Forerunners of Fascism. *Ethics* 50 (1): 16–34. https://doi.org/10.1086/290117.
Rawls, Anne Warfield. 1996. Durkheim's Epistemology: The Neglected Argument. *American Journal of Sociology* 102 (2): 430–482. https://doi.org/10.1086/230952.
Rawls, Anne Warield. 2004. *Epistemology and Practice: Durkheim's* The Elementary Forms of Religious Life. Cambridge UK: Cambridge University Press.

Richman, Michèle. 1997. Rejection and Renewal: Durkheim in the 20th Century. *Modern & Contemporary France* 5 (4): 409–419. https://doi.org/10.1080/09639489708456395.
Ricoeur, Paul. 1965. *De l'interpétation*. Paris: Éditions du Seuil.
Riley, Alexander. 2002. Durkheim Contra Bergson? The Hidden Roots of Postmodern Theory and the Postmodern 'Return' of the Sacred. *Sociological Perspecrives* 45 (3): 243–265. https://doi.org/10.1525/sop.2002.45.3.243.
Riley, Alexander. 2026. Against the Caprice of Wills: Durkheim on Sex, Marriage, and Family. In: Fournier and Carls (2026).
Rosati, Massimo. 2004. Forms of Radicalism: Theoretical and Social Radicalism in Durkheim. *Durkheimian Studies/Études durkheimiennes* 10: 10–18. https://doi.org/10.3167/136202404780459311.
Rosati, Massimo. 2009. *Ritual and the Sacred: A Neo-Durkheimian Analysis of Politics, Religion and the Self*. London: Routledge.
Salmon, Gildas. 2016. Foucault et la généalogie de la sociologie. *Archives de philosophie* 1 (79): 79–102. https://doi.org/10.3917/aphi.791.0079.
Sawyer, Keith. 2002. Durkheim's Dilemma: Toward a Sociology of Emergence. *Sociological Theory* 20 (2): 227–247. https://doi.org/10.1111/1467-9558.00160.
Schmaus, Warren. 1994. *Durkheim's Philosophy of Science and the Sociology of Knowledge: Creating an Intellectual Niche*. Chicago, IL: University of Chicago Press.
Schmaus, Warren. 2004. *Rethinking Durkheim and His Tradition*. Cambridge UK: Cambridge University Press.
Schmaus, Warren. 2010. Understanding and Explanation in France: From Maine de Biran's *Méthode Psychologique* to Durkheim's *Les Formes Élémentaires de la vie Religieuse*. In *Historical Perspectives on Erklären and Verstehen*, ed. U. Feest, 101–120. Dordrecht: Springer.
Searle, John. 2006a. Social Ontology: Some Basic Principles. *Anthropological Theory* 6 (1): 12–29. https://doi.org/10.1177/1463499606061731.
Searle, John. 2006b. Culture and Fusion: Reply to D'Andrade. *Anthropological Theory* 6 (1): 40–44. https://doi.org/10.1177/1463499606061733.
Searle, John. 2006c. Searle versus Durkheim and the waves of thought: Reply to Gross. *Anthropological Theory* 6 (1): 57–69. https://doi.org/10.1177/1463499606061735.
Smith, Philip. 2020. *Durkheim and After: The Durkheimian Tradition 1893-2020*. Cambridge UK: Polity Press.
Steinmetz, George. 2026. *Durkheim and the Post-Colonial Critique*. In: Fournier and Carls (2026).
Stedman-Jones, Susan. 1996. What Does Durkheim Mean by "Thing"? *Durkheimian Studies/Études Durkheimiennes* 2: 43–59.
Stedman-Jones, Susan. 2001. *Durkheim Reconsidered*. Cambridge, UK: Polity Press.
Susen, Simon. 2020. *Sociology in the Twenty-First Century: Key Trends, Debates, and Challenges*. London: Palgrave Macmillan.
Takla, Tendzin, and Whitney Pope. 1985. The Force Imagery in Durkheim: The Integration of Theory, Metatheory, and Method. *Sociological Theory* 3 (1): 74–88. https://doi.org/10.2307/202175.
Taylor, Charles. 1971. Interpretation and the Sciences of Man. *The Review of Metaphysics* 25 (1): 3–51.
Tilly, Charles. 1981. Useless Durkheim. In *As Sociology Meets History*, 95–108. New York, NY.
Tilly, Charles. 1984. *Big Structures, Large Processes, Huge Comparisons*. New York: Russell Sage Foundation.
Tiryakian, Edward. 1962. *Sociologism and Existentialism: Two Perspectives on the Individual and Society*. Hoboken: Prentice Hall.
Thomson Reuters. 2009. Most Cited Authors of Books in the Humanities, 2007. *Times Higher Education*, Mar 29.
Turner, Stephen. 1995. Durkheim's *The Rules of Sociological Method*: Is it a Classic? *Sociological Perspectives* 38 (1): 1–13. https://doi.org/10.2307/1389258.

Turner, Stephen. 2007. Defining a Discipline: Sociology and its Philosophical Problems, from its Classics to 1945. In *Philosophy of Anthropology and Sociology*, ed. Stephen Turner and Mark Risjord, 3–70. Oxford UK: Elsevier.

Weiss, Raquel. 2012. From Ideas to Ideals: Effervescence as the Key to Understanding Morality. *Durkheimian Studies/Études Durkheimiennes* 18: 81–97. https://doi.org/10.3167/ds.2012.180106.

Chapter 6
Max Weber's Contribution to the Philosophy of Science

Basit Bilal Koshul

Abstract In this chapter I build on John Dewey's observation that the emergence of modern science is a watershed in the history of philosophy and argue that Max Weber's methodological reflections show that the emergence of the social sciences is (at least) as significant an event in this regard. For Dewey the scientific study of Nature marks a new stage of the development of philosophy because it provides the logical warrant to dignify ontically and epistemically that which Classical philosophy denigrated (i.e. matter, time, and the individual). Weber's reflections on the social scientific study of Culture provide the logical warrant to deepen and broaden this shift on two counts. First, his reflections on the methodology of the *Giesteswissenshaften* (soft sciences) in comparison to the *Naturwissenschaften* (hard sciences), confers ontic dignity on social, historical, and psychological phenomena. Second, his observation that science is one cultural activity among other such activities shows that in addition to being the pursuit of knowledge about "objective" facts of Nature, scientific inquiry is the self-conscious pursuit to understand the "subjective" values of Culture that underpin this inquiry. The ontic and epistemic shift implicit in Weber's methodological reflections; (a) deepen post-Newtonian science's critique and reconstruction of the Classical and Enlightenment conceptions of "reality" and "knowledge," (b) initiate the process of value-neutral (scientific) inquiry into the value-laden (cultural) claims of the universal validity and meaningfulness of scientific knowledge, and (c) reveal heretofore unseen interrelations between Nature and Culture—thus making the emergence of the social sciences an historical landmark for the philosophy of science.

Keywords Weber · Dewey · Nature · Culture · Methodology of social science

B. B. Koshul (✉)
Lahore University of Management Sciences, Lahore, Pakistan
e-mail: basitb@lums.edu.pk

Y. Shan (ed.), *History and Philosophy of the Social Sciences*,
https://doi.org/10.1007/978-3-032-08167-4_6

6.1 Dewey on the History of Philosophy and the Philosophy of Science

John Dewey argues that with the help of historical hindsight, we see philosophy passing through three phases of development; (a) classical, (b) early modern, and (c) modern. Classical philosophy's conception of "reality" and "knowledge" (in its Aristotelian form) shaped the outline of the philosophy of science for about two millennia. Scientific discoveries in the early modern period led to calls for revising the classical conceptions—with Francis Bacon being the trailblazer in this regard. But due to the socio-cultural environment and the relatively underdeveloped state of science, the early modern "development represents merely a transitional stage. It tried, after all to put the new wine in old bottles" (Dewey 1920, 51). It is only with the emergence of modern science in the late nineteenth and early twentieth centuries that the intellectual resources and cultural environment became conducive to forging a viable alternative to the "old bottles" and providing the "new wine" with a suitable vessel. I will use Dewey's description of the "old bottles" and the "new wine" to set the stage for describing the historical and philosophical significance of Max Weber's reflections on the methodology of the social sciences.

Dewey's use of the term "old bottles" refers to the characteristics that classical philosophy attributed to "reality" and "knowledge." On the first point, "[a]ll philosophies of the classic type have made a fixed and fundamental distinction between two realms of existence" (Dewey 1920, 22)—ultimate reality and apparent reality. The former was the "absolute and noumenal reality which could be apprehended only by the systematic discipline of philosophy itself" and the latter was "the ordinary empirical, relatively real, phenomenal world of everyday experience" to which "positivistic science referred" (Dewey 1920, 23). Three characteristics distinguished ultimate reality from apparent reality. First, the ultimate "True Being is pure Immaterial Mind" (Dewey 1920, 112), while that which appears to be real (matter) is "something threatening to mind and opposing it" (Dewey 1920, 72). Second:

> ...[C]lassic philosophy maintained that change, and consequently time, are marks of inferior reality, holding that true and ultimate reality is immutable and eternal. Human reasons, all too human, have given birth to the idea that over and beyond the lower realm of things that shift like the sands on the seashore is the kingdom of the unchanging, of the complete, the perfect. (Dewey 1940, 217).

Third, "ultimate reality" was also completely homogeneous, undifferentiated, and indivisible. Such "Being" is completely devoid of personality—if the latter is understood as a unique individual interacting with other individuals and purposefully pursuing an "ought" in the face of an "is." In short, while ultimate reality is characterized by immateriality, eternity, and homogeneity, "apparent reality" is characterized by materiality, temporal flux, and multiplicity.

The "relationship" between the two realms of reality is reflected in the classic conception of a fixed, completed, and hierarchical universe. At the bottom of the hierarchy is the earth which is; (a) composed of matter, (b) the site of constant flux, and (c) the dwelling place of many different species and individuals (engaged in incessant conflict and strife among themselves because of their differences). Above the earth are the celestial spheres. While they are not completely immaterial, they are composed of a finer type of "matter" (i.e. fire and air) than the earth. Similarly, while flux and change are seen in the heavens (as evidenced by the orbits of planets and moons) this movement is of a different kind than earthly movement—it is a perfect circle. At the top of the hierarchy is the completely immaterial, immutable, undifferentiated "divine" ether—which hems in and encompasses the entire universe. Being completely homogenous and undifferentiated, this outermost realm is free of all materiality, multiplicity, and diversity. Consequently, the "being" of ether is free of all friction, struggle, and conflict—which are the defining characteristics of all "becoming" on earth.

The absolute distinction between ultimate reality and apparent reality was carried over into classical conception of "knowledge." Philosophy claimed that it "was in possession of a higher organ of knowledge than is employed by positive science and ordinary practical experience, and that it was marked by a superior dignity and importance…" (Dewey 1920, 23). The use of this "higher organ of knowledge" led the philosophers to posit that "the world as it stood was an embodiment of a fixed and comprehensive Mind or Reason" (Dewey 1920, 50). The knowledge of this "comprehensive Mind or Reason" was called *episteme*. The method of its attainment was solitary contemplation—which would yield an unmediated, intuitive vision of Pure Being. The defining characteristics of *episteme* was that it is "knowledge for its own sake" and it was certain, exact, and universal.

The knowledge of different aspects of apparent reality came under the headings of *emperia*, *historia*, and *doxa*—pursued and produced by a motley group of people: politicians, artisans, bards, and common folk. It differed in terms of dignity, method, and kind (not just degree) than the *episteme* of the philosophers. In terms of dignity, (strictly speaking) it does not deserve to be called "knowledge" because it is knowledge of material, temporal phenomena and related to mundane needs of individuals and society. These three are "a low and untrue type" of knowledge because (collectively) it "is not disinterestedly for its own sake" (Dewey 1920, 110). While their goals differ, the knowledge of the politician, artisan, bard, etc. "has an ulterior aim, and one which itself testifies to imperfection. For want, desire, affection of every sort, indicate lack. Where there is need and desire—as in the case of all practical knowledge and activity—there is incompleteness and insufficiency" (Dewey 1920, 110).

In terms of method, this lower type of knowledge is attained by some sort of mediation:

- *Emperia* (knowledge of the natural world based on observed similarities),
- *Historia* (knowledge of the past based on memory).
- *Doxa* (knowledge of mundane affairs based on common sense).

This means that the knower plays a dual role in the attainment of this knowledge. The knower's observations, memory, and common sense play a role in gathering information and the knower's needs and desires determine the end to which the knowledge will be used. Given the role of the knower in the act of knowing, *emperia*, *historia*, and *doxa* are a different kind of knowledge than *episteme* (which is unmediated, direct knowledge). In contrast to the certainty, exactitude, and universality of the latter, the former is probabilistic, vague, and contextual—and hardly worthy of being called "knowledge" according to canons of classical philosophy.

Second order reflections on the classic understanding of reality and knowledge allow us to describe the defining characteristics of the "old bottles" with greater precision. Firstly, the classic conception of ultimate reality rules out any possibility of relations. Being completely homogenous and undifferentiated, Ultimate Reality is devoid of relations—both internally and externally. Internally, it is an immaterial, immutable, homogenous monad. Externally, this self-enclosed monad has no possible relationship with apparent reality because the latter is composed of many different individuals/phenomena whose alliances/relationships constantly change with the passage of time. Secondly, the classic conception of reality rules out the possibility of growth or novelty—which are examples of change and a sign of irredeemable deficiency. It is perfectly reasonable that change is a most prominent feature of apparent reality because it is the realm of perpetually deficient "becoming." But it is utterly unreasonable to even consider the possibility of growth and novelty being attributed to ultimate reality because it is eternally perfect "Being." Thirdly, the classical conception of reality does conceive of something like individuality because "Being" is "itself and not something else" (Dewey 1940, 220). But this perfect individual Being "had no development, no history, because it had no potentialities…Time did not enter into its being either to corrode it or to develop it" (Dewey 1940, 220). In other words, the Perfect Individual of classical philosophy is the exact opposite of the imperfect human individuals of history. The latter are creatures whose lives are a process of becoming in which their (conflicting and changing) desires, needs, and purposes play a central role. This life process is shaped (and reshaped) by decisions made to pursue certain ends. While the former is completely devoid of all such "personal" and "subjective" elements. For our purposes, the "old bottles" refers to the philosophical position which considers "reality" and "knowledge" of it to be free of relations, growth, and novelty—or (more generally) free of social, temporal, and psychological elements. Even more precisely the "old bottles" is a metaphor for an unbridgeable divide between the Natural World and Human Culture.

Dewey argues that this division between the Natural and the Human begins to be challenged at the dawn of modern science. Dewey contrasts the Aristotelean "logic of demonstration" with Baconian "logic of discovery" to show how the classical conception of "reality" and "knowledge" to describe the most general outlines of this development. Aristotle's "logic of demonstration" aspired to reach eternal, immutable Truth that has been known since eternity—because it is Knowledge of Eternal Being:

> It was an axiom of Aristotle that only which was already known could be learned, that growth in knowledge consisted in simply bringing together a universal truth of reason and a particular truth of sense which had been previously noted separately. (Dewey 1920, 31)

From this perspective "learning" does not mean "*growth* of knowledge" because "growth belongs to the region of becoming and change" but rather the "*possession* of knowledge in the syllogistic self-revolving manipulation of what was already known—demonstration" (Dewey 1920, 30). In contrast to Aristotle's detached contemplation of the Unitary Being that is veiled by the apparent diversity of the material world, Bacon proposes a new "technique of inquiry" that will lead to the discovery of harmony and relations:

> Bacon eloquently proclaimed the superiority of discovery of new facts and truths to the demonstration of the old. Now there is only one road to discovery, and this is penetrating insight into the secrets of nature. Scientific principles and laws do not lie on the surface of nature. They are hidden, and must be wrested away from nature by an active and elaborate technique of inquiry. (Dewey 1920, 31f.)

The "discovery of new facts" begins with close observation of the material, heterogenous, changing diversity that one finds in the natural world. Then this naturally given is subjected to human control and manipulation with the goal of discovering characteristics, patterns, and relationships that were obscured by the immediately observable. This is the method of "active experimentation" and its fruit is the discovery of new knowledge: "Active experimentation must force the apparent facts of nature into forms different to those in which they familiarly present themselves; and thus make them tell the truth about themselves…" (Dewey 1920, 44). Since it cannot be determined a priori how many facts are known and unknown, Bacon's method entails the continuous and perhaps unending growth of knowledge. Because the goal of "active experimentation" is "the invasion of the unknown, rather than the repetition in logical form of the already known" (Dewey 1920, 34) a "logic of discovery" is implicit in Bacon's method. This logic gives material reality, temporal flux, and human actions, decisions, and purposes a central role in the pursuit of knowledge—a role that is principally ruled out by the Aristotelean "logic of demonstration."

It is not just with respect to the origins of knowledge that early modern developments challenge classical philosophy—it is also with respect to the final goal. For the former "Knowledge is Power" means "power over natural forces in the common interest of all" and for the latter it means "power over other men in the interest of some class or sect or person" (Dewey 1920, 30). For Bacon the growth of knowledge should produce "works, fruits, and consequences of value to humankind through power over natural forces" (Dewey 1920, 34) rather than facilitating "victory of man over man in law courts, diplomacy, and political administration" (Dewey 1920, 34). The social character of the goal of knowledge is not just in its use but also in its "ownership": "Power over nature was not to be individual but collective; the Empire, as he says, of Man over Nature, substituted for the Empire of Man over Man" (Dewey 1920, 37). Dewey notes:

> [S]o far as concerns the spirit and atmosphere of the pursuit of knowledge, Bacon may be taken as a prophet of a pragmatic conception of knowledge. Many misconceptions of its spirit would be avoided if his emphasis on the social factor in both the pursuit and the end of knowledge were carefully observed. (Dewey 1920, 38)

Dewey's metaphor of the spider, ant, and bee succinctly contrasts the classical and emergent modern positions:

> Pure reasoning as a means of arriving at truth [of Eternal Being] is like the spider who spins a web out of himself. The web is orderly and elaborate, but it is only a trap. The passive accumulation of experiences—the traditional empirical method—is like the ant who busily runs about and collects and piles up heaps of raw materials. True method, that which Bacon would usher in, is comparable to the operations of the bee who, like the ant, collects material from the external world, but unlike that industrious creature attacks and modifies the collected stuff in order to make it yield its hidden treasure [which will be used for nurture its own future generation and prove to be of immense value to other creatures in the natural environment]. (Dewey 1920, 32)

Here the spider represents the classical, the ant the early modern, and the bee the emergent modern position. In contrast to the classical conceptions of "reality" and "knowledge" the emergent modern position considers relations, growth, and novelty to be as much a part of "reality" as social, temporal, and psychological factors are part of the human knowledge of reality. This means that at its very birth the modern position is at odds with the classical divide between the Natural World and Human Culture.

In Bacon we see the raw material from which the "new wine" begins to emerge—but this "new wine" was eventually poured back into "old bottles." This can be illustrated by examples from the "hard" and "soft" sciences. The best example from the former is Isaac Newton's "natural philosophy." On the surface it breaks with the classical tradition by doing away with the absolute distinction between two realms of reality and posits the claim that there is only one realm—that of "Nature" (which would later come to be termed "empirical" reality). But a closer look shows that the classical distinction between the absolute and apparent is not just maintained but forcefully affirmed in the modern conception of "empirical" reality. Newton's scientific method posits an absolute distinction between the "primary" and "secondary" qualities of phenomena. While materiality, mutability, and heterogeneity are the observable characteristics of "secondary qualities," all such phenomena is underpinned by more fundamental, unobservable "primary qualities." And the latter have traits that classical philosophy attributed to "ultimate" reality—devoid of relations, growth, and novelty. This means that while there was a change in important details regarding the understanding of "reality" (it was not understood in terms of two distinct realms) there was no change in the basic principles (absolute distinction between two domains in one "empirical" reality).

Dewey offers perceptive remarks regarding the historical origins of Newton's distinction between primary and secondary qualities—and the impact of this distinction upon "Enlightenment" philosophy. Regarding Newton's claim that the observable multiplicity and flux in secondary qualities is underpinned by certain eternal, immutable primary qualities, Dewey notes:

> [T]he minds of men, including physical inquirers, were still possessed by the old notion that reality in order to be solid and firm must consist of those things which philosophy calls substances. Changes could be known only if they could be somehow reduced to recombinations of original unchanging things... Thus a popular metaphysics, given rational formulation

> by the Greeks, and taken over into the intellectual tradition of the western world, controlled at first the interpretations placed upon the procedures and conclusions of experimental knowing. (Dewey 1929, 119)

Consequently, "Newtonian philosophy allowed itself to become entangled in Greek metaphysics according to which the immutable is the truly real" (Dewey 1929, 119)—and from Newtonian philosophy this metaphysics found its way into the "empiricism" of John Locke:

> It thus appears that Newton retained a part of the qualitative equipment of the objects of Greek science, in spite of their irrelevance to both mathematics and experiment. When one searches through philosophical commentary and discussion (based mainly on Locke's version of Newton's results), one finds a great deal of discussion about the fact that the so-called secondary qualities, color, sound, odor, taste, were eliminated from "reality." But not a word as far as I can discover is said about the fact that other sensible qualities under the name of primary were retained in defining the object of science. And yet this retention is the fons et origo malorum [source and origin of the malaise]. (Dewey 1929, 120)

The distinction between observable secondary and unobservable primary qualities had a direct impact on the Newtonian conception of knowledge—as illustrated by his well-known saying: "I do not invent hypotheses." When viewed alongside his lesser known saying—"Whatever is not derived from phenomena is to be called a hypothesis and hypotheses…have no place in experimental philosophy" (Dewey 1929, 116)—it means Newtonian "experimental philosophy" sees unmediated sense perception as both the "origin" and "warrant" (or "justification") of knowledge claims—with the knower and the act of knowing playing no role in the construction of the knowledge claim. The fact that this is but Aristotle's definition of *episteme* in a new guise is almost self-evident—almost as self-evident as the fact this Newtonian supposition is the shared premise of rationalist idealism and empiricist sensationalism:

> In spite of the polar opposition between the two schools, they depend upon a common premise. According to both systems of philosophy, reflective thought, thinking that involves inference and judgement, is not originative. It has its test in antecedent reality as that is disclosed in some non-reflective immediate knowledge. (Dewey 1929, 109)

And it is not only these two schools of modern philosophy that share this premise:

> In breaking away from antique and medieval thought,…, early modern thought continued the older tradition of Reason that creates and constitutes the world, but combined it with the notion that this Reason operates through the human mind, individual or collective. This is the common note of idealism sounded by all the philosophies of the seventeenth and eighteenth centuries, whether belonging to the British school of Locke, Berkeley and Hume or the Continental school of Descartes. In Kant as everybody knows the two strains came together; and the theme of the formation of the knowable world by means of a thought that operated exclusively through the human knower became explicit. Idealism ceased to be metaphysical and cosmic in order to become epistemological and personal. (Dewey 1920, 50f.)

From the foregoing, it is clear that key elements of the classical heritage are found at the core of the "Enlightenment" understanding of not just "reality" but also of "knowledge." The Greek term *episteme* was translated to *scientia* in Latin and "scientific" knowledge came to be associated with "true knowledge of reality." Early modern science did reject the notion of "certain" knowledge and replaced it with the notion of "objective" knowledge. But the difference between "certain" and "objective" is one of degree not of kind because both claim to be free of all psychological, temporal, and social influences. And it is by virtue of being free of these "subjective" traits that the certainty of episteme and the objectivity of science attain their (self-declared) universal validity. In contrast *emperia*, *historia*, and *doxa* were (to varying degrees) simultaneously knowledge of psychological, temporal, and social phenomena and infected by these personal and subjective elements. Consequently, they produce only probabilistic and context bound claims—claims hardly worthy of being called "knowledge."

In the "soft sciences" August Comte and Wilhelm Dilthey's ideas serve as an example of the far-reaching impact of classical conception of "knowledge." Comte put forward the idea of "social physics" as a way to study and understand social reality. He argued that that just as all natural phenomena was "governed" by "laws of nature," so were human societies, institutions, and individuals. The discovery of the "laws" that "govern" socio-cultural phenomena will lead to knowledge of social reality that was as precise and predictive as knowledge of natural phenomena. In other words, scientific knowledge of psychological, historical, and social phenomena was (in principle) possible. At face value, Comte's conception appears to challenge the classical position. In the late nineteenth century, a different challenge emerged from a totally opposite direction—with the classical position being turned on its head. Dilthey argued that genuine knowledge could only be attained of unique, individual, particular phenomena but not "universal" natural phenomena. This is because the subject and object of socio-cultural inquiry share and are linked by Geist (spirit). Analysis using inner intuition yields *Versthen* (unmediated understanding) of the object of inquiry. In contrast, there will always be an unbridgeable gap between the human observer and natural phenomena because nature is devoid of *Geist*. As a result, scientific study of nature can only yield "causal explanation" which is expressed in the terms of conceptual laws. For Dilthey "understanding" of socio-cultural phenomena constitutes genuine knowledge and "explanation" of natural phenomena only the appearance of knowledge.

On the surface Comte and Dilthey are challenging the classical position regarding knowledge of social reality—both are claiming that such knowledge is possible. But in making their arguments, (like Newton before them) both thinkers offered "new wine" but ended up putting it in "old bottles." Dewey notes that these "old bottles" are aptly described in Henri Bergson's critique of mechanistic and teleological theories:

> [B]oth are guilty of the same error, although from opposite points. Fixed laws, which govern change and fixed ends towards which changes tend are both products of a backward look, one that ignores the forward movement of life. (Dewey 1940, 219).

For Bergson, these theories do describe what has been left behind by life but not life's "ongoing vital creative course…whose behavior and outcome are unpredictable both mechanically and from the standpoint of [fixed] ends" (Dewey 1940,

219). While the past, present, and future interact dynamically in the "ongoing vital course" that is life, mechanistic theories (like those of Comte) and teleological theories (like those of Dilthey) negate the significance of the present and future. They consider the experienced present and yet-to-be future as givens from the distant past—for Comte-like mechanism the predetermined effects of eternally fixed laws, for Dilthey-like teleology the necessary pursuit of eternally fixed ends. The observation of William James shows the two types of theories to share another important trait:

> Both sacrifice individuality and all the values, moral and aesthetic, which hang upon individuality. For according to absolute idealism, as to mechanistic materialism, the individual is simply a part determined by the whole of which he is part. (Dewey 1940, 219)

In such a conception, there is no possibility "for the emergence of the genuinely new" individual in the universe (Dewey 1940, 219). By drawing attention to the importance of the study of earthly material phenomena, social reality, and the possibility of discovering new facts about the universe, early modern philosophy was able to produce "new wine." But the socio-cultural conditions of the seventeenth to nineteenth centuries and available intellectual resources forced it to put this "new wine" into the "old bottles" inherited from classical philosophy.

Dewey argues that it is only with the birth of modern science that the conditions and material needed to forge "new bottles" becomes available. This material comes in the form of the findings of modern physics and chemistry which not only show the shortcomings of the classic conceptions of reality and knowledge but also provide the resources for articulating an alternative. Dewey argues that the philosophical lessons to be drawn from the pathbreaking discoveries of modern physicists illustrate this point. Einstein's discoveries show that "relations" are a "fundamental" characteristic of "reality"—not idea, substance, essence, or other such abstract concoction. For example, mass (or "substance") is not "an inherent property which cause[s] inertia and momentum" but rather "mass is inertia-momentum and these are strictly measures and relations" (Dewey 1940, 221). The same holds (to the same degree) for other "fundamental" objects in physics as well: "[T]he very method of physical science, with its primary standard units of mass, space, and time is concerned with measurements and relations of change" (Dewey 1940, 221). Special relativity has shown that space is not an empty vacuum (or an absolute void) unaffected by the bodies of mass it "contains" but rather a relationship of space-time that impacts and is impacted by the movement of these bodies. Consequently, "a physical law states correlations of changes or ways and manners of change. The law of gravitation, for example, states a relation which holds between two bodies with respect to distance and mass" (Dewey 1940, 22). And (as is well-known) Einstein's theory of general relativity showed knowledge to have a characteristic that was principally ruled out by classical philosophy: the position of an observer directly impacts the observer's knowledge claim. In a sentence, special relativity showed reality to be characterized by dynamic relationships and general relativity showed knowledge claims to be impacted by the perspective of the observer.

The philosophical implications of Heisenberg's discoveries on the classic conception of reality and knowledge are as far reaching as Einstein's. Up till the early twentieth century it was widely believed that precise knowledge of the past, present, and future could be had. In the light of Newton's discoveries, it was argued that since the velocity and location of a body could be determined at any given point in time, the past and future of the entire system could be determined based on the present knowledge of its individual components. Heisenberg's discoveries showed this claim to be absolutely fantastical for not just a system composed of many bodies but for even a single element in the system. For an electron orbiting the atoms, there is an inverse relationship between the precision with which its velocity and its position can be known: "[G]iven the determination of position, [the particle's] velocity can be stated only as of a certain order of probability, while if its velocity is determined the correlative factor of position can be stated only as of a certain order of probability" (Dewey 1940, 222). Since the velocity and location of a single particle cannot be determined simultaneously, "it follows necessarily that the future of the whole collection cannot possibly be foretold except in terms of some order of probability" (Dewey 1940, 222). Dewey argues that Heisenberg's indeterminacy principle is "a way of acknowledging the pertinence of real time to physical beings. The utmost possible regarding an individual is a statement as to some order of probability about the future" (Dewey 1940, 222). From this perspective, the "actual force and significance" of Heisenberg's discoveries is "generalization of the idea that the individual is a temporal career whose future cannot be *logically* deduced from its past" (Dewey 1940, 222). The discoveries of quantum physics challenged the classical conception of "reality" and "knowledge." In terms of the former they showed that to be actually real which was considered only apparently real—time. In terms of the latter they showed scientific knowledge to have a trait which classical philosophy claimed invalidated a claim's status to "knowledge": probability.

The philosophical lessons to be learned from the work of the physicists are affirmed and deepened by those to be learned from a philosophical engagement with the "laws of gasses." Looking at the relationship between the individual molecule, the gas in which this molecule is an element, and the "law of gasses" Dewey observes:

> A single molecule is not and cannot be a gas. It is consequently absurd to suppose that the scientific law is about the elementary constituents of a gas. It is a statement of what happens when a very large number of such constituents interact with one another under certain circumstances. (Dewey 1940, 221)

Considering laws of nature as "statistical statements" shows them to be "of the nature of probability formulations" in contrast to the "so-called dynamic laws that are abstract and mathematical, and disguised definitions" (Dewey 1940, 221). Viewing laws as statistical statements has far-reaching implications for conceiving the relationship between the individual and the group:

> No statement is made about what will take place in the case of an individual. The application to scientific formulations of the principle of probability statistically determined is thus

> a logical corollary of the principle already stated, that the subject matter of scientific findings is relational, not individual. (Dewey 1940, 222)

The fact that a "law" is a statistical statement about the probable behavior of a system, has direct implications about our conception of the relationship of law to the individual:

> Laws do not 'govern' the activity of individuals. They are a formulation of the frequency-distributions of the behavior of large numbers of individuals engaged in interaction with one another. (Dewey 1940, 222)

Laws are statistical descriptions of the probable behavior of a system (or group of individuals)—they "govern" neither the behavior of the whole nor the individual parts of the whole.

The discoveries of modern physics also challenge the early modern conception of "potential" which was conceived as "a *fixed* end which the individual endeavors by its own nature or essence to actualize" (Dewey 1940, 223). A close study of the philosophical implications of the process by which potential energy is converted into kinetic energy and the chemical processes by which some products are turned into other (i.e. milk into plastic) shows that "potentialities must be thought of in terms of consequences of interaction with other things. Hence potentialities cannot be *known* till *after* interactions have occurred" (Dewey 1940, 223). This means that "potentialities are not fixed and intrinsic, but are a matter of an indefinite range of interactions in which an individual may engage" (Dewey 1940, 224). Revising the conception of "laws of nature" and the "nature" or "essence" of individuals in light of the findings of modern chemistry shows that to be real which classic philosophy ruled out in principle: the unique individual, indeterminate future possibility, and genuine novelty. Combining this point with the earlier reflections on the "relational character" of reality, it is easy to see that "temporal quality and historical career are a mark of everything, including atomic elements, to which individuality may be attributed" (Dewey 1940, 220). For Dewey this is a landmark in the history and philosophy of science:

> The conclusion…is that the principle of a developing career applies to all things in nature, as well as to human beings—that they are born, undergo qualitative changes, and finally die, giving place to other individuals. The idea of development applied to nature involves differences in forms and qualities as sure as it rules out breaches of continuity. (Dewey 1940, 222f.)

By highlighting this commonality, Dewey points to a relationship between the Natural World and Human Culture that was ruled out—in principle—by the classical conception of reality and knowledge.

6.2 Weber's Methodological Reflections and the Philosophy of (Social) Science

In the previous section we saw Dewey analyzing the discoveries of modern physics and chemistry to draw philosophical lessons for the purpose of critiquing and repairing the classical and early modern conceptions of "reality" and "knowledge." In this section, I will follow Dewey's lead and build on it—with one important modification. Instead of the discoveries of modern (social) science, I will analyze Weber's reflections on the methodology of the "soft sciences" for the same purpose. The philosophical lessons to be derived from Weber's methodological reflections will show that his contribution to the history and philosophy of science begins where the contribution of Einstein, Bohr-Heisenberg, and thermodynamics ends.

For Dewey, the most important philosophical lessons to be derived from the discoveries of the "hard sciences" are:

- Relationality, temporality, and individuality are defining characteristics of reality.
- Probability, perspective, limitation are defining characteristics of knowledge.
- Because "the principle of a developing career applies to all things in nature, as well as to human beings," there is continuity between Nature and Culture—with both sharing key characteristics.

Even a plain sense reading of Weber's 1904 article "The 'Objectivity' of Knowledge is Social Science and Social Policy" shows each of these three points being explicitly discussed in (sometimes painstaking) detail. He wrote this article after becoming co-editor (along with Werner Sombart and Edgar Jaffe) of the most important journal in the social sciences of the day (and arguably ever). The purpose of the article was to outline the editorial policy of the journal. Weber uses this forum and format to offer his reflections on the subject matter, goals, and methodology of the "soft sciences." He does this not only with the "hard sciences" in view but also with a keen eye on the debates in the areas of philosophy, logic, and methodology. At the very start of the article he states: "The social science that *we* want to pursue is a *science of reality*" (Weber 2012d, 114—all emphasis in original). Shortly after describing social science to be a "*science of reality*," he describes the "real" in more precise terms:

> [T]he interest of social science no doubt focuses on the real (that is to say, the individual) configuration of the social [and] cultural life that surrounds us, in its universal context (which is, however, no less individual in character), and as it has evolved from other social and cultural circumstances, which in turn had their individual character. (Weber 2012d, 115).

From this passage it is clear that whatever other characteristics the "real" may have, "individuality" is the most important of these characteristics. Right after Weber gives this description of the "real," he notes that this is "the same situation that we have just described in relation to astronomy" (Weber 2012d, 115). The "object" of inquiry for the "hard science" is not "laws" *per se* but an "individual constellation" (whether that constellation is a "planet," "solar system," or "galaxy") that "laws" describe:

> The "astronomy" itself is interested in the following question: what is the individual result of the action of those laws on an individual constellation, since those constellations are significant for us. (Weber 2012d, 114)

He gives a more detailed description of the "individual constellation" that astronomy studies and the limitations of "laws" in this regard:

> Naturally, any individual constellation which is "explained" or predicated by [this "astronomy"] can only be causally explained as the result of another, equally individual one that preceded it; and, however far we go back into the grey mist of the most distant past, the reality for which the laws are valid retains just the same individual character, and it remains just as impossible to deduce it from those laws. (Weber 2012d, 114)

In other words, knowledge of "laws" is "extremely important—indeed indispensable!—[as a] *preliminary* investigation" (Weber 2012d, 117) for the sciences of empirical reality. But even in a hard science like astronomy, these "laws" cannot be accorded exclusive explanatory power and even less so any causal agency.

While the "real" is equally "individual" for the hard and soft sciences, the soft sciences ask certain questions about the individual and the interrelation of individuals that the hard sciences are incapable of asking. Social science aspires to:

> [U]nderstand the distinctive character of the reality of the life in which we are placed and which surrounds us—on the one hand: the interrelation and the cultural significance and importance of its individual elements as they manifest themselves today; and, on the other, the reasons why the[se elements] historically developed as they did and not otherwise. (Weber 2012d, 114).

Weber's description of "sociology" pointedly describes the unique perspective from which the soft sciences can study the unique individual. At the very start of *Economy and Society*, Weber describes "sociology" in these terms:

> Sociology…is a science concerning itself with the interpretive understanding of social action and thereby with causal explanation of its course and consequences. We shall speak of "action" insofar as the acting individual attaches a subjective meaning to his behavior—be it overt or covert, omission or acquiescence. Action is "social" insofar as its subjective meaning takes into account the behaviour of others and is thereby oriented in its course. (Weber 1968, 4)

For sociology human "social action" is as much an empirical fact as the orbit of the planets for physics and photosynthesis is for biology—all are observed "occurrences." While human behavior shares certain characteristics with occurrences in the natural world, it has some unique features:

> The course of human ("external" or "inner") behaviour exhibits interrelations and regularities, like all occurrences. But what is, at least in its fullest sense, peculiar to human behaviour is [that it exhibits] interrelations and regularities whose course can be interpreted so that it can be understood. (Weber 2012a, 273)

In other words, while "interrelations and regularities" are characteristic of both Nature and Culture, human behavior is unique because its "interrelations and regularities" are meant to be and capable of being "understood" by the different individuals in terms of "some (*subjective*) *meaning* that it 'has' or that is 'meant'." Furthermore, the investigator studies the "interrelations and regularities" among the different individuals because the investigator finds them to be of "cultural *significance and importance*." Weber's description of "social action" in the context of his description of social science shows interrelation to be the second important characteristic of "reality."

A third characteristic of reality is implicit in Weber's description of social science—temporality. Every individual that is studied by any science of empirical reality is the product of a historical process—a process virtually synonymous with "time" and "temporality." This is not just the case in the soft sciences, it is (first and much more so) the case in the hard sciences. While the classical notions of spontaneous generation and the eternity of the universe considered the "essence" of all existents to be eternal and immutable, from the perspective of modern physics, chemistry, and biology every individual in the natural world (i.e. planet, solar system, galaxy, chemical, compound, phyla, genus, species, etc.) is the result of a historical process. This means that the scientist can identify a point in time in the past when this individual was not present in its present form. Modern science leaves no doubt that the individual emerges at a certain point in time as the result of a novel set of interrelations being established among previously existing elements—which themselves were the products of an historical process. In other words, the individual is an emergent phenomenon in history—and such phenomena interest us "because they developed in the way they did, and not differently" (Weber 2012d, 122). Once the individual emerges it becomes the "cause" of moving the historical process in certain directions and not in others. It is precisely because of their ability to recognize the reflexive relationship between the individual and time that the soft sciences are able to ask an investigative question that the hard sciences are incapable of imagining: Why did these individuals "historically develop as they did and not otherwise"?

The foregoing summary of Weber's reflections of the character of "reality" as understood by social science shows that it is characterized by relationality, individuality, and temporality. A closer look at the intersection of "cultural significance and importance" and the unique individual shows reality to be characterized by an additional factor. Speaking as a social scientist, Weber notes that the individual is "a complex of relationships" that we unite "into a conceptual" whole "from the vantage point of their *cultural significance*" (Weber 2002, 13). Put a little differently, the individual is a phenomenon "which was 'evaluated' by us as having *universal* significance by virtue of its *individual*, distinct character" (Weber 2012c, 152, fn. 2). Because the individual "denotes a phenomenon that is of qualitative importance as a consequence of its *uniqueness*," this concept cannot be defined according to how it is 'demarcated' vis-à-vis other concepts (*genus proximum, differntia specifica*)" (Weber 2002, 13). But rather, it "must be gradually *put together* from its single component parts, each of which is taken out of historical reality" (Weber 2002, 13)—and this is always done from the investigtor's "particular vantage points of interest" (Weber 2002, 13). Weber emphasizes this point in these words:

> The must be acknowledged, namely, the central role played by the researcher's particular vantage point in identifying what is essential to each case, as it belongs to the very essence of the formation of historical concepts. (Weber 2002, 13)

These reflections on the process behind the construction of historical concepts shows that reality is invariably characterized by "subjective" elements—the perspective of the investigator which plays a determining role shaping the concept.

Weber's description of "culture" allows us to get an "objective" view of the "subjective" elements that are invariably a part of scientific inquiry. Weber describes "culture" as follows: "'Culture' is a finite section of the meaningless infinity of events in the world, endowed with meaning and significance from a *human* perspective" (Weber 2012d, 119). This description of culture suggests that scientific inquiry is one type of cultural activity among others—because it is a specific and unique way to endow "meaning and significance" on the "meaningless infinity of events in the world" by human beings. In very pointed terms Weber notes that "even our knowledge of the most established theoretical propositions—in the exact natural sciences and mathematics, for instance—is a product of culture, in just the same way our conscience is developed and refined by culture" (Weber 2012d, 103). This interrelation between science and the reality which science investigates is deepened by Weber's description of the presupposition of cultural science:

> The transcendental precondition of every cultural science is not that we find a particular, or indeed any, "culture" valuable, but that we are cultural beings, endowed with the capacity and the will to adopt a deliberate position with respect to the world, and to bestow meaning upon it. (Weber 2012d, 119)

From the perspective of the cultural sciences, the object and subject of scientific inquiry are first and foremost cultural beings—the former engaged in the activity of conferring meaning on a particular segment of reality and the latter engaged in the activity of studying the activity of meaning conferral (with it being "understood" that the study of meaning conferral it itself a meaningful and significant activity.) Weber gives a more detailed description of the "ground" that is shared by the "object" of inquiry and the "subject" conducting the inquiry:

> The cultural problems that move humankind constantly assume new forms and colourings; within that ever-infinite stream of individual events, the boundaries of the area that acquires meaning and significance for us—which becomes a "historical individual"—therefore remain fluid. The intellectual framework within which it is considered and scientifically comprehended shifts over time; thus, the points of departure of cultural sciences remain subject to change in the limitless future…(Weber 2012d, 121)

For Weber, implications of taking the historical dimension of culture seriously are far reaching for the health and growth of scientific inquiry: "[S]ome sciences are fated to remain eternally youthful, namely all historical disciplines; all those that are constantly confronted with new questions by the ever-advancing flow of culture" (Weber 2012d, 133). While Weber's reflections on social science and sociology highlighted the relational, temporal, and individual character of reality, his reflections on the cultural sciences show that psychologically meaningful, historically contingent, and socially significant factors are always present in all human knowledge claims—including those of science.

The role of these "subjective" elements in scientific knowledge claims can be shown by looking at the similarities and differences between the hard and soft sciences. Just as there is no principle difference between the hard and soft sciences in terms of the defining characteristics of reality, there is no principle difference in terms of the means to pursue knowledge—since both are pursuing "empirical" knowledge. But this similarity has been obscured by the "enthusiastic spirit of naturalistic monism" which breathed new life to,

> the hope—inherited from Antiquity and later developed further that it would be possible by means of generalizing abstraction and of an analysis of empirical [reality] aimed at uncovering law-like regularities, to arrive at a monistic knowledge of total reality… (Weber 2012d, 122)

More specifically, this knowledge would be expressed.

> [I]n the shape of a conceptual system with metaphysical validity and in mathematical form—knowledge which would be completely "objective" (which, in this connection, would mean: detached from all values) and at the same time completely "rational" (that is to say: free from all individual "contingent elements"). (Weber 2012d, 122)

A closer look at the type of knowledge that all sciences of empirical reality aspire for shows that this type of "objectivity" and "rationality" is not logically possible. To begin with, before there can be anything like "objective" knowledge, there must be an affirmation of the claim that such knowledge is valuable and worth having—this precondition must be fulfilled to make the pursuit of "objective" knowledge possible:

> The objective validity of all empirical knowledge rests, and rests exclusively, upon the fact that the given reality is ordered according to categories that are in a specific sense subjective, in that they form the precondition of our knowledge, and that are based on the presupposition of the value of that truth which empirical knowledge alone is able to give us…(Weber 2012d, 137)

This basic definition shows empirical knowledge to be composed of interrelations between "objective" and "subjective" elements—the objectivity being the grounds for its "validity" and the subjectivity being the "preconditions" for attaining it.

Empirical knowledge is attained through "categories" or "concepts and judgements that are neither empirical reality, nor reproductions of empirical reality, but that allow empirical reality to be *ordered intellectually* in a valid manner" (Weber 2012d, 137). In this regard the difference between the hard and soft sciences "does *not* concern differences of causality, or differences in the significance and character of the concept formation" (Weber 2012b, 81)—the difference is based only on the perspective of the inquirer. The hard sciences employ "concepts and judgements" to study the unique features that are shared among a selected group of individuals—with the goal of "discovering" the "law" under which these features can be grouped/subsumed. The soft sciences employ "concepts and judgements" to study that constellation of features which make an individual unique—especially the needs, desires, and purposes of individuals that cause them to behave in a particular way when they could have done otherwise. Despite their differing ends, the means used by the hard and soft sciences is the same—abstract concepts (with the former employing generic concepts and the latter ideal type concepts). The fact that all scientific knowledge is attained by means of concepts that are subject to critique and revision means that all scientific is probabilistic.

Given the fact that scientific inquiry is a unique type of cultural activity—that of conferring meaning, from some perspective, one some activity, for some reason, at some point in history—the concluding paragraph of *OSSP* is equally applicable to the knowledge produced by soft and hard sciences:

> [A]t some point, the colouring changes; the significance of points of view that have been applies unreflectingly grows uncertain, the way forward fades away in the twilight. The light shed by the great cultural problems has moved on. Then science, too, prepares to find a new standpoint and a new conceptual apparatus, and to contemplate the stream of events from the summits of thought. It follows those stars that alone can give meaning and direction to its work…(Weber 2012d, 138)

The fact that scientific is ready to "find a new standpoint and a new conceptual apparatus" so that it is free to follow "those stars that alone can give meaning and direction to its work" means that the inner urge of scientific knowledge is to grow, evolve, and expand.

The casual reader would have noted that the philosophical lessons derived from Dewey's analysis of the discoveries of the hard sciences are practically the same as those derived from Weber's reflections on the methodology of the soft sciences: relationality, temporality, and individuality are defining characteristics of reality. The careful reader would have sensed that Weber's insights go much further than those of Dewey:

- Einstein's special relativity shows the relationship between two individuals (time and space) to be a "fundamental" characteristic of reality. But his theory erases the uniqueness of time and reduces it to a dimension in space. Weber's sociological inquiry focuses on the interrelation of two (or more) individuals— and shows that individuals and relations are (simultaneously) the products and producers of history.
- Einstein's general relativity shows that knowledge claims are impacted by the position of the observer but his theory is formulated from the perspective of a third "neutral" or "objective" position. Given the fact that all scientific inquiry is a cultural activity, Weber's description of the transcendental presupposition of the cultural sciences shows that such "neutrality" is not possible—and his own scientific research is an attempt to determine the validity and significance of not just his activity but of the culture and cultural values that made this activity possible:

  > Any heir of modern European culture will, unavoidably and justifiably, address universal-historical themes with a particular question in mind: What combination of circumstances led in the West, and only in the West, to the appearance of a variety of cultural phenomena that stand—at least as we like to imagine—in a historical line of development with *universal* significance and empirical validity? (Weber 2002, 149)

- Quantum mechanics affirmed the reality of time by showing "the individual is a temporal career whose future cannot be *logically* deduced from its past" and that all scientific knowledge claims are probabilistic. Weber gives a "thicker" description of the interrelation of time and the individual by showing the impact that the activity of meaning conferral has on not just determining the future of an individual but also its impact on understanding (and shaping) its past. By stating that the transcendental presupposition of cultural science is that human beings to be "endowed with the capacity and the will to adopt a deliberate *position* with respect to the world, and to bestow *meaning* upon it," Weber puts every human being in the position of Schrödinger's observer—someone who has to capacity to not only have a say in the trajectory of his own future but also a say in the character and future of the world around him.

- The "law of gasses" shows "law" to be a statistical description of the probable behavior of a system—which says nothing about the behavior of the individual components of the system. Weber's conception of social science (more specifically sociology) makes it possible to attain probable knowledge of not just the behavior of different individuals interacting in the course of social interaction but also of the reason why they are behaving so.[1] Weber does not bemoan the fact that "laws" can only take us so far in terms of knowing empirical reality. For Weber this awareness only heightens his consciousness of the possibilities that this opens up:

> The fate of a cultural epoch that has eaten from the tree of knowledge is that it must realize that we cannot read off the *meaning* of events in this world from the results—however complete they may be—of our scrutiny of those events, but that we ourselves must be able to create that meaning. (Weber 2012d, 104)

In the first section of this chapter, we saw that Dewey challenges the classic and early modern conceptions of reality and knowledge by drawing philosophical lessons from the findings of modern physics and chemistry. He concludes by bringing Nature and Culture into relationship by showing that "the principle of a developing career applies to all things in nature, as well as to human beings." In the second section, we see Weber going much further than Dewey in challenging the "ineradicable monist trend" (Weber 2012d, 112) which was "inherited from Antiquity and later developed further" (Weber 2012d, 122) based on his reflections on the methodology

[1] The work of Ilya Prigogine and Dilip Kondepudi in far from equilibrium thermodynamics shows "inert, dead" matter to have characteristics almost always associated with human beings. Since the field is dominated by equilibrium thermodynamics, we remain in the dark about "some essential features of the behavior of matter and energy" (Kondepudi and Prigogine, 2015, x). But when matter and energy are studied "in far from equilibrium situations we begin to observe new properties of matter that are hidden at equilibrium" (Kondepudi and Prigogine, xviii). Among these "new properties" are; (a) matter as a dynamic system interacting with its environment (i.e. dissipating structure), (b) the system "may in general 'choose' between various states" at bifurcation points (Kondepudi and Prigogine, xviii), and (c) irreversible processes which lead "to new space-time structures. They therefore play a basic and constructive role" (Kondepudi and Prigogine, xvii). In simpler terms, "inert, dead" matter has displayed the characteristics of "responding" to its environment, "choosing" between states, and "evolving" with the passage of time. These discoveries in one of the hardest of the hard sciences provide experimental evidence that supports the conceptions of "reality" and "knowledge" which are implicit in Weber's description of the methodology of the social and cultural sciences.

(and practice of) the social and cultural sciences. A cursory reading of these reflections shows that Weber begins with conferring ontic dignity on concrete relations, history, and the unique individual and goes on to show that empirical reality (as conceived by the soft sciences) is composed of psychological, historical, and social elements. A closer reading of his methodological reflections reveals that meaning conferral, value affirmation, and a purposive attitude are indispensable elements in all cultural activity—most importantly the activity of scientific inquiry. Consequently, for Weber while scientific inquiry starts with the conscious study of "objective" facts in Nature, its ultimate goal is to arrive at a self-conscious understanding of the "subjective" Cultural values that generate and sustain this inquiry. By showing scientific study of Nature to be a unique type of cultural activity and by offering a methodology that makes scientific investigation of cultural values possible, Weber goes far beyond just demonstrating that Nature and Culture share important characteristics (as noted by Dewey)—thus making his reflections on the methodology of the social sciences a watershed event in the history of the philosophy of science.

References

Dewey, J. 1920. *Reconstruction in Philosophy*. New York: Henry Holt and Company.

Dewey, J. 1929. *The Quest for Certainty: A Study of the Relation of Knowledge and Action*. New York: Minton, Blach & Company.

Dewey, J. 1940. Time and Individuality. In *The Essential Dewey: Pragmatism, Education, Democracy*, ed. Larry A. Hickman and Thomas M. Alexander, 217–225. Bloomington and Indianapolis, IN: Indiana University Press.

Kondepudi, D., and Ilya Prigogine. 2015. *Modern Thermodynamics: From Heat Engines to Dissipative Structures*. London: Wiley.

Weber, M. 1968. *Economy and Society*, ed. Guenther Ross and Claus Wittich. New York: Bedminster.

Weber, M. 2002. *The Protestant Ethic and the Spirit of Capitalism*. Trans. Stephen Kalberg. Los Angeles, CA: Roxbury Publishing Company.

Weber, M. 2012a. "On Some Categories of Interpretive Sociology." *Max Weber: Collected Methodological Writings*. Eds. H.H. Brun and S. Whimster. Trans. H.H. Bruun. New York, NY: Routledge. 273-301.

Weber, M. 2012b. Roscher and Knies and the Logical Problem of Historical Economics *Max Weber: Collected Methodological Writings*, ed. H.H. Brun and S. Whimster. Translated by H.H. Bruun, 3–93. New York, NY: Routledge.

Weber, M. 2012c. Critical Studies in the Logic of the Cultural Sciences *Max Weber: Collected Methodological Writings*, ed. H.H. Brun and S. Whimster. Translated by H.H. Bruun, 139–184. New York, NY: Routledge.

Weber, M. 2012d. The 'Objectivity' of Knowledge in Social Science and Social Policy. *Max Weber: Collected Methodological Writings*. Eds. H.H. Brun and S. Whimster. Translated by H.H. Bruun, 100–138. New York, NY: Routledge.

Chapter 7
The Critical Philosophy of Social Science of Left-Wing Logical Empiricism

Thomas Uebel

Abstract The understanding of logical empiricist philosophy of social science is bedeviled by misrepresentations which continue to dominate how it is characterized by its opponents across the philosophical and political spectrum. This chapter sets out how typical misunderstandings can be avoided and how the critical dimension of the logical empiricist philosophy of social science is readily recovered. Focussing on its main exponent Otto Neurath, the discussion surveys and responds to the main objections which pertain to how science itself is conceived of by him and what it legitimately can pronounce upon. It is shown that Neurath's conception of physicalism and his encyclopedic version of the unity of science are non-reductive and that his understanding of value-neutrality only bars unconditional value judgments and allows research in the service of a transformative agenda.

Keywords Otto Neurath · Physicalism · Unity of science · Value-neutrality · Critical agenda

7.1 Introduction

Logical empiricist philosophy of social science is not widely discussed these days.[1] To be sure, while important work was done on the periphery of the Vienna Circle, philosophy of social science was not as central a concern for logical empiricism

[1] Exceptions to be noted are Nemeth (2007), occasional pieces by the present author built upon and extended here (see various footnotes below), and the wide-ranging volume Tuboly (2022).

T. Uebel (✉)
University of Manchester, Manchester, UK
e-mail: thomas.uebel@manchester.ac.uk

Y. Shan (ed.), *History and Philosophy of the Social Sciences*,
https://doi.org/10.1007/978-3-032-08167-4_7

generally (certainly not in the inter-war years to which I must limit myself here) as were philosophy of logic and mathematics or philosophy of physics.[2] That said, social science *was* central to Otto Neurath's understanding of the "scientific world-conception" that was propagated by the Vienna Circle (one of two points of origin of logical empiricism).[3] What this understanding meant for social science, however, is less than readily apparent and getting the measure of this philosophy of social science is not without difficulties. Many of them trace back to the exigencies of the dialectical situation in which Neurath and the Vienna Circle found themselves. On the one hand, discussions of their philosophy with opponents were anything but collegial, for their challenge to the tradition was recognized by all to be more than academic: the "metaphysics" they opposed served as obfuscating ideologies of an ever more reactionary political order.[4] So some of (not only) Neurath's pronouncements exhibit an anti-idealist militancy whose lack of subtlety is readily exploited.[5] On the other hand, the Circle's general philosophy of science also underwent extremely rapid and sometimes divergent development from its inception through the later 1920s and 1930s. In consequence, some early statements of Neurath's philosophy of social science share the shortcomings of the then contemporary state of the art: to get the proper measure of his conception the trajectory of its development must be borne in mind.

What makes such hermeneutic efforts worthwhile is that Neurath's program for social science presents a highly original combination of elements that normally are thought to stand in irresolvable tension. Notably, Neurath affirmed both of the flagship doctrines that characterize his logical empiricism, the unity of science and physicalism, *and* the irreducibility and relative autonomy of the special sciences, doctrines dear to contemporary philosophers of social science that seem endangered by the former. In addition, Neurath wished to pursue critical social science, the apparent bars set against it by the Weberian doctrine of value neutrality and his own (and the Circle's) value noncognitivism notwithstanding. Whether all this amounted to a budget of paradoxes or whether these ambitions cohere is investigated here.

---

[2] For a brief discussion of works by Felix Kaufmann, Karl Menger and Edgar Zilsel, with further references, see Uebel (2007). In the 1940s the *International Encyclopedia of Unified Science* tried to broaden the canvas a bit. The 1950s saw Philipp Frank's explorations of the relevance of the sociology of science (see his 1956 but also Reisch and Tuboly 2025) and important work by Ernest Nagel and C.G. Hempel (briefly discussed in Uebel 2003).

[3] See the collectively produced "manifesto" Verein Ernst Mach (1929); for its production history and early reception, see Uebel (2008). For the critical interest that some fellow members and associates of the Circle (Karl Menger, Felix Kaufmann, Edgar Zilsel) showed in Neurath's physicalist social science, see Uebel (2007).

[4] The reaction of Vienna's right-wing press to the murder of the Circle's nominal leader Moritz Schlick in 1936—that he got what he deserved—is emblematic; see the documentation in Stadler (1997/2015, Ch.13).

[5] One striking example: "The problem of 'matter' or 'mind' is solved by the disappearance of the theory of the 'mind', leaving nothing but the theory of 'matter': only physics remains. *All genuine science can only be physics*." (1931a/1973, 360) As argued below, Neurath's actual view is badly misrepresented by this formulation: careful reading reveals far more subtle and insightful ones.

7.2 Philosophy in Science and Science in Philosophy

Indispensable for understanding Neurath's philosophy of social science is a grasp of the overall philosophical program pursued by him and some of his colleagues in the Vienna Circle, their revisionist metaphilosophy. For all Circle members, the advances of what they still called "modern science" called into question the kinds of philosophy that had legitimated, by appeal to supposedly higher insight, the long-held certainties now revealed as false. Philosophy's position vis-à-vis the sciences had to be radically reconsidered: philosophy itself had to become scientific—just how remained under discussion.

For Neurath—alongside Rudolf Carnap and Philipp Frank with whom he formed the Circle's "left wing"—philosophy was to join science as its metatheory. Science affirmed its independence from traditional philosophy by recognizing that reflection about scientific procedures and principles belongs to science itself—as its metatheory. Importantly then, neither did science remain merely positive, nor was all philosophy discarded. Philosophy was retained as second-order reflection—for the price that its former claim to a separate source of knowledge was rejected while all the tools of the formal and empirical sciences remained available for it.

Philosophy became either what Carnap called the "logic of science" or what Neurath misleadingly called its "behavioristics" (more on this below) and Frank more appropriately renamed as "pragmatics of science".[6] Both versions of scientific metatheory were meant to complement each other: the logic of science pursues *a priori* formal inquiries into the structure of scientific theories and their relations to possible evidence, whereas the pragmatics of science pursues empirical inquiries into the history, sociology and psychology of doing science, both proceeding with the intent of improving the understanding of science and its performance. This practical orientation and the strategy of trying to effect either improvement "from within" indicates the basic outlook of philosophical naturalism that was shared with American pragmatists.[7] No ultimate justifications were sought; to allude to Neurath's famous simile, the point was to keep science afloat in choppy waters.[8]

To avoid possible confusion due to Quine having made naturalism the motto of the fight against "empiricist dogma", note that the wholesale rejection of the distinctions between a priori and a posteriori reasoning and analytic and synthetic truths was made a mark of naturalism only with Quine. Both distinctions were retained by Carnap and Neurath without damage to their naturalism because they stripped them of their pretension of old. It was no longer apodictic truths of reason that a priori thought discovered and analytic statements only told of logical and conceptual

[6] See Carnap (1934/1937, §§72–73) and (1942, §39), Neurath (1936a/1983, 160, 168) and Frank (1957, 360).

[7] This affinity did not go unnoticed; see Uebel (2015).

[8] "We are like sailors who have rebuild their ship on the open sea, without ever being able to dismantle it in dry-dock and reconstruct it from the best components." Neurath (1932b/1983, 92). For the history and significance of Neurath's simile, first used in 1913, see Uebel (1996).

truths that came with and followed from the adoption of one or another language with which to reason. (Kant's synthetic a priori was rejected as a scientifically refuted over-reach of philosophical ambition.)

So far, so nearly contemporary; other aspects of Neurath's thought appear more rebarbative nowadays. In the spirit of Mach's dictum that "where neither confirmation nor refutation is possible, science is not concerned" (1883/1960, 587), the Vienna Circle adopted a criterion of meaningfulness by which assertions were rejected as "metaphysics" and opposed not as false but "cognitively meaningless" when they were incapable of being evaluated for their truth or probability, either by methods of formal proof or by empirical tests. (Importantly, no general theory of meaning was offered and the existence of associative and emotive meanings—and their relevance for retaining metaphysical beliefs—was not denied.) Except by the Wittgensteinian wing, it was held that for empirical statements to be cognitively significant only the possibility of fallible testing was required, neither its actuality nor its conclusiveness. As is well-known, determining a formally precise formulation of the criterion of empirical significance proved difficult, but efforts are ongoing; interestingly enough, Frank and Neurath appeared to favor a contextually informed and exemplar-based understanding of it.[9]

Adopting the criterion of meaningfulness was part of raising philosophy to scientific standards (for it applied to philosophical discourse too). This is well-illustrated by how the charge of self-refutation was undercut. Consistent with not offering a general theory of meaning, the criterion was formulated by Carnap not as a descriptive assertion but as a proposal for an empiricist language *L* that is focused entirely on cognitive meaning: "Every synthetic statement must be confirmable." (1936–37, 34) Once accepted, the proposal turns into an analytic statement of the metalanguage *ML*: Every statement of *L* is confirmable. Arguably, adopting the criterion makes good sense for the language of science—and philosophy.

7.3 The Challenges to Logical Empiricist Philosophy of Social Science

So much for the philosophical background of Neurath's conception of social science. The controversial doctrines that concern us pertain to how science itself is conceived of (the unity of science and physicalism) and what it legitimately can pronounce upon (value neutrality).

The unity of science doctrine holds that all the sciences are connected or connectable with each other and that there is no radical divide that separates *Naturwissenschaft*, natural science, from *Geisteswissenschaft*, the human or social sciences, as was customarily held in 19th and much of twentieth century German philosophy. The great diversity among the sciences was recognized, of course, but

[9] For more on Vienna Circle verificationism see Uebel (2019a).

it was denied that behind it lurked an unbridgeable difference of an epistemological sort, let alone an ontological one. All scientific knowledge claims faced the test by intersubjective evidence and for this any difference in the mode of putative discovery was utterly irrelevant. It follows that the sciences so conceived share a common language, namely that in which any results were to be formulated in order to be tested.

Disregarding some niceties of formulation, we can say that physicalism holds that this basic language ensuring the unity of science speaks of physical objects, properties and processes. Needless to say, different versions of both the unity of science and physicalism were held in the Vienna Circle (at various times even on its left wing), and it will be important to distinguish them to see whether a very general charge holds up against Neurath's. That is the charge that the doctrines of the unity of science and physicalism entail a form of reductionism that makes social science impossible for it denies its disciplines the conceptual autonomy from physics that they need to be practiced at all.[10]

Another difficulty that faces contemporary observers is presented by the doctrine of scientific value neutrality. Famously introduced by Max Weber in a foundational dispute in early German social science, it finds no explicit endorsement in the Vienna Circle's writings, but something much like it follows from their doctrine of value noncognitivism (unconditional value statements are cognitively meaningless). Both Weber and the Circle held that the office of science concerns the investigation of facts, of what is, but not the determination of values, of what should be. From this, critics have concluded that any social science pursued accordingly must be of an inherently conservative, if not politically conformist orientation and cannot offer criticism of established states of socio-political affairs.[11]

That the Circle's quasi-Weberian stricture offers no bar to critical social science discourse (for it allows differentiation that is typically neglected) will be argued below. First I must turn to the doctrines of physicalism and unity of science to investigate whether they have the deleterious consequences often alleged.

7.4 Physicalism

Three different meanings of the term "physicalism" can be discerned in Neurath's writings. Physicalism as a "comprehensive attitude" or "total conception" amounts to his metaphilosophical position of anti-foundationalist epistemological

[10] See, e.g., Horkheimer (1937), Hayek (1941–44/1979, 78–79), Adorno (1969/1976, passim), Habermas (1970/1990), Fay (1975, 13), Keat and Urry (1975/1982, 25–26), Dallmayr and McCarthy (1977, 8). More recently, such complaints are either repeated, like Hayek's in Caldwell (2003, 428), or catchily rephrased, like as "system closure" in Steinmetz (2005, 34).

[11] Evidently, this charge is raised by the left-leaning authors listed in the previous note.

naturalism: I touched on it briefly above.[12] Sometimes Neurath means by "physicalism" a doctrine about the relation of the theories of the different scientific disciplines to each other and physics in particular: this will be discussed under the more appropriate heading of "unity of science" in the next section. Then there's the physicalism to be discussed here: a metalinguistic doctrine which concerns the language of the sciences.

Before turning to Neurath's version, we must note the difference between him and Carnap on this matter. For Carnap early on, "physicalism" meant that the languages of all scientific disciplines can be translated into the language of physics (1932a, 1932b).[13] With all the terms of the special sciences held to be definable in physical terms and their statements translatable into physical statements, it followed that the laws of the special sciences were translatable into physical terms as well. (Whether they were or were derivable from physical laws was yet another question.) By the mid-1930s, due to the indefinability of dispositional predicates in terms of observational ones, Carnap changed to holding that all scientific terms stood in a relation of now non-eliminative "reducibility" to the elementary terms of the basic language of science, newly conceived as a "thing language" which spoke of intersubjectively observable objects, properties and processes (1936–37). This liberalization complicated the logico-linguistic relations between the terms and statements of different sciences, but they remained surveyable: "the class of observable thing-predicates is a sufficient reduction basis for the whole of the language of science" (1938, 60). Soon fully irreducible "theoretical terms" were recognized to feature in theories which faced observational evidence only in their entirety (1939, 1956), but Carnap did not discuss the unity of science further.[14]

Two points are important to note here. First, that Carnap never intended physicalism to make an ontological claim: this distinguishes Vienna Circle physicalism sharply from the kind of physicalism that was debated in analytical philosophy of mind from the 1960s onwards.[15] The second point is that what ultimately mattered for Carnap was not the eliminability by definition or even only the non-eliminable reduction of certain types of terms, but that their use was responsive in principle, and detectable as such due to logical relations obtaining, to intersubjectively discernable evidence.[16]

---

[12] Neurath's term at (1936a/1983, 159). For discussion of Neurath's epistemological naturalism, first diagnosed by Hempel (1982) and Haller (1982), then Koppelberg (1987), see Uebel (1991) and now also Bentley (2023).

[13] "The physicalistic language contains not only physical terms (in the narrow sense) but also all the various special terminologies (of biology, psychology, sociology, etc.) understood as reduced by definitions to their basis in physical determinations." Carnap (1932a/1934, 95–96)

[14] That Carnap in effect rested with the broad pragmatism of his conclusion that "the unity of the language of science is the basis for the practical application of theoretical knowledge" (1938, 62) for the rest of his career is argued in Crewe and Richardson (2024).

[15] The shift to contemporary usage may have begun with Feigl's mid-1950s concern with early forms of mind-body identity theory; see his (1958) and (1963).

[16] Carnap never was embarrassed to call earlier formulations "not entirely adequate" (1936–37, 464) or a previous theses "formulated too narrowly" (1963, 56). That way progress lay.

Neurath was in full agreement that fallibilist epistemology and not ontology was their overriding concern, but his metalinguistic physicalism nevertheless was centered differently. For him from the start, the basic physical language was that language in which ordinary middle-sized objects and their behavior were described and also results of experiments were reported. This was the everyday language, albeit "physicalistically cleansed" of metaphysical phrases but inevitably retaining a certain indefiniteness—what he also called the "universal jargon" or "universal slang" (e.g., 1932b/1983, 92, 1933/1987, 9). Importantly, Neurath also allowed into this basic language of science, unlike Carnap, psychological predicates that were intersubjectively confirmable but only subjectively observable, like the perception terms that figure in his preferred form of so-called protocol sentences recording scientific evidence.[17] But he also admitted statements like "Otto felt so strange yesterday" once it was granted that they possessed an as yet unspecifiable translation in terms of "'brain changes' or 'total bodily states'" (1933/1987, 8). Neurath's conception of the basic physicalist language, this suggests, was geared from the start to the needs of empirical social science. As long as intersubjective testability was assured in principle, he saw no reason to bar first-person reports of psychological states from having an evidential function. He was not limited, even less committed to exclusively quantitative methods of investigation.[18]

For Neurath, physicalism expressed the condition of empiricism. Physicalistic statements are statements about "spatio-temporal structures" (1931a/1981, 425). The motivation was epistemological. "Physicalism … only makes pronouncements about what can be related back to observation statements *in some way or other*." (Ibid., emphasis added) Only those statements are admissible that can be tested—or, as he put it, "controlled"—by direct or indirect reference to intersubjectively discernable states of affairs. It follows that social scientific theories also must allow for derivations that can be formulated in the everyday language speaking of spatio-temporal structures and can be tested.

> Physicalism encompasses psychology as much as history and economics; for it there are only gestures, words, behaviour, but no 'motives', no 'ego', no 'personality' beyond what can be formulated spatio-temporally. It is a separate task to ascertain what part of the traditional material can be expressed in the new strict language. Physicalism does not hold the thesis that 'mind' is a product of 'matter', but that everything we can sensibly speak about is spatio-temporally ordered. (1931a/1973, 325)

Note that even though he spoke less carefully than Carnap, Neurath also sought to avoid ontological claims. When Neurath claimed that "physicalism" represented the "heir" and "logically consistent development of materialism" (1932c/1981, 568; cf.

[17] "A complete protocol statement might be worded like this: 'Otto's protocol at 3:17 o'clock: [Otto's speech-thinking at 3:16 o'clock was: (at 3:15 there was a table in the room perceived by Otto)].'' (1932b/1983, 93. note the dark anticipation at 1932a/1983, 66–67, written in 1931). For interpretation, of Neurath's proposal, see Uebel (2009) and Bentley (2022).

[18] For demonstration that Neurath's rejection of the methodology of *Verstehen* pertained only to its metaphysically misunderstood version and of his awareness of the hermeneutical dimension of interpersonal understanding and historical research, see Uebel (2019b).

1933/1987, 1), he similarly meant a materialism purified by the linguistic turn. His forceful rejection of talk of *Volksgeist* etc. is not to be read as an ontological pronouncement, but as a characterization of admissable languages by their domain. Expressions that speak of supernatural forces or supra-individual agents are not be admitted into unified science on account of their making assertions untestable.[19]

The differences from Carnap's physicalism underscore that Neurath's was also an anti-eliminativist, even though he did not from start oppose a reductive hierarchy of concepts with physics at its base.[20] Unfortunately, however, Neurath's chosen terminology is all too easily misunderstood. Like Carnap (and before that Russell in *Analysis of Mind* [1921]), Neurath helped himself to the language of behaviorism (he even coined the term "social behaviorism") to counteract dualist mentalism without intending thereby to endorse the eliminative claims that we nowadays associate with such talk. Disclaimers like the following were all too easily overlooked: "From the standpoint advocated here it does not matter whether certain individual tenets of Watson, Pavlov or others are maintained or not. What matters is that only physicalistically formulated correlations are used in the description of living things, whatever is observed in these beings." (1932a/1983, 73) At bottom, behaviorism for Neurath meant that in the description and explanation of human activities we limit ourselves to the attribution of states that could in principle be "controlled" by evidence (concerning behavior or neurophysiological events).[21] This allowed talk of many of the "intervening variables" which for the psychologists mentioned had become illegitimate.[22]

By the mid-30s Neurath called Watson's form of behaviorism "too narrow", unable to "withstand physicalist criticism", and preferred the term "behaviouristics" for its less restrictive associations (1936a/1983, 164; cf. 1944, 17, 51). Now he stated:

---

[19] Hempel rightly noted that "Neurath put mentalistic terms like 'mind' and 'motive' on the Index on the grounds that they tended to be construed as standing for immaterial agencies." (1969, 169).

[20] Three publications from 1931 testify to Neurath's uncertainty about Carnap's early physicalist reductionism. In one he seems to have endorsed it: "This unified language of unified science, which by and large can be derived from everyday language by certain alterations, is the language of physics." (1932a/1983, 62; this was written in 1931) In another, he endorsed as still applicable the "pyramidal" structure of the genealogy of concepts of the *Aufbau* (1931a/1973, 390), though in yet another he envisaged its revision so as to accord with physicalism (1931b/1983, 56; this is echoed at 1933/1987, 274). Later on, however, Neurath's rejection of "pyramidism" was part of his encyclopedic understanding of the unity science (1938/1983, 203).

[21] This contradicts Hempel's earlier reading that "[i]n Neurath's science of behavioristics, statements about phenomena of consciousness and about mental processes would be replaced by statements about spatio-temporally localizable occurrences such as macroscopic behaviour (including gestures and speech acts) and about physiologically and physiochemically described processes in the brain and the central nervous system" (1969, 170–1), Neurath did not require such replacements.

[22] Already in the early 1930s Neurath headed a working group dedicated to the "physicalization" of Freud's texts; for Neurath on psychoanalysis see Neurath (1932a/1983, 80), (1939/1983, 210), Frenkel-Brunswick (1954) and Carnap (1963, 58).

> One can use this term "conscious" always if, for example, one makes the hypothesis that the scholar, if asked, would have said: "I have made the following observation …" We then say: the scholar has *formulated*: "I have made the following observation." That is, this statement is "coordinated" to him to characterize the state of consciousness. … Besides the statements that we pronounce we thus get in addition the statements that we take to be coordinated to certain "states of consciousness". Explicit control of a prediction thus would amount to a confrontation of one statement with another statement, whether this be one that the subject actually made or one coordinated to him as suggested. We can logically compare all statements with each other, whether they are pronounced or only coordinated. (1936a/1983, pp. 162–163, orig. emphasis, trans. altered)

> While avoiding metaphysical trappings physicalism can in principle predict future human action to some degree from what people 'plan' or 'intend' ('say to themselves'). (1936b/1981, 714)

In later years, Neurath made his anti-eliminationist intention increasingly explicit and he denied "exclud[ing] what is called 'inner experience'" (1939/1983, 209). He again noted that "[s]tatements of the type 'this entrance hall of a building thrills me' can be regarded as physicalist ones because they are observation statements" (1941/1983, 221) and pointed out that "[h]istorians of human social life are highly interested in descriptive terms such as deal with the feeling-tone of persons, their devotion, their fear and hopes" (1944, 15). It was in this inclusive sense that Neurath already early on expounded his social behaviorism that "ultimately comprehends all sociology, political economy, history etc." (1932c/1981, 565).

In light of this trajectory it is justified to interpret Neurath's physicalism, despite its sometimes unsophisticated formulation, as sharing the resistance of late twentieth century non-reductive physicalism to the eliminative designs that motivate radical behaviorism and certain forms of materialism.[23] When Neurath ill-advisedly joked that terms like "mind" or "motive" were on his "index verborum prohibitorum", he meant their employment in the service of dualism: as we see clearly now, use of the intentional idiom in social scientific matters was by no means ruled out, only its metaphysical misuse. If it were objected that Neurath lacked what nowadays undergirds versions of non-reductive physicalism (an argument for supervenience), we must remember that his metalinguistic version abstained from ontology and owed no answer on how the objects of science were materially constituted. This does not mean, however, that he lacked any argument against conceptual reducibility that entailed eliminability. Since his student days he was familiar with the functional definitions of social kinds that characterize Simmel's sociology.[24] Neurath

[23] Even Neurath's initial sympathy for Carnap's early physicalist reductionism does not tell against this once it is taken into account that he shared Carnap's dismissal of ontological concerns. Without such concerns, no elimination threatens due to the claimed translatability of special science terms and laws into those of physics; if anything, it would help to legitimate special science pronouncements.

[24] Attendance of Simmel's seminars is noted in a letter from Neurath to Ferdinand Tönnies, an early mentor his, of 31 October 1903, Tönnies Nachlass, Schleswig-Holsteinische Landesbibliothek, Kiel. Thus already early in the physicalist phase Neurath appealed to a consonant argument: see the quotation from (1931a/1973, 333) in the next section.

was aware of the functional and structural analyses and explanations that were being explored at the time in Durkheimian sociology and anthropology and that had long been a mainstay of Marxist analysis.[25]

To be sure, Neurath did not provide an explicit argument spelling out the phenomenon of variable realization (and its widely presumed anti-reductionist consequences), but he did write: "One can understand the working of a steam engine quite well on the whole without surveying it in detail. And indeed, the structure of a machine may be more important than the material of which it consists." (1931a/1973, 333) Neurath's concern was epistemological and he concluded that social scientific explanations were independent of precisely what social kinds were made of as long as their spatio-temporal nature was assured. It was due these considerations that, still before Carnap conceded the unworkability of physicalist reductionism, Neurath granted to the special sciences all the *conceptual autonomy* that was compatible with the empiricist mission that animated his physicalism.

7.5 Encyclopedism

That Neurath granted such conceptual autonomy to the social sciences becomes still clearer when his concept of the unity of science is considered. Turning to it we move from consideration of the terminology of the language of unified science to that of the structure of unified science and the nature of the statements that make it up.

Here we must first take another step back and distinguish the Vienna Circle's understanding of the unity of science from that common in analytic philosophy since the 1950s, before focusing again on how Neurath's must be distinguished from Carnap's. Paul Oppenheim and Hilary Putnam's "Unity of Science as a Working Hypothesis" developed a conception that became standard enough to supply the contrary subtitle for a paper by Jerry Fodor which two decades later (1974) argued the nowadays widely accepted anti-reductionist case (with reference to variable realization), but left unchallenged any presumption (perhaps shared it) that the unity position it opposed was of much longer standing in "positivist" circles.

Oppenheim and Putnam argued for a conception of the unity of science as a hierarchy of reductively related disciplines with six successively more basic levels of scientific theory: theories of social groups, of (multicellular) living beings, of cells, of molecules, of atoms, of elementary particles. Notably this conception demands not only the unity of the language but also the unity of laws such that "the laws of science become reduced to the laws of some one discipline" (1958, 4). Oppenheim and Putnam noted that this requires so-called micro-reductions: the objects of the theory to be reduced are to be regarded as decomposable so that their behavior becomes explainable in terms of their components at the next lower level.

[25] How precisely the legitimate forms of such explanations would go was another matter, of course. Food for thought is given by Neurath's elaborations to a Viennese discussion group including Carnap in 1930–31; see his (1963, 58).

With micro-reductions being transitive, irreflexive and asymmetric, it becomes possible "in principle to dispense" with the laws of the special sciences and "explain the relevant observations" by the laws of the fundamental science alone (1958, 9). One of their prize exhibits was methodological individualism in economics according to which regularities of group behavior in the market are derived as aggregative consequences of the workings of the laws of individual psychology (1958, 17).

As we saw, Carnap held the laws of the special sciences to be translatable into physical terms but always distinguished from this the question of the derivability of the laws of the special sciences from the laws of physics. That, he held, was an empirical question, not to be decided on logical grounds. While holding that no argument ruled out its possibility, he concluded: "there is at present no unity of laws" (1938, 61). Just such a unity, however, Oppenheim and Putnam sought to establish as a working hypothesis. So Carnap's and their ideas about the unity of science differ deeply for their notion of reduction is understood differently. Oppenheim and Putnam explicitly distanced themselves from the "epistemological uses of the terms 'reduction', 'physicalism', and 'Unity of Science'" (1958, 5). (They did not mention Carnap or Neurath by name, but they are clearly the target.) Instead, Oppenheim and Putnam based their understanding of reduction on the concept of theory reduction developed by Kemeny and Oppenheim (1956) to which they added as their innovation the notion of micro-reduction.[26] By contrast, while the later Carnap may well have considered conceptions of theory reduction favorably, he never entertained the ontological question of micro-reduction and saw no reason to abandon his approach to the unity of science as concerned with its language alone: the derivability of laws did not follow from it on purely logical grounds.

Now consider Neurath's conception of the unity of science. That he would have shared Carnap's disdain for Oppenheim-Putnam's micro-reductions and ontologizing is clear. What to him was of overriding importance from the beginning was the connectability of the laws or generalizations of the different sciences. "Unified science contains all scientific laws; these can be connected without exception." (1932a/1983, 62) Such connectability had been unproblematic given their translatability on Carnap's early reductionist model, but that was abandoned by 1935. Yet this change on Carnap's part only enforced what could be called Neurath's "encyclopedic turn", even if it did not prompt it outright.[27]

[26] Oppenheim and Putnam summarized Kemeny and Oppenheim's conception (1956, 15) as follows: "given two theories T_1 and T_2, T_2 is said to be reduced to T_1 if and only if: (1) The vocabulary of T_2 contains terms not in the vocabulary of T_1. (2) Any observational data explainable by T_2 are explainable by T_1. (3) T_1 is at least as well systematized as T_2" (1958, 5) Note that Ernest Nagel's concept of theory reduction (see 1961, Ch.11, esp. 345–366)—which differs from Kemeny and Oppenheim's by introducing bridge-laws connecting the concepts of T_1 and T_2—likewise did not feature micro-reductions. Note also that Fodor's counter (1974) did not distinguish between these conceptions of theory reduction and used Nagel's as his model.

[27] Carnap's non-eliminative reduction sentences were first introduced in his (1936), read at the Paris Congress in September 1935, whereas Neurath first announced his "encyclopedism" first at the Preperatory Conference in Prague in September 1934 (see his [1935a] quoted below).

To begin to see what Neurath's encyclopedic unity of science comes to, first consider his conception of the laws established by social science in particular.

> The development of physicalistic sociology does not mean the transfer of the laws of physics to living things and their groups, as some have thought possible. Comprehensive sociological laws can be found as well as laws for definite narrower social areas, without the need to be able to go back to the microstructure, and thereby to build up these sociological laws from physical ones. (1932a/1983, 75)

Here Neurath is best understood to reject the derivability of the laws of the special science from those of physics; again no difference from Carnap obtained.[28] Yet Neurath recognized a much greater variety of types of scientific laws, or better: systematic generalizations, and this is crucial for social science. All along Neurath accepted both deterministic and probabilistic or stochastical laws: "We are always searching for correlations between magnitudes that occur in the physicalist description of events. It makes no difference in principle whether the descriptions are statistical or non-statistical." (1932a/1983, 68) On this point, of course, he was in agreement with Carnap and other logical empiricists. But Neurath did not require that explanatory generalizations across the sciences be the same in their scope.

When he spoke of "laws for definite narrower social areas" as above, he endorsed what Robert Merton much later was to call "theories of the middle range" (1967). Thus Neurath wrote: "Let us not start from what one tends to call a 'law of nature', but those less demanding generalizations which are common in the social sciences. Results gained from a rather restricted range of examples are extended to a further set of cases that also are fairly restricted." (1937/1981, 788) This allows for non-universal regularities that are limited to certain spatio-temporal domains: "To be sure, most of the sociological regularities that help us in deriving predictions are formulated in such a way that they are valid only for relatively complex structures in certain spatio-temporal areas." (1936c/1981, 771; cf. 1932a/1983, 85) When he expressed this view in the early 1930s—e.g., sociologists "can speak of 'social laws' which are valid for distinct social formations" (1931a/1973, 371)—he added: "We cannot yet indicate precisely on what certain correlations depend: 'historical period' = non-analysed complex of conditions." (1932a/1983, 85) Given the association of laws with exceptionless strictness, Neurath preferred the generic "correlations". That this feature of the social sciences frustrated hopes for the derivation, not only the translation of "laws", is clear.

In arguing for these non-universal regularities, Neurath's familiarity with the practical exigencies of the social sciences evidently trumped the authority of Carl Menger and the school of Austrian economics in the spirit of whose strict deductivism he was rudely criticized by Karl Popper.[29] Thus it is important to note Neurath's principled reasons. Applying regularities always meant to presume as constant

[28] As here: "Another question is to what extent one can reduce the statements or laws of biology, behaviouristics or sociology to physical statements or laws." (1938, 19, trans. changed)

[29] See Popper (1944–45/1961, 103); for some discussion see Uebel (2000). Popper's criticism presupposed his strict falsificationism which Neurath had criticized as "pseudorational" in (1935b).

certain "total conditions"; as Neurath noted, even astronomical laws presume the constancy of certain boundary conditions (1936c/1981, 772). It followed that the social sciences cannot claim the same degree of universality for its generalization as physics or even biology for its boundary conditions are more likely to undergo alteration. This difference however, does not lessen the participation of the social sciences in unified science (the connectability of its generalisations with those of other disciplines). With full generality therefore Neurath stated that "our scientific practice is based on local systematizations only, not on overstraining the bow of deduction." (1946/1983, 232)

Neurath, this suggests, argued from the variety of explanatory principles and the heterogeneity of laws to the *nomological irreducibility* of social science. He did not only invoke the distinction between the contexts of discovery and justification such that only in the latter inter-theoretic reductions of laws are required (as may be suspected), for he noted: "The sociological laws found without the help of physical laws in the narrower sense must not necessarily be changed by the addition of a physical substructure discovered later." (1932a/1983, 75) Whatever Neurath's encyclopedic unity was, it was not a reductive hierarchy of laws.

Yet Neurath's nomological antireductionism also has a still more specific dimension of relevance in social science, namely the rejection of methodological individualism in one of its many senses. The strenuous opposition to metaphysical social science in his *Empirical Sociology* (1931a) clearly documents that Neurath opposed the invocation as causal agents of the supra-individual entities that populated the then rising *völkisch* ideologies: in this sense he was undoubtedly a methodological individualist. Yet concerning sociological laws Neurath also wrote: "Naturally certain correlations result that cannot be found with individuals, with stars or machines. Social behaviourism establishes laws of its own kind." (1932a/1983, 75) Here he appears to be stressing the nomological autonomy of sociology or any other social science by pointing to the irreducibility of their laws to those of psychology.[30] Yet even within the social sciences pluralism was expected: "It would be wrong to assume that all the subdisciplines of so-called Social Science are based on a universal sociological terminology." (1938c/1983, 203) Hempel once reported that Neurath "refrained from making any general claims on the realizability of the program of methodological individualism" (1969, 174). As we can see, he passed a differentiated judgement on different aspects of methodological individualism.

So much for the content that the unity of science turned on. Let's turn to its encyclopedic nature. Neurath's first proclamation of encyclopedism occurred in an address programmatically entitled "Unity of Science as a Task":

> After the removal of traditional metaphysics, in constant struggle against metaphysical leanings, positive work could be our occupation, namely the creation of an *encyclopedic* synthesis of the sciences on uniform logical foundations. We would establish the 'cross-connections' from science to science and thus create a structure that knows no 'philosophy', no 'epistemology' with special propositions—whichever one of these two is

[30] On this point there obtained (independent) agreement even with Popper (1945/1966, 90).

> applicable has found its place either in the 'logic of science' or in 'behavioristics': the program of unified science. (1935a/1983, 115, orig. emphasis)

To the question of what such an encyclopedic synthesis amounts to Neurath at this point gave a short positive and a longer negative answer. Let's begin with the latter.

> One may try to accomplish all these programs [of unified science, empiricism, physicalism TU] as systematically as possible and to carry out the logical construction as carefully as possible—we do *not* arrive at *'one'* system of science that could take the place of the 'real world' so to speak; everything remains ambiguous and in many ways uncertain. *'The' system is the great scientific lie. ... Multiplicity* and *uncertainty* are essential. From the data at our disposal we can, in more than one way, deduce predictions that are in harmony with science; the multiplicity of predicting cannot be excluded by any method; no degree of systematic procedure can alter this. One can, so to speak, not agree on a 'machine' that unambiguously produces 'inductions' in the wider sense. The progress of science consists, as it were, in constantly changing the machine and in advancing on the basis of new decisions. Still, the result in fact is far-reaching unity that can *not* be deduced logically. (1935a/1983, 116, orig. emphases)

There is perhaps little that need to be added to the vehemence with which Neurath here denounced the idealization of science that characterizes it as one determinate comprehensive system of theories individually formalized as axiomatized sets of highly abstract sentences that nevertheless have universal applicability due to complex but logically precise correspondence rules. It would not have pacified Neurath to be told that such systems are put forward as conscious idealizations:

> Not even as an anticipated goal is it a useful guiding thought as it takes us close to Laplace's spirit which, it is thought, with knowledge of all the equations of the sciences, constantly makes correct predictions: an assumption that serves no prediction, an assumption that cannot be verified in any way; metaphysical formulations, useless for the purposes of science ... (Ibid.)

Not only the "behavioristics" of science, Neurath urged, should take account of the messy reality of science for over-systematicizing idealization does not help science perform better. That his vocal opposition to "pyramidism" (1938c/1983, 203–204)—ruffled feathers among fellow logical empiricists could not be helped.

> Not a few friends of Scientific Empiricism will criticize this attitude as a lack of systematization. But the anticipation of a pyramid of non-overlapping sciences and subsciences hinders the free evolution of the sciences. We should regard the Social Sciences as a collection of a great many scientific units *which can be combined in very different ways*. That is real '*Encyclopedism*' within the Unity of Science Movement. (1939/1983, 211)

The short answer to what encyclopedic unity of science comes to positively speaking we have encountered by now. The actual unity which is being created is not one comprehensive system but consists in a multitude of specific "'cross-connections' from science to science". Note the architectural difference: unity is established by horizontal links, not a vertical hierarchy. The point is not to unify entire theories into ever greater wholes, but to be able to make connections between individual statements belonging to different theories and combine them to predict and manage courses of events. Neurath's favorite example are forest fires (e.g., 1932a/1983, 59; 1933/1987, 3; 1936e/1983, 132). Towards furthering this aim the logic of science was also to be directed (1938b/1987, 136–137). Neurath was no luddite then, but a hard-nosed pragmatist. Nancy Cartwright dubbed his version of the unity of science a "patchwork conception" and, with Jordi Cat and Hasok Chang, observed that Neurath—contrary to his misleading rhetoric—was a de facto representative of today's "disunity of science" movement.[31]

Once this is borne in mind, it is readily seen that the similes and metaphors by which Neurath conveyed why an encyclopedia should serve as the model for the unity of science are far from empty. That "the" system is replaced by the "mosaic of science" (1938b/1987, 132) characterizes how work on encyclopedic unity proceeds, and the claim that an encyclopedia is not the "logical product" of the various contributions to it (their intersection), but instead the "'logical sum' of barely reconcilable views" (their union) well characterizes the outcome of this work (1938b/1987, 140).

The structure of encyclopedic unity thus allows for great qualitative difference between the sciences. Neurath did not require that social science be conducted just like natural science. "The programme of unified science does not presuppose that physics can be regarded as an example for all the science to follow." (1937/1981, 788) As noted above, social science has laws of varying scope and may have distinctive principles of explanation. Research procedures oriented to non-universal generalizations may be of a different kind than those aiming for universal laws. Neurath's recognition of the need for methodological pluralism finds expression not only in stressing that it would be mistake to hold social science to the standard achieved by physics, but also in his admonition to also investigate sciences that do yet meet those exacting standards.

> Sometimes one tends to prefer handling precise terms to such a degree that certain problems are avoided which are still structured less clearly. Certain phrases characteristic of the appreciation of art or of sociological considerations are thus discarded too quickly as being too vague and still too indeterminate and containing potentially metaphysical terms. But such incomplete reflections often contain all the scientific results so far achieved in this field and one should rather try to build on them. (1936b/1981, 712)

In sum, the encyclopaedic unity envisaged by Neurath did not require a reductive hierarchy of homogeneous laws with physics at its base. Nor did he require a strong form of methodological monism—beyond the basic testability of scientific assertions. Consequently, for Neurath the claim of unified science was minimalist: "all laws of unified science must be capable of being linked with each other, if they are to fulfill the task of predicting as often as possible individual events or groups of events." (1932a/1983, 68) The laws and generalizations established in its various disciplines are "linked" in unified science neither by translation nor by derivation from a privileged set, but by sharing the "universal slang" represented by the "thing language", now broadened to include psychological terms, in which their test conditions are formulated. It was the sharing of this specific and more or less clearly delimited, by no means all-encompassing, core "universe of discourse" (to use

[31] See Cartwright (1994) and Cat et al. (1996).

modern terminology) that for Neurath constituted both the physicalist unity of language and the encyclopedic unity of science. Both developed in line with the slogan "No system from above, but systematisation from below" (1936d/1983, 153).

7.6 The Critical Dimension

Readers who followed me so far may wonder what makes this "normalization" of Neurath's logical empiricist philosophy of social science newsworthy. Wasn't it precisely the preposterous reductive claims seemingly characteristic of his physicalism that made it stand out and distinguish it in its field? To think so is to overlook what it meant to advocate the unity of science in the 1920s and 1930s in Central Europe. Then the most unsavory social ideologies hid behind the pretense of *Geisteswissenschaft* supposedly being freed from the strictures of scientific probity that *Naturwissenschaft* observed. To insist on the unity of science was to insist on such probity also in social science, namely to issue only testable knowledge claims, and, by implication, to reject and unmask as nothing but inegalitarian and racist ideologies much that paraded as conservative social thought at the time.[32] The point of Neurath's physicalism was to ensure the legitimacy of social science in principle and to de-legitimize false pretenders.

Neurath was not alone in this estimation of the state of social scientific scene of the day. In numerous respects it agrees with that expressed by Max Horkheimer. In his inaugural address as Director of the Frankfurt Institut für Sozialforschung in 1931, Max Horkheimer called for the development of an interdisciplinary materialism, a research program held together by a form of verificationism which he characterized no less crudely than Neurath at times: "Materialism has in common with positivism that it acknowledges as real only what is given in sense experience. … Every assertion requires its confirmation in sense experience." (1933/1972, 42, second sentence restored)[33] Likewise, at the time Horkheimer rejected the principled separation of *Geistes*—from *Naturwissenschaft* (1932, 197). To be sure, any appearance of compatibility vanished with the second phase of the articulation of Critical Theory initiated in 1937 when philosophy was granted a distinct authority to pronounce on science and Horkheimer rode his ill-judged attack on logical empiricism.[34] My point now is simply that, largely unbeknownst to them, a large *de facto*

[32] One entire chapter, "Metaphysical Countercurrents", of Neurath's *Empirical Sociology* (1931a) was dedicated to this, with Othmar Spann and Werner Sombart's theories serving as representatives of the socio-politically regressive tendency. (Neurath's partial misunderstanding of Weber's methodology there is a separate matter; for that and the "de-transcendentalizing" that Weber's conception nevertheless requires, as hinted at below, see Uebel 2022.)

[33] For unknown reasons, the second sentence (in the orginal at 1933, 30), was dropped from the translation.

[34] For analysis of the exchange between Horkheimer (esp. his 1937) and Neurath see O'Neill and Uebel (2004).

agreement obtained about how to go about doing engaged-and-objective social science between Frankfurt and Vienna prior Horkheimer and Neurath's first meeting in January 1936. In many respects Neurath's program for unified science was compatible, even agreed with what Horkheimer envisaged. Unified science left plenty of room for interdisciplinary research into the economical-political and psychological-cultural determinants of late capitalism. Neurath's call for "sociology in the framework of physicalism" (as one his titles had it: 1932a) must be understood much like Horkheimer's for interdisciplinary materialist research—both of them call for the radical reorientation of the still fledgling social sciences along anti-metaphysical, anti-idealist lines.

What they did not agree on was the normative itself. As elsewhere, logical empiricists reject all ontologizing and point out that in ethical and political matters all we can do as scientists is describe behavior, seemingly rule-based or random, and take account of participants' reports. But it's not up to social or any other scientists to set the rules for behavior. With this position, of course, classical critical theorists disagree sharply and denounce positivists for their professed noncognitivism. Others may be tempted simply to charge that by nevertheless advocating critical social science logical empiricists violate their own strictures on scientific discourse. Since their noncognitivism in this instance equates to Max Weber's doctrine of scientific value neutrality, the charge is that such critical engagement must remain out of bounds for them. This worry is not groundless, but it is readily disarmed.

Viennese verificationism does indeed rule out from scientific discourse unconditional value statements as noncognitive: whether a given value "ought" to be realized unconditionally cannot be established the basis of what "is" (as Hume argued long ago). On this point the Circle was in agreement with Weber, as on the point that conditional value statements were straightforwardly cognitive in expressing means-ends statements. Where they differed was, first, in that Weber appeared to reserve the determination of the truth of unconditional value statements for philosophy whereas the Circle granted it no such office; and, second, that the Circle (to be precise, its left wing) explored and exploited margins of Weber's doctrine that not only had been left undiscussed by Weber himself but typically also by critical theorists who dismissed Weber's doctrine as "positivistic".

What are those margins? Consider, to begin with, a key demand of critical theory: to call into question "ideologically frozen relations of dependence that can in principle be transformed" (Habermas 1965/1972, 310). Does calling such social relations into question require the proclamation of unconditional values? Would it even make for an effective route towards their transformation? It certainly would not do so by itself, nor is it needed for social scientists to further such transformation by their work. Lest they are employed by a totalitarian state whose leadership can simply implement measures of social reorganization, anyone aiming for social transformation must first of all make plain for all to see what the facts of the case are. This means that, to begin with, they have to make available to the public answers to the following questions. What are the relations that are under discussion? How do they create or sustain relations of dependence and by whom on whom? Are the relations at least changeable in principle? What prevents such a change taking place?

Who gains from such a change being prevented? That these are social scientific questions is clear, as should be that no appeal to unconditional value statements (or absolute values) is needed to ask them. (Whether certain relations are relations of dependence is wholly independent of whatever value one places on independence.)

Still the worry may be that the very act of asking ethically or politically critical questions in science is out of bounds for logical empiricists and Weberians. To see that, on the contrary, social scientists are not prevented from pursuing potentially transformative agendas by the proscriptions issued by either logical empiricist non-cognitivism or Weberian value neutrality, note what these doctrines actually say. Weber stated explicitly and with full generality that social scientific research is "value-relevant" (1917/1949, 21) and reflects the interests and values of its practitioners (1904/1949, 76). Neurath repeatedly stressed that science and in particular sociology does not enjoy "extraterritoriality" and stand outside the socio-political struggles of the day (even if it is not directly concerned with them), and he advised the pursuit of a "sociology of sociology", a science reflective of the place of science among these struggles (1931a/1973, 406; 1944, 46–47). Putting both Weber's and Neurath's insights together (and de-transcendentalizing Weber's conception of values), we find that recognition of social science's inevitable value-relevance and inescapable non-extraterritoriality entails the legitimacy of potentially transformative research programs (as long as the strictures on unconditional value statements are observed).

Of course, the scientists in question better believe in the values at issue themselves, lest they are prepared to suffer cognitive dissonance, but this does not have to affect the scientific nature of their work. Note therefore that the presentation of the results of their scientific research to the public also does not require affirmation of unconditional value statements (or absolute values). It suffices to present the results as a combination of factual description and conditional policy advice: state p obtains under conditions $p_1 \ldots p_n$; if it is thought desirable that *not-p* obtain, then counter-measures q conducive to *not-p* obtaining should be taken. (In our example: in certain circumstances, social relations of dependence obtain that appear natural but are changeable in principle; if it is thought desirable to increase the personal independence of citizens, then certain types of social relations should be discouraged if not outlawed.)

To be sure, presenting such results falls short of unconditional advocacy of social change. But to require this of scientists in their role as scientists, Weber and the logical empiricists held, would be to confuse and conflate different roles that persons can and must play in society. *As citizens* social scientists like everyone else are much encouraged to advocate outright the realization of the values they believe in. *As scientists* they are limited to pointing out the facts that pertain to their case, the numerous consequences that are to be expected from maintaining the status quo and from changing it.[35]

I noted above that the point of Neurath's physicalism was both to ensure the legitimacy of social science and to de-legitimize false pretenders. So it may also be asked: what was his norm of legitimacy and what legitimated its choice? Neurath's norm of legitimacy was the criterion of intersubjective accountability which finds expression in the Circle's verificationist cognitivism. Intersubjective accountability, giving reasons for one's assertions, was never more than a necessary condition for science, however fundamental, so faced with theoretical choices we will have to call on a variety of further criteria to guide us—but a necessary condition it is. What's important here is that like all norms adduced by philosophical naturalists, this norm cannot be absolute but can only be conditional one—and that it is too. The justification of the norm of intersubjective accountability is instrumental and readily intelligible as such: on it as a presupposition of scientific communication depends the demand for the testability, even the mere intelligibility of scientific claims.

7.7 Conclusion

Left-wing logical empiricist social science was neither unduly positivist in a reductionist or eliminativist sense nor limited to quantitative fact gathering. In also serving critical social science it violates neither Weberian nor naturalist-physicalist strictures. Note how it broadens social scientific horizons. From its canons it follows that social scientists are not faced by the choice of either (i) remaining good empiricists by refraining from placing any ethical or political demands of relevance on their scientific work or (ii) becoming good activists by having their scientific work affirm ethical or political values whose own legitimation lies outside of science. Neurath's philosophy of social science shows that researchers do not have to choose between personal conviction and professional obligation: it is possible to be both scientific and critical.

References

Adorno, Theodor. 1969/1979. Einleitung In *Der Positivismusstreit in der deutschen Soziologie*. Wiesbaden: Luchterhand. Trans. "Introduction". In D. Frisby (ed.), The Positivist Dispute in German Sociology, London: Heinemann, 1–67.

Bentley, Joseph. 2022. Protocol Statements, Physicalism and Metadata: Otto Neurath on Scientific Evidence. *Studies in History and Philosophy of Science* 96:125–134.

[35] Carnap made this point very clearly in his (1934) and Neurath's numerous writings on socialist economics convey a variety of stances, from dispassionate analysis to outright advocacy, depending on the audience and his role with regard to them; see Uebel (2020) for discussion and Neurath (1920a) and (1920b) as two examples.

Bentley, Joseph. 2023. *Logical Empiricism and Naturalism. Neurath and Carnap's Metatheory of Science*. Cham: Springer.

Caldwell, Bruce. 2003. *Hayek's Challenge: An Intellectual Biography of F. A. Hayek*. Chicago: University of Chicago Press.

Crewe, Bianca, and Alan Richardson. 2024. Carnap on Unity of Science. In *Interpreting Carnap*, ed. A. Richardson and A. T. Tuboly, 252–269. Cambridge: Cambridge University Press.

Carnap, Rudolf. 1932a. Die physikalische Sprache als Universalsprache der Wissenschaft. *Erkenntnis* 2:432–465. Trans. *The Unity of Science*, London: Kegan, Paul, Trench and Teubner & Co., 1934.

Carnap, Rudolf. 1932b. Psychologie in physikalischer Sprache. *Erkenntnis* 3:107–142. Trans. "Psychology in Physicalist Language", in A. J. Ayer (ed.), *Logical Positivism*, New York: Free Press 1959, 165–198.

Carnap, Rudolf. 1934. *Logische Syntax der Sprache*. Wien: Springer. Rev. ed. trans. *The Logical Syntax of Language*, London: Kegan, Paul, Trench Teubner & Cie, 1937, repr. Open Court, Chicago, 2004.

Carnap, Rudolf. 1936a. Über die Einheitssprache der Wissenschaft. Logische Bemerkungen zum Project einer Enzyklopädie. *Actes du Congres Internationale de Philosophie Scientifique, Sorbonne, Paris 1935*, Facs. II, "Unité de la Science. Paris: Herman & Cie, 60–70.

Carnap, Rudolf. 1936b–37. Testability and Meaning. *Philosophy of Science* 3: 419–471, 4: 1–40.

Carnap, Rudolf. 1939. *Foundations of Logic and Mathematics*. Chicago: University of Chicago Press.

Carnap, Rudolf. 1956. The Methodological Character of Theoretical Concepts. In *The Foundations of Science and the Concepts of Psychology and Psychoanalysis*, ed. H. Feigl and M. Scriven, 38–76. Minneapolis: University of Minnesota Press.

Carnap, Rudolf. 1963. Intellectual Autobiography. In *The Philosophy of Rudolf Carnap*, ed. P. A. Schilpp, 3–84. La Salle: Open Court.

Cartwright, Nancy. 1994. Fundamentalism vs. the Patchwork of Laws. *Proceedings of the Aristotelian Society* 94:279–292.

Cat, Jordi, Nancy Cartwright, and Hasok Chang. 1996. Otto Neurath: Politics and the Unity of Science. In *The Disunity of Science*, ed. P. Galison and D. Stump, 347–359. Stanford: Stanford Univeristy Press.

Fay, Brian. 1975. *Social Theory and Political Practice*. London: Allen & Unwin.

Feigl, Herbert. 1956. The 'Mental' and the 'Physical'. In Feigl, Scriven and Maxwell 1958, 370–497. Repr. *The "Mental" and the "Physical": The Essay and a Postscript*, Minneapolis: University of Minnesota Press, 1967.

Feigl, Herbert. 1963. Physicalism, Unity of Science and the Foundations of Psychology. In Schilpp 1963, 227–268.

Feigl, H., M. Scriven, and G. Maxwell. 1958. *Concepts, Theories, and the Mind-Body Problem*. Minneapolis: University of Minnesota Press.

Frank, Philipp. 1957. *Philosophy of Science. The Link Between Science and Philosophy*. Englewood Cliffs, NJ: Prentice Hall. Repr. Mineola, NY: Dover, 2004.

Frank, Philipp, ed. 1956. *The Validation of Scientific Theories*. Boston: Beacon Press.

Frenkel-Brunswik, Else. 1954. Psychoanalysis and the Unity of Science. *Proceedings of the American Academy of Arts and Sciences* 80:271–350.

Habermas, Jürgen. 1965. Erkenntnis und Interesse. *Merkur* 213 (December). Trans. "Appendix. Knowledge and Human Interests: A General Perspective" in Habermas, *Knowledge and Human Interests*, London: Heinemann 1972, 301–317.

Habermas, Jürgen. 1970. *Zur Logik der Sozialwissenschaften*. Frankfurt: Suhrkamp. Trans. *On the Logic of the Social Sciences*, London: Polity, 1990.

Hajek, F.A. 1941–1944. Scientism and the Study of Society. *Economica* 9–11, repr. as Part One of *The Counter-Revolution of Science. Studies in the Abuse of Reason*. Free Press, New York, 1952, 2nd ed. Liberty Fund, Indianapolis, 1979.

Haller, Rudolf. 1982. Zwei Arten der Erfahrungsbegründung *Grazer Philosophische Studien* 16/17: 9–33. Trans. "Two Ways of Experiential Justification" in Uebel (ed.), *Rediscovering the Forgotten Vienna Circle*, Dordrecht: Kluwer, 1991, 191–202.

Hempel, C. G. 1969. Logical Positivism and the Social Sciences. In *The Legacy of Logical Positivism*, ed. P. Achinstein and S. F. Barker, 163–194. Baltimore: John Hopkins Press.

Hempel, C.G.. 1982. Schlick und Neurath: Fundierung vs. Kohärenz in der Wissenschaftlichen Erkenntnis. *Grazer Philosophische Studien* 16 (7): 1–18. Translated by Hempel, *Selected Philosophical Essays* (ed. by R. Jeffreys), Cambridge: Cambridge University Press 2000, 181–98.

Horkheimer, Max. 1932. Marx, Hegel und die Metaphysik. In *Festschrift für Carl Grünberg*, ed. Anon, Leipzig, 185–197.

Horkheimer, Max. 1933. Materialismus und Metaphysik. *Zeitschrift für Sozialforschung* 2: 1–34. Trans. "Materialism and Metaphysics" in Horkheimer 1972, 10–46.

Horkheimer, Max. 1937. Der neueste Angriff auf die Metaphysik. *Zeitschrift für Sozialforschung* 6:4–51. Trans. "The Latest Attack on Metaphysics" in Horkheimer 1972, 132–187.

Horkheimer, Max. 1972. *Critical Theory*. New York: Seabury Press.

Keat, Russell, and John Urry. 1975. *Social Theory as Science*. London: Routledge, Kegan & Paul, 2nd ed. 1982.

Kemeny, J. G., and P. Oppenheim. 1956. On Reduction. *Philosophical Studies* 7:6–19.

Koppelberg, D. 1987. *Die Aufhebung der Analytischen Philosophie*. Frankfurt: Suhrkamp.

Mach, Ernst. 1883. *Die Mechanik in ihrer Entwicklung historisch-kritisch dargestellt*. Leipzig: Brockhaus. Trans. *The Science of Mechanics*, Chicago: Open Court, 1960.

McGuinness, Brian, ed. 1987. *Unified Science*. Dordrecht: Kluwer.

Merton, Robert K. 1967. On Sociological Theories of the Middle Range. In Merton, *On Theoretical Sociology*, 39–72. New York: Free Press.

Nagel, Ernest. 1961. *The Structure of Science. Problems in the Logic of Scientific Explanation*. New York: Harcourt, Brace and World.

Nemeth, Elisabeth. 2007. *Logical Empiricism and the History and Sociology of Science*. In Richardson and Uebel 278–302.

Neurath, Otto. 1920a. Ein System der Sozialisierung. *Archiv für Sozialwissenschaft und Sozialpolitik* 48:44–73. Trans. "A System of Socialisation" in Neurath 2004, 345–370.

Neurath, Otto. 1920b. *Vollsozialisierung*, Diederichs, Jena. Trans "Total Socialisation" in Neurath 2004, 371–404.

Neurath, Otto. 1931a. *Empirische Soziologie. Der wissenschaftliche Gehalt der Geschichte und Nationalökonomie*. Vienna: Springer. Repr. in Neurath 1981, 423–527, excerpts trans. "Empirical Sociology" in Neurath 1973, 319–421.

Neurath, Otto. 1931b. Physikalismus. *Scientia* 50:297–303. Trans. "Physicalism" in Neurath 1983, 52–57.

Neurath, Otto. 1932a. Soziologie im Physikalismus. *Erkenntnis* 2: 393–431. Trans. "Sociology in the Framework of Physicalism" in Neurath 1983, 58–90.

Neurath, Otto. 1932b. Protokollsätze. *Erkenntnis* 3: 204–214. Trans. "Protocol Statements". In Neurath 1983, 91–99.

Neurath, Otto. 1932c. Sozialbehaviorismus. *Sociologicus* 8, repr. in Neurath 1981, 563–570.

Neurath, Otto. 1933. *Einheitswissenschaft und Psychologie*. Vienna: Gerold, Trans. "Unified Science and Psychology" in B. McGuinness 1987, 1–23.

Neurath, Otto. 1935a. Einheit der Wissenschaft als Aufgabe. *Erkenntnis* 5:16–22. Trans. "The Unity of Science as a Task" in Neurath 1983, 115–120.

Neurath, Otto. 1935b. Pseudorationalismus der Falsifikation. *Erkenntnis* 5: 353–365. Trans. "Pseudorationalism of Falsification" in Neurath 1983, 121–131.

Neurath, Otto. 1936a. Physikalismus und Erkenntnisforschung. *Theoria* 2: 97–105, 234–237. Trans. Physicalism and the Investigation of Knowledge. In *Neurath* 1983, 159–171.

Neurath, Otto. 1936b. Mensch und Gesellschaft in der Wissenschaft. In *Actes du Congres International de Philosophie Scientifique, Sorbonne, Paris 1935*, Fasc. II, Unite de Science, Herrmann, Paris, 32–40. Repr. in Neurath 1981, 711–718.

Neurath, Otto. 1936c. Soziologische Prognosen. *Erkenntnis* 6:398–405. Trans. "Sociological Predictions" in Neurath, *Economic Writings. Selections 1904-1945* (ed. by T. Uebel and R.S. Cohen), Dordrecht: Kluwer, 2004, 507–512.

Neurath, Otto. 1936d. L'encyclopedie comme 'modele'. *Revue de Synthese* 12: 187–201. Trans. "Encyclopedia as 'Model'. In Neurath 1983,145–158.

Neurath, Otto. 1936e. Einzelwissenschaften, Einheitswissenschaft, Pseudorationalismus. *Actes du Congres Internationale de Philosophie Scientifique, Sorbonne, Paris 1935*, Facs. I "Philosophie Scientifique et Empirisme Logique. Paris: Herman & Cie, 57–64. Trans. "Individual Sciences, Unified Science, Pseudo-Rationalism" in Neurath 1983, 132–138.

Neurath, Otto. 1937. Prognosen und Terminologie in Physik, Biologie, Soziologie. In *Traveaux du IXe Congres International de Philosophie*, Herrmann, Paris, 77–85. Repr. in Neurath 1981, 787–794.

Neurath, Otto. 1938a. Unified Science as Encyclopedic Integration. In *Encyclopedia and Unified Science*, ed. Neurath et al., 1–27. Chicago: University of Chicago Press.

Neurath, Otto. 1938b. Die neue Enzyklopadie. In Neurath et al, *Zur Enzyklopadie der Einheitswissenschaft*, TheHague: Van Stockum & Zoon, 6–16. Trans. The New Encyclopedia. in McGuinness 1987, 132–141.

Neurath, Otto. 1938c. The Departmentalization of Unified Science. *Erkenntnis* 7:240–246. Repr. in Neurath 1983, 200–5.

Neurath, Otto. 1939. The Social Sciences and Unified Science. *Journal of Unified Science* (*Erkenntnis*) 9: 244–248. Repr. in Neurath 1983, 209–212.

Neurath, Otto. 1941. Universal Jargon and Terminology. *Proceedings of the Aristotelian Society* New Series 41: 127–148. Repr. in Neurath 1983, 213–229.

Neurath, Otto. 1944. *Foundations of the Social Sciences*. Chicago: University of Chicago Press.

Neurath, Otto. 1946. Prediction and Induction In Neurath 1983, 243–246.

Neurath, Otto. 1973. *Empiricism and Sociology*, ed. M. Neurath and R. S. Cohen. Dordrecht: Reidel.

Neurath, Otto. 1981. *Gesammelte philosophische und methodologische Schriften*, 2 vols, ed. R. Haller, H. Rutte. Vienna: Hölder-Pichler-Tempsky. Repr. Vienna: LIT, 2021.

Neurath, Otto. 1983. *Philosophical Papers 1913-1946*, ed. R.S. Cohen and M. Neurath. Dordrecht: Reidel.

Neurath, Otto. 2004. *Economic Writings. Selections 1904-1945*, ed. T. Uebel, R.S. Cohen. Kluwer, Dordrecht.

O'Neill, J., and T. Uebel. 2004. Horkheimer and Neurath: Restarting a Disrupted Debate. *European Journal of Philosophy* 12:75–105.

O'Neill, J., and T. Uebel. 2008. Logical Empiricism as Critical Theory? The Debate Continues. *Analyse und Kritik* 30:379–398.

Oppenheim, Paul, and Hilary Putnam. 1958. The Unity of Science as Working Hypothesis In Feigl, Maxwell and Scriven 1958, 3–36.

Popper, Karl. 1944–45. The Poverty of Historicism. *Economica* 11–12. Rev. repr. *The Poverty of Historicism*, London: Routledge & Kegan Paul, London 1957, 2nd corr. ed. 1961.

Popper, Karl. 1945. *The Open Society and Its Enemies*, Vol. 2. London: Routledge. Repr. 1960.

Reisch, George, and Adam Tamas Tuboly. 2025. A Forgotten Chapter in the History of Sociology of Science - Philipp Frank's 'Good Nose'. In *The Vienna Circle – History and Legacy*, ed. G. Schiemer. Cham: Springer.

Richardson, Alan, and Thomas Uebel. 2007. *Cambridge Companion to Logical Empiricism*. Cambridge: Cambridge University Press.

Russell, Bertrand. 1921. *Analysis of Mind*. London: Allen & Unwin.

Steinmetz, George. 2005. Introduction: Positivism and Its Others in the Social Sciences. In *The Politics of Method in the Human Sciences*, ed. Steinmetz, 1–56. Durham: Duke University Press.

Stadler, Friedrich. 1997. *Studien zum Wiener Kreis. Ursprung, Entwicklung und Wirkung des Logischen Empirismus im Kontext.* Frankfurt a.M.: Suhrkamp. Trans. *The Vienna Circle. Studies in the Origins, Development, and Influence of Logical Empiricism.* Vienna: Springer, 2001, 2nd rev. ed. Cham: Springer, 2015.

Tuboly, Adam Tamas, ed. 2022. *The History of Understanding in Analytic Philosophy. Around Logical Empiricism.* London: Bloomsbury

Uebel, T. 1991. Neurath's Programme for Naturalistic Epistemology. *Studies in the History and Philosophy of Science* 22:623–646.

Uebel, T. 1996. On Neurath's Boat. In *Otto Neurath: Philosophy between Science and Politics*, ed. N. Cartwright, J. Cat, L. Fleck, and T. Uebel, 89–166. Cambridge: Cambridge University Press.

Uebel, T. 2000. Some Scientism, Some Historicism, Some Critics: Hayek's and Popper's Critiques Revisited. In *The Proper Ambition of Science*, ed. M. Stone and J. Wolff, 151–173. London: Routledge.

Uebel, T. 2003. 20th Century Philosophy of Social Science in the Analytical Tradition. In *The Blackwell Guide to the Philosophy of Social Science*, ed. P. A. Roth and S. T. Turner, 84–116. Oxford: Blackwell.

Uebel, T. 2007. Philosophy of Social Science in Early Logical Empiricism. In Richardson and Uebel, 250–277.

Uebel, T. 2008. Writing a Revolution. On the Production and the Early Reception of the Vienna Circle's Manifesto. *Perspectives on Science* 16:70–102.

Uebel, T. 2009. Neurath's Protocol Statements Revisited: Sketch of a Theory of Scientific Testimony. *Studies in History and Philosophy of Science* 40:4–13.

Uebel, T. 2015. American Pragmatism and the Vienna Circle: The Early Years. *Journal for the History of Analytical Philosophy* 3 (3): 1–35.

Uebel, T. 2017. The Development of Logical Empiricism. In *The Routledge Companion to the Philosophy of Social Science*, ed. A. Rosenberg and L. Macintyre, 41–55. Abingdon: Routledge.

Uebel, T. 2019a. Verification and (Some of) Its Discontents. *Journal for the History of Analytical Philosophy* 7 (4): 1–37.

Uebel, T. 2019b. Neurath on Verstehen. *European Journal of Philosophy* 27:912–938.

Uebel, T. 2020. Intersubjective Accountability: Politics and Philosophy in the Left Vienna Circle. *Perspectives on Science* 28:35–62.

Uebel, T. 2022. More on Neurath on Verstehen: The Rejection of Weber's Ideal Type Methodology. *In Tuboly* 2022:103–134.

Verein Ernst Mach. 1929. *Wissenschaftliche Weltauffassung. Der Wiener Kreis*, Vienna: Wolf. Translated by The Scientific Conception of the World. The Vienna Circle. In Neurath 1973, pp. 299–318; rev. trans. (with orig. annotated bibliography) "The Scientific World-Conception. The Vienna Circle," in F. Stadler, T. Uebel (eds.), *Wissenschaftliche Weltauffassung. Der Wiener Kreis. Hrsg.* vom Verein Ernst Mach (1929). Vienna: Springer, 2012, 75–116.

Weber, Max. 1904. Die 'Objektivität' sozialwissenschaftlicher und sozialpolitischer Erkenntnis. *Archiv für Sozialwissenschaft und Sozialpolitik* 19. Trans. "'Objectivity' in Social Science and Social Policy" in Weber, 1949, 49–112.

Weber, Max. 1917. Der Sinn der 'Wertfreiheit der soziologischen und ökonomischen Wissenschaften. *Logos* 7:40–88. Trans. "The Meaning of Ethical Neutrality" in Weber 1949, 1–47.

Weber, Max. 1949. *The Methodology of the Social Sciences*, ed. E. A. Shils. New York: The Free Press.

Chapter 8
Thomas Kuhn and the Social Sciences

William Outhwaite

Abstract This chapter examines Kuhn's views on the social sciences and the impact of his work on the self-understanding of social scientists, from 1968 to the present. It traces the multifarious aspects of the relation between Kuhn's work on natural science and the activities of social scientists. Kuhn's knowledge of social science was uneven, and his own practice of it left him unsatisfied, but his model of paradigms and scientific change greatly strengthened post-positivist trends in the philosophy of science, including that of social science. His holistic approach, shared with French traditions, shaped structuralist Marxism in the English and German language areas. The term paradigm has become used in increasingly metaphorical ways, notably in the so-called 'paradigm wars' around social science methodology. More generally in the social sciences, however, the relations between opposed *theoretical* 'paradigms', or whatever we choose to call them, have been largely pacified, as the rest of the world becomes more warlike and dangerous.

Keywords Kuhn · Social science · Paradigm · Fleck · Marxism

This chapter examines Kuhn's views of the social sciences and also the impact of his work on the self-understanding of social scientists, from the 1968 years (when it was used among other things to mark the difference between Marxist and 'bourgeois' social science) to the present, when it continues to shape discussions of what used to be called the classification of the sciences.[1]

[1] I addressed this issue in a chapter in a memorial volume for Hermínio Martins (Outhwaite 2018). The present chapter benefits from my belated reading of the comprehensive discussion of Kuhn by Steve Fuller (2000), especially chapter V.

W. Outhwaite (✉)
Newcastle University, Newcastle upon Tyne, UK
e-mail: william.outhwaite@newcastle.ac.uk

Y. Shan (ed.), *History and Philosophy of the Social Sciences*,
https://doi.org/10.1007/978-3-032-08167-4_8

In a letter to Karl Deutsch, Kuhn wrote: 'I know a great deal less than I should, and in any case virtually nothing at all, about the social sciences and I will not create confusion by bluffing it.' (Deutsch et al. 1986: 278). Deutsch had invited him to a conference in which *The Structure of Scientific Revolutions* was to be celebrated as a major advance in contemporary social science. Kuhn's slightly pleonastic explanation of his unwillingness to accept the invitation suggests a certain embarrassment, and it does seem that his knowledge of the social sciences was inversely related to his impact upon them worldwide.

As well as a background in physics, Kuhn had in a sense home ports in history and philosophy, and if we count history as a social science and philosophy as intimately related to it, these two fields deserve special attention. Fuller (2000: 10) particularly emphasizes the context of US science after WW2 and Kuhn's teaching at Harvard in James Conant's General Education in Science programme: 'Professionally ill equipped and ill disposed to pursue research in the area of "Big Science", Kuhn […] took refuge in the history of science.'[2] As an undergraduate at Harvard he had been attracted by philosophy, as he stated in the 1995 interview in *The Road Since Structure* (Kuhn 2000: 264): 'I gave a presentation on Kant and the notion of preconditions for knowledge…it's an important story because I go round explaining my own position saying I am a Kantian with moveable categories.'

Kuhn's academic identity could thus be described as a philosophical historian of science, shaped by an early 'ah' experience which he described retrospectively as his 'Aristotle experience'. Reading Aristotle through lens of Newtonian mechanics made no sense.

> Might not the fault be mine rather than Aristotle's, I asked myself. Perhaps his words [such as motion WO] had not always meant to him and his contemporaries quite what they meant to me and mine.
>
> …all at once Aristotle seemed a very good physicist indeed, but of a sort I'd never dreamed possible. Statements that had previously seemed egregious mistakes, now seemed at worst near misses within a powerful and generally successful tradition. That sort of experience—the pieces suddenly sorting themselves out and coming together in a new way—is the first general characteristic of revolutionary change…(Kuhn 2000: 16–17)

This rethink shaped Kuhn's approach to history of science, stressing (unlike Conant) 'the need to say what people had believed *before.*' (Kuhn 2000: 275)

> I used to think […] I could read texts, get inside the heads of the people who wrote them, better than anybody else in the world. […] So, being a historian of *that* sort was something I was quite willing to be […] But my objectives in this, throughout, were to make philosophy out of it. […] And I thought of *Structure*, when I finally got to it, as being a book for philosophers.

As with Foucault, then, the label 'historian of ideas' does not quite fit. For Kuhn, there was the additional problem that a post poised between philosophy and history did not fit the sharply disciplinary set-up in US academia, both at Harvard and at Berkeley. This was his strategy, as he described it in retrospect: 'My notion was […] that there was important philosophy to come out of it; but I needed first […] to establish myself professionally as a historian before I let the cat out of the bag.'

[2] See also Nathan Reingold (1991), esp. chp. 17, pp. 379–409.

(Kuhn 2000: 281) Not surprisingly, perhaps, Kuhn focussed on French *épistémologie*, with its combination of historical and philosophical analysis.[3] He was quite close to Alexandre Koyré, met Bachelard, and was impressed by the work of Emile Meyerson, the historian of chemistry Hélène Metzger, who had been murdered in the Holocaust, and the historian of medieval science Anneliese Meyer. (Kuhn 2000: 284–7) He also read, with some linguistic difficulty, Ludwik Fleck's *Entstehung und Entwicklung einer wissenschaftlichen Tatsache*, published in the unpromising year of 1939. '…it was very important that I read that book because it made me feel, all right, I'm not the only one who's thinking about things the way I was, thinking about the historical material the way I was.' (Kuhn 2000: 283) Kuhn later wrote a foreword to the US translation.[4]

His engagement with strands of philosophy prominent in the US seems more aleatory, though during his time in California he was close to Stanley Cavell, bilingual in so-called continental and analytic philosophy. They were appointed to Berkeley at the same time and shared a desire to shake up their respective fields—Cavell with Austin and Wittgenstein. (Cavell 2010: 352–5). Cavell (2010: 354) describes Kuhn's response to a philosophy discussion.

> 'I wouldn't have believed it. You people don't behave like academics in any other field. You treat each other as if you are all mad.' This seemed right to me, but normal enough, and because normal, suddenly revelatory. The immediate implication I drew is that philosophy is not (a) science. Later Kuhn will say that philosophy displays a wider disagreement over measures and standards […] than any other academic discipline.

The social sciences as conventionally defined seemed to Kuhn rather similar to philosophy in this respect. As has often been pointed out, Kuhn had abundant opportunities to study them in exchanges with colleagues, but initially contented himself with the constatation which opens *Structure*. Referring to his stay in 1958–9 at the Stanford Center for Advanced Studies in the Behavioral Sciences,

> …I was struck by the number and extent of the overt disagreements between social scientists about the nature of legitimate scientific problems and methods. Both history and acquaintance made me doubt that practitioners of the natural sciences possess firmer or more prominent answers to such questions than their colleagues in social science. Yet, somehow, the practice of astronomy, physics, chemistry or biology normally fails to evoke the controversies over fundamentals that today often seem endemic among, say, psychologists or sociologists. Attempting to discover the source of that difference led me to recognize the role in scientific research of what I have since called "paradigms". (Kuhn 1970: viii)

Seen, by himself or others, as a historian, there is nothing surprising in Kuhn's focus on natural science. (Some historians restrict themselves to a few years in the course of, for example, the English Civil War.) But given his philosophical concerns, it is more surprising that he did not pursue the comparison. As a psychologist pointed out rather grumpily, 'To discover differences between two entities one needs to make a comparison which, of course, involves a detailed study of both entities.'

[3] As discussed later, this formed an important link with English-language work in philosophy and social science which referred in particular to the work of Louis Althusser and the notions of problematic and rupture or break. See for example Benton (1977, 1984), Outhwaite (1983).

[4] On the later reception of Fleck's book, see Schnelle (1982).

(O'Donohue 1993: 269) In a suggestion which will recur a good deal later in this chapter, O'Donohue (1993: 285) writes:

> Perhaps psychologists like the relativism that is sometimes associated with this notion [paradigm WO]. Psychologists might believe that Kuhn offers them a 'feel good' account of science in which everyone wins and all the diverse schools within psychology can claim their own truth and legitimacy.

Kuhn had himself directed a very substantial research project involving interviews with the surviving pioneers of quantum theory. This three year project, including a year based in Copenhagen at Niels Bohr's invitation, drew on contemporary work in social science interviewing and the study of life histories, anthropological field work and the emergent area of oral history. At the beginning of the project Kuhn travelled from California to New York to consult the oral historian Saul Benison.[5] Although Kuhn was disappointed with the results of the interviews and disliked the process of interviewing, te Heesen (2022: 177) notes that the project foreshadowed later work in history and social studies of scientific research.

Having reviewed Kuhn's evaluation of the social sciences as he experienced them, I turn now to examine his impact on them. According to Kuhn (2000: 307), 'I gradually realized that a lot of the response [to *Structure*] was coming from social scientists, and I was quite unprepared for that; I thought of the book as directed to philosophers.' Having already cited O'Donohue's critical account of the response in psychology, I shall begin there. There are perhaps two reasons why psychologists should have been particularly attracted to the argument of the book. First, Kuhn's focus on the scientific community, as opposed to more 'externalist' explanations in terms of the broader social context of scientific developments,[6] suggested a social psychological angle. Kuhn, who engaged closely with psychologists (Kaiser 2016), described what he had produced as a social psychology of science: 'the explanation must, in the final analysis, be psychological or sociological.' (Kuhn 1970: 21) In *Structure* itself, Kuhn (1970: 62–4) described in some detail a social psychological experiment by Bruner and Postman (1949) in which subjects attempted to make sense of anomalously decorated playing cards, trying to assimilate them to the usual categories.

---

[5] An excellent book on this project, which was named 'Sources for History of Quantum Physics' (SHQP), reprints a page of notes which Kuhn took of this meeting. (te Heesen 2022: 85) Benison also drew Kuhn's attention to Henry Stack Sullivan's book on *The Psychiatric Interview.*

[6] Kuhn (2000: 286) cites the last letter he received from Koyré after he had read *Structure.* '"…you have brought the internal and external histories of science, which in the past have been very far apart, together." Now, I hadn't thought of that at all as what I was doing. I saw what he meant, and coming from him it was particularly agreeable because he had been so anti-external history; his gifts were as an analyst of ideas.' For reasons which are not clear, Kuhn (2000: 283) was suspicious of Ludvik Fleck's concept of the 'thought collective' (Gedankenkollektiv). 'It was clear it was a group, since it was collective, but [Fleck's] model was the mind and the individual.' Authors often perceive differences between their work and similar work by others which are not obvious to outsiders. Psychologists have a word for this: narcissism; Freud ([1930] 1961: 114) wrote of the 'narcissism of small differences'.

> In science, as in the playing card experiment, novelty emerges only with difficulty, manifested by resistance, against a background provided by expectation. Initially, only the anticipated and usual are experienced even under circumstances where anomaly is later to be observed. (Kuhn 1970: 64)

Second, psychology, like geography and anthropology, has a physicalist or natural scientific dimension and, unlike the other two disciplines, also an experimentalist culture, and this may have increased the perceived relevance of *Structure*.[7] As O'Donohue shows, mentions of Kuhn in the US psychological journals were frequent and usually favourable. More recently, a standard handbook (Sternberg and Pickren 2019) includes a chapter (Benjafield 2019) which uses the term paradigm as an organising theme, while concluding, following Driver-Linn (2003) and Green (2015), that it is not directly applicable in its Kuhnian sense(s). Green is more emphatic about this: 'Kuhn's model of paradigmatic science is inapplicable to psychology, whether or not it is true elsewhere in the sciences.' Driver-Linn stresses the rhetorical use of Kuhn's model and argues for a more cautious and pluralist approach in a 'splintered discipline'.

This raises a broader point about the social sciences. The Kuhnian paradigm, in the more precise sense of what he later called a 'disciplinary matrix', is monopolistic: you can't have more than one at a time. Kuhn, in his 'Reflections on my Critics', backtracked on the sharp division between what he had originally called the pre-paradigm period and the paradigm period.

> …paradigms have throughout been possessed by any scientific community, including the schools of what I previously called the 'pre-paradigm period'. My failure to see that point earlier has certainly helped to make a paradigm seem a quasi-mystical entity or property that, like charisma, transforms those infected by it. (Kuhn 1970: 272)

Kuhn however retained a somewhat 'whiggish' conception of the advance of natural science.

> The early stages in the development of most sciences are characterized by the presence of a number of competing schools. Later, usually in the aftermath of a notable scientific achievement, all or most of these schools vanish, a change which permits a far more powerful professional behaviour to the members of the remaining community. (Kuhn 1970: 272 n.1)

To anticipate my later discussion, this rather cumbersome locution points towards a more explicitly sociological account of science than Kuhn offered (except in passing hints such as this one). The Portuguese sociologist Hermínio Martins (1972: 51), living in exile in Britain, pointed to this ambiguity.

> There is something paradoxical in Kuhn's appeal to sociology, for Kuhn's work *combines sociologism with anti-sociology*…It is sociologistic in that it presents something like a social theory of natural-scientific knowledge…On the other hand, it can be used as an ideological tool of anti-sociology, in so far as sociology appears to be lacking in the diagnostic criteria of scientific maturity—paradigmaticness and revolutions.

[7] On the other hand, as we shall see, many social sciences without an experimentalist dimension were quick to adopt a Kuhnian approach to the history of their disciplines.

Before looking at sociology and the sociology of scientific knowledge in more detail, it is worth considering some other social sciences, notably economics which, along with psychology, tends to display more 'scientific' pretensions than the others. I suspect that a stroll down a corridor of most economics departments in the 'West' would reveal a large majority of 'orthodox' or 'neoclassical' economists, along with one or two Marxists and/or feminists and perhaps some more specialised colleagues such as development economists. Here then seems a possible candidate for a Kuhnian paradigm. On closer inspection, however, this seems less plausible. Tony Lawson (2013: 950), who has challenged the economic orthodoxy for decades in the name of a more sociological alternative, questions the identity of a neoclassical 'school'.

> Certainly the contemporary discipline is dominated by a mainstream tradition. But whilst the concrete substantive content, focus and policy orientations of the latter are highly heterogeneous and continually changing, the project itself is adequately characterised in terms of its enduring reliance, indeed, unceasing insistence, upon methods of *mathematical modelling*.

The details of Lawson's critique of this orthodoxy's atomistic ontology and covering-law conception of causality, long criticised by him and other realists,[8] are not relevant here. It converges however with that by Arne Heise (2020: 116), who calls it 'the dynamic stochastic general equilibrium (DSGE) model'.

> All other paradigms, scientific research programmes or Denkstile,[9] such as Marxian theories, the various post-Keynesianisms, behavioural economics, complexity economics, evolutionary economics, feminist economics and the Austrian School.[10]

Heise sees the dominance of the DSGE in terms of ideology and argues in favour of pluralism.

After psychology, it was sociology which responded most fully to *Structure*, generating the second highest volume of correspondence to Kuhn in the two decades following its publication (Kaiser 2016: 84). It was also one of the most fraught relationships. Norton Wise (2016: 39) reports:

> I have [a] vivid memory of walking innocently by his open office door sometime about 1972 when I heard, 'Norton, come in here. Did you see this? Just look at what they're saying', as he stabbed at an article on his desk by a sociologist.

The sociology of knowledge was inaugurated by a largely forgotten figure, Wilhelm Jerusalem (1854–1923). It was Jerusalem who formulated the idea of what he called *soziale Verdichtung* (social condensation), the gradual reinforcement of beliefs and memories. (Huebner 2013: 436).[11] This anticipated the idea of the social construc-

---

[8] See for example Outhwaite (1987).

[9] He frequently uses Fleck's term.

[10] He refers here to Erlei and Haucap (2019), who compare these with the orthodoxy. The editors list evolutionary economics, the Austrian School, neo-Marxist political economy, Ordoliberalism, Neo-Ricardian and feminist economics, agent-based modelling, classical institutional economics, behavioural economics and postkeynesianism. (Erlei and Haucap 2019: 404–5)

[11] The term *Verdichtung* had been used by Freud in his study of dreams.

tion of reality, formulated by Peter Berger and Thomas Luckmann (1966), which in turn really took off much later with the vogue of postmodernism. The sociology of knowledge had been consolidated in the 1920s by Karl Mannheim (1929) and Max Scheler (1926) but was still a rather marginal activity within sociology. There was also a more institutional sociology of science, which was really a sociology of 'scientific communities'—though Robert Merton covered both, as in his 1945 'Paradigm' essay (Merton 1945). This was in fact the first of three paradigm essays, followed by another for functional analysis (Merton 1945) and a third for structural analysis (Merton 1976). sociology of science.

Mannheim was sceptical about extending his sociology of knowledge to mathematics and the natural sciences, though he had always been interested in a sociological analysis of epistemologies and other intellectual products. The sociology of scientific knowledge relied substantially on Kuhn in taking up this challenge, notably in the 'strong programme' of the 'Edinburgh School', with its insistence that sociologists should adopt an attitude of methodological neutrality and symmetry in attempting explanations of the emergence of 'true' and 'false' beliefs, though an early paper by David Bloor (1973) does not in fact cite Kuhn. For his part, Kuhn (1992) rejected this relativism as a reductio ad absurdum of his approach to scientific change. Wise (2016: 39) comments:

> No doubt this public hostility represented his pressing concern that SSK opened up his project [...] to renewed charges of relativism [...] But more profoundly, in my view, he had been working for years to eliminate social considerations from having any purchase on the exquisitely technical function of paradigms. SSK threatened the heart of his enterprise.

Hermínio Martins made a substantial contribution to this new field and the broader context of social studies of science, bringing to bear an awareness of the historical dimension of the question of the classification of the sciences. He co-edited an early volume in the *Sociology of the Sciences* series (Elias et al. 1982) and his papers in this field are reprinted in Principe 2015. The long essay on 'Images and Imagining in Science' is a brilliant tour d'horizon of a whole range of topics related to the theme. One is the controversy, in physics and elsewhere, over the role of images in the formation of concepts and theories. The physicist Ludwig Boltzmann, as he notes, spoke of concepts and images or pictures (*Bilder*) more or less interchangeably (Martins, in Principe 2015: 79–80; see also Blackmore 1999), but there was a clear opposition, partly mapped onto the English Channel, between Kelvin and other British physicists and the French approach of Duhem, following Lagrange, who presented algebra without diagrams. Kelvin (1904) was 'never content until I have constructed a mechanical model of the subject I am studying. If I succeed in making one, I understand; otherwise I do not.' For Duhem (1906: 100; my translation), the 'English mind' has 'an extraordinary skill at imagining very complicated ensembles of concrete facts and an extreme difficulty in conceiving abstract notions and formulating general principles.'

These variations in what Ludwik Fleck (1935) was to call 'thought styles' are relevant both to the reception of physics[12] and of the social sciences—the latter torn between the imitation of natural science and alternative antinaturalistic approaches. Martins was concerned to head off a tendency in the social sciences to pursue notions of paradigm and epistemic consensus, shaped by positivism and the growth of technical methodology, arguing instead for the study of controversy, 'cognitive agonistics', in both natural and social science, Social science is more pluralistic, as he pointed out: 'the same piece of work may conjoin the Marxist theory of alienation, the Durkheimian theory of anomie, the Parsonian pattern variables…etc.'[13]

It may however be the case that a watered-down and plural notion of paradigm shaped the social sciences, especially sociology, in the following decades, and continues to compete with, to put it paradoxically, a more unitary image of fragmented research linked by a veneration of high-tech methodology.[14] Mark J. Smith (1998: 197–8) has suggested that, although the dominance of functionalism in the US in the 1950s offers 'something approaching' a paradigm, it is more appropriate to see the social sciences as in what Kuhn called a 'pre-paradigmatic' state. (There remain the questions whether this is likely to be a permanent state—if the social sciences are, in the unfortunate phrase of a book title, 'Völker ewiger Urzeit', permanently primitive (Breysig 1939)—and whether a paradigm phase would in any case be beneficial.) Smith (1998: 198) suggests distinguishing three categories of paradigm:

Paradigm 1—the strict sense identified by Kuhn […].
Paradigm 2—where the intellectual foundations of the community are such that to move from this paradigm to another would require reorientation of the type identified by Kuhn as a gestalt shift. […] This use acknowledges the presence of competing paradigms in the same field of knowledge but requires them to be incommensurable.
Paradigm 3—used to designate a school of thought, theoretical perspective or set of problems. This is the loosest but most common use […].

A number of commentators have pointed out the appeal of Kuhn's model to social scientists concerned to legitimate their activity to themselves and others. As Fuller (2000: 244) puts it,

> Although Kuhn himself clearly wanted to restrict his account to the natural sciences, his tendency to identify a paradigm with a self-selecting, self-governing community of inquirers helped persuade many humanists and social scientists that they too could enjoy scien-

[12] Paul Forman (1970) suggested that quantum theory was welcomed in part in Weimar Germany because it counted against popular critiques of scientific determinism. In the German-language discussion, the key term was the perceptual accessibility (Anschaulichkeit) of physics—particularly that of quantum theory.

[13] See also Sharrock and Read (2002: 126–9).

[14] A book published in 1997 describes its impact as 'explosive' (Fischer and Hoyningen-Huene 1997: 2) and traces the notion of paradigm through a number of fields, including law, music and moral philosophy as well as the natural and social sciences.

> tific status—or at least raise their standing in the university—by doing things that look like normal science.[15]

Fuller (2000: 258–9) suggests an interesting parallel between Kuhn's model and Walt Rostow's contemporary modernisation theory in his *The Stages of Economic Growth: A Non-Communist Manifesto* (Rostow 1960). Rostow's 'take-off' to 'high mass consumption' parallels the scientific revolution initiating a period of normal science, with communism in Rostow's narrative seen not as the culmination of capitalist development but as a 'disease of the transition' to advanced industrialism. As in aviation, take-off is a moment of danger, while there is little risk after climb-out to cruising altitude.

> Intellectually speaking, Kuhn and Rostow faced largely the same challenges. Both were mounting a historical argument in fields that had been dominated by static, formalist conceptions of their respective subject matter (i.e., positivistic conceptions of science and neoclassical conceptions of the economy. Both emphasized that over the long term, growth was not continuous, but subject to spurts from one sector…that eventually reconstituted the entire organism, be it a national economy or a field of inquiry.

In a related move, I shall suggest, sociologists, like nationalists, can use the weaker conception of paradigm as a protection against criticism from outside: 'my paradigm right or wrong'.[16] The Marxism of the 1970s, which came closest to constituting a paradigm in the social sciences (Bernstein 1976; Fuller 2000: 249), was particularly prone to this approach. It rejected, for good reasons, Popperian falsification but went on to argue for a paradigmatic incompatibility between Marxist and bourgeois economics and, with rather more difficulty, between Marxist and non-Marxist sociology. For Marxists who wanted something more sophisticated than partisanship as a way of marking the difference, Althusserian structuralism provided the answer. In the UK, Barry Hindess and Paul Hirst argued, in the context of their critique of history,

> What is specified in theoretical discourse can only be conceived through the form of that discourse (or another, complementary, discourse). It cannot be specified extra-discursively.

Though no Marxist, Foucault also argued in terms of an 'écart énonciatif' or, as Roger Chartier (2009: 228; my translation) summarises it, the 'regime of enunciation which is specific to the production of this discourse; and a "conceptual network" defined by the rules of concept formation of a given discursive formation.'

In his first book, shaped by the Althusserian approach in which sciences are defined by their 'object', Göran Therborn (1976: 428) argued that the division between economics, dominated by an individualistic model of self-interested action in a market, and sociology, focussed on values and what he called 'the ideological community', was the underlying problem of non-Marxist social science.

---

[15] Among these are the fetishisation of essentially meaningless metrics of the 'impact factor' of journals, especially in the social sciences but also in the humanities.

[16] Barkin (2010: 1) suggests the image of castles.

> Economics and sociology allot no proper space to the object of historical materialism, which is constituted outside their common problematic of self-interest and social values. Economists may speak of capitalism and sociologists of classes and class conflicts, but the former do not see capital as a relation of production and the latter do not relate classes and class struggle to particular historical combinations of the forces and relations of production. On the other hand, in historical materialism, the market and the ideological community have a space as irreducible social determinants.

Therborn (1976: 429) argued at that time for 'a *transcendence of sociology,* similar to Marx's transcendence of classical economics, and the development of historical materialism as *the* science of society.'

One of the other underlying themes was the relation between social science and normative analysis, or between fact and value. Here Tom Bottomore, following the Austro-Marxists, answered his former co-editor[17] Maximilien Rubel's question: 'Is there a Marxist sociology?' by arguing that Marxism contained both a sociology and a normative programme and that these were independent, although Marx's 'theoretical analysis and his allegiance to the labour movement were congruent and, in a sense, mutually supporting'. (Bottomore 1975a: 83.) Bottomore was unusual in combining his political commitment with an equally strong commitment to value-free science; more usually, both in 'Frankfurt' critical theory and in more standard versions of Marxism, the concept of critique served to bridge the gap. The same was true of other forms of theory such as feminism and, more recently, what came to be called postcolonial theory.

A third issue revived in the second half of the twentieth century, and particularly with the revival of social theory in the 1970s, was that of the relation between the natural and the social sciences. The conjunction of analytic philosophy of language, 'phenomenological' sociology and philosophical hermeneutics reinforced earlier strands of symbolic interactionism, itself reinvigorated by Erving Goffman's influential work. The question remained open whether naturalistic or antinaturalistic paradigms of social science were simple alternatives or whether they could somehow be combined, for example in neo-Weberian theory, Habermasian critical theory or in Roy Bhaskar's realist naturalism.

The topic of sociological paradigms was a prominent theme in the 70s and after. Garry Gutting (1980) brought together some of these contributions in his edited reprint collection of 'Appraisals and Applications of Thomas Kuhn's Philosophy of Social Science'. Eckberg and Hill (1979) usefully presented 'Twelve Sets of Sociological "Paradigms"' from the previous decade, in an order shaped by the number of paradigms identified. Michael Carroll (1972) discussed a single paradigm candidate, the methodological approach centred on the analysis of variance, but was more concerned to criticise this than to analyse its role as a prospective paradigm. Next came a number of dualistic models, of which the most lasting, here represented by Walsh (1972), contrasted positivistic and interpretive or

[17] Bottomore and Rubel (1956).

'phenomenological' approaches.[18] A triadic variant (Sherman 1974) followed Habermas (1968/1971) in adding a third, 'critical' approach, though he was mainly concerned to stress the importance of the (then only recently canonised) trinity of Marx, Durkheim and Weber (Sherman's preferred classic). Tom Bottomore (1975b) provided a characteristically insightful and judicious overview of four approaches: structural-functional, historical, structuralist and phenomenological.

The most systematic discussion of the concept of paradigm in the books and papers reviewed by Eckberg and Hill is that by Lehman and Young. They identify in US sociology a 'pre-paradigm-prophetic period' (1890–1920), the 'quest for a scientific sociology (1920–1940) in positivist and functionalist approaches, 'the consolidation of the functionalist paradigm' (1940–1960) and a move, which they welcome, 'toward a conflict paradigm' (1960–1973). This polarisation has generated 'a revolutionary crisis' and 'a state of paradigmatic anarchy' (Lehman and Young 1974: 15), to be remedied by the further development of the conflict approach. Friedrichs (1970: xxi, 3) also proclaimed his indebtedness to Kuhn's model and presented it at some length in the first chapter, but it rather drops out of sight in the rest of the book, after being used to frame the 'revolutionary' break-down of the functionalist orthodoxy and the free-for-all which followed in the 1960s. This is partly because he concluded that 'The system paradigm appeared to be a product of a prior commitment to a paradigmatic image of the sociologist as *value-free*; the conflict paradigm an outcome of the paradigmatic vision of the sociologist as *engagé*. (Friedrichs 1970: 290–1) This is of course a very different sense of paradigm from any of those in Kuhn's conception.

The other commentators discussed by Eckberg and Hill are more cautious, suggesting that 'it is quite problematic whether there are any "paradigms" in sociology as yet' (Effrat 1972: 3n) or that 'Sociology (and this applies also to other social and human sciences) has always been…a *multi-paradigm science*' (Bottomore 1975b: 191). Eckberg and Hill (1979: 130) cite approvingly Hermínio Martins' observation that 'Paradigms pertain to fields like the study of heat, optics, mechanics etc.; *there are not and cannot be* paradigms of physics or chemistry.' (emphasis added) They conclude (p. 132) that 'If paradigms (exemplars) exist in the discipline of sociology, they are difficult to find.' One of the other works they discuss, Nicholas Mullins' very substantial analysis of 'Theory Groups' and social networks in sociology, contrasts his own approach with those of Friedrichs, Edward Shils (1970) and Alvin Gouldner (1965, 1970); see Mullins 1973: 306–9). Mullins, who had earlier worked on molecular biology (Mullins 1973), offered a more fine-grained analysis of the growth of theory groups from 'networks' through 'clusters' to 'specialties', usefully distinguishing between theories as a whole and specific works which have a status as 'critical landmarks', initiating further research traditions in the same way as paradigmatic theories or, in Kuhn's terminology, 'exemplars' (Mullins 1973: 6).

[18] Two other dualisms were conflict/consensus (Lehman and Young 1974) and priestly/prophetic (Friedrichs 1970).

Mark Blaug (1975: 151), in an essay reprinted in the same volume as Eckberg and Hill's, argues that Imre Lakatos's model of research programs and progressive versus degenerative problem-shifts fits the Keynesian Revolution, as it gained ground against a background of unemployment and depression, better than Kuhn's model:

> ...the Keynesian theory succeeded because it produced the policy conclusions most economists wanted to advocate anyway, but it produced these as logical inferences from a tightly knit theory and not as endless epicycles on a full-employment model of the economy.

Be that as it may, it is the Kuhnian paradigm model, rather than Lakatos's alternative, which has mostly been drawn on in the past third of a century.[19] The paradigm candidates come and go (no-one now talks about conflict theory, while evolutionary social theory is resurgent), but the paradigm metaphor remains an attractive one.

As so often, it was in Germany that these issues were most fully debated, notably in the *Theorievergleichsdebatte* in (West) German sociology, which recognised a theoretical pluralism encompassing Habermasian communicative theory, Luhmann's system theory, interactionist and phenomenological action theory, neo-behaviourism and historical materialism. Unlike the contemporaneous and extremely acrimonious debate around the 'finalisation' of natural science discussed later, this was conducted at sociological congresses in 1974 and 1976 and in the professional journals. As Rainer Lepsius (2008: 125) wrote in a retrospective, politicised currents gave way to alternative paradigms: 'Aus politisierten Richtungen wurden alternative Paradigmen.' (See Greshoff 2010: 185.) In Germany, the revival of sociological theory and the philosophy of social science began with the discussion around Ralf Dahrendorf's *Homo Sociologicus* (1958), the 'Positivismusstreit' in the early 1960s (Adorno et al. 1969) and the Habermas-Luhmann exchange at the end of the decade (Habermas and Luhmann 1971) (See Greshoff 2010: 183). Greshoff himself founded a discussion journal in 1990, *Ethik und Sozialwissenschaften*, later (2001–2015) *Erwägen Wissen Ethik* (*EWE*) with an open door publishing policy.[20] He and colleagues also proposed a 'conceptual framework' to 'mediate between paradigms' (Greshoff et al. 2007). A thoughtful, if rather negative explanation of the failure of such initiatives was offered by Hans-Joachim Giegel (1977: 228), who suggested that for the social sciences, in the absence of clear cognitive advances, innovation simply means originality and that this leads to 'irrational pluralism':

> The self-sufficient search for new methods, the deviation towards side-issues and irrelevancies, the flight into speculation with no scientific basis, the protection from criticism, the instrumentalisation of scholarly communication for non-scholarly purposes, all this encourages a development which leads to the indifferent coexistence of diverse theories.

This cap is probably a partial fit, though it is also possible to argue, as Claus Offe (1977: 251) did in the same volume, that theories as different as classical sociology, Marxism and contemporary system theory are linked by a common focus on crises

---

[19] For a fairly early response, see Bayertz (1981).

[20] I once sent a contribution with the suggestion that they should leave it out if they felt it didn't fit, and was reminded that that would be contrary to the policy of the journal.

of different kinds and a problematisation of unreflective common-sense or theoretically limited conceptions.

> The critical potential of empirical social research lies not so much in the critique of answers as the critique of questions. Its practical relevance derives from the collection of evidence for the practical irrelevance of the criteria of validity to which it is subjected—whether by state or private research administration or by its own theoretical tradition.

If we have to live with pluralism in the social sciences, we may at least escape some of the naiveties which pervade discussions of the social organisation of natural science. Here again, of course, Kuhn was a crucial source of enlightenment in the philosophy and sociology of science.[21] Kuhn's intervention was rightly perceived as a threat to what had come to be called the 'standard view' in the philosophy of science: a logical empiricist model, more or less modified with Popperian epicycles. Although it was possible for Popper and his followers to claim that puzzle-solving normal science was simply not good science by the standards of critical rationalism,[22] this conception of philosophy as a slightly *weltfremd* judge pronouncing on a massively expanding and stupendously successful social practice was not very convincing. A broadly Kuhnian or Lakatosian account of scientific practice therefore fitted into the gradual process of modification and unravelling of the original logical empiricist model (and later its Popperian variant) which began as soon as it was articulated. Like Stalinist economic planning, or the latest innovation by Microsoft, the positivist model was no sooner in place than there were suggestions for what we would now call updates to make it more sophisticated and sensitive. As Peter Halfpenny (1982: 120) summarised the process, positivism 'lives on philosophically, developed until it transmutes into conventionalism or realism'.

Nevetheless, as Steven Shapin (1995) pointed out, 'scientistic North American sociological traditions and, to a lesser extent, traditions in Britain and Europe continue actively to disseminate a picture of scientific "method" and scientific knowledge radically at variance with those offered by SSK [sociology of scientific knowledge].' This latter field seems in retrospect one of the most innovative and impressive features of English-language, and particularly British sociology, yet one which remains marginal in the profession, even within the interdisciplinary area of science and technology studies. Like Hermínio Martins in the passage quoted earlier (Martins 1972: 51), Shapin (1995: 301–2) points to a paradox:

> From a sociological point of view, Kuhnian SSK is at once conservative and radical. On the one hand, it seeks inter alia to answer traditional questions about the grounds of a communal order, and it does so by pointing to the regulative role of norms. While the regulatory relevance of social maxims ("Be sceptical", "Be disinterested") is doubted, the significance of norms for ensuring order and for marking the boundaries of communities is vigorously respecified and reaffirmed in a new idiom. The solidarity of specialist communities—or

[21] On the disjunction between philosophy of science and historical and sociological studies of scientific knowledge, see, for example, Golinsky (2011).

[22] As Popper (1970: 57) put it, 'to me the idea of turning for enlightenment concerning the aims of science, and its possible progress, to sociology or to psychology (or…to the history of science), is surprising and disappointing.'

> such solidarity as is found to exist—is coordinated through their specialist knowledge. Good and bad, proper and improper, interesting and banal scientific behaviour is recognized and sanctioned by members' knowledge of the natural world. On the other hand, by arguing that the relevant norms are made of the same stuff as the community's technical knowledge, the Kuhnian move overturns the existing sociology-of-knowledge scheme that asks how "society might influence knowledge"...[Thus] SSK's insistence upon a quite elementary feature of the sociological sensibility has seemed to acquire a shockingly radical, even subversive, character.'

This may partly explain why mainstream sociology in the UK and US was unwelcoming to the sociology of scientific knowledge, though other factors, such as the marginality of the sociology of knowledge as a whole and the entry costs of acquiring even a minimal reading knowledge of the relevant areas of natural science provide a more obvious explanation. More significant is the wider social hostility to any social insight into or oversight of scientific progress. Whereas a historian or sociologist of religion would be laughed at for suggesting that Judaism was succeeded by Christianity and in turn by Islam because that was what God wanted, any social explanation of scientific advance was suspected of 'explaining it away', as if the social explanation somehow undermined an alternative and perhaps complementary explanation in terms of 'internal' criteria of success.[23]

A particularly striking example of hostility to social approaches to science was a controversy in West Germany in the late 1970s around the 'finalisation thesis'. This concerned the practical applications of 'mature' science and was developed by a team of young researchers at the Max-Planck Institute for the Study of Life in Social-Technical World at Starnberg, just south of Munich, working with the physicist Friedrich von Weizsäcker. The thesis, presented in two articles in the *Zeitschrift für Soziologie* (Böhme, van den Daele and Krohn 1972, 1973), at a conference I attended in 1974 which brought together researchers from Paris and Sussex as well as elsewhere in Germany[24] and in another article a little later (van den Daele and Weingart 1975) and a subsequent book publication in German (van den Daele et al. 1979) and English (Schäfer 1983) drew on the history of science to suggest a three-stage model of scientific development. Following a pre-paradigmatic phase of largely amateur science and a second phase characterised by the consolidation of Kuhnian paradigms or, better, Lakatosian research programmes, there follows a third phase in which the mature science, as Kuhn called it, is applied in a variety of specialist areas, such as steady state physics drawing on quantum theory. (In other areas of physics, such as particle physics and cosmology, the search for new fundamental theories continues (Böhme, van den Daele and Krohn 1973: 135).) Whereas

[23] Kuhn addressed this issue in a number of places, for example in a contribution to symposium on Hempel in 1983 and published that year in *The Journal of Philosophy* (Kuhn 2000: 208–15). See also Bourdieu (2004); I am grateful to Bridget Fowler for this reference.

[24] Habermas was co-director of the Institute. Having been invited to comment on the finalisation thesis, we tried to earn our keep by making critical comments, to an extent which rather upset the researchers. Don't worry, Habermas said to them, as I recall the discussion. Stick to your guns and say what you want; it doesn't matter if it's wrong—if it's criticised, that's how scholarly progress works.

the emphasis in the second phase is on the development of the research programme, in the third phase, as in the first, science tends to be open to external goals—as, for example, in environmental science with its bases in chemistry, biology, meteorology and so forth. This orientation to external purposes is 'conditioned [bedingt] by the theoretical development itself' (Böhme, van den Daele and Krohn 1973: 136 n.15). 'We have conceived the third phase in our model not as the application of results of the fundamental theory, but as a concretisation of this theory in a particular developmental direction.' (Böhme, van den Daele and Krohn 1973: 137)

Steve Fuller (2000: 251–9) discusses the finalisation thesis and Francis Fukuyama's conception of the end of history as prolongations of Kuhnian themes. My focus here is on the controversy which finalisation unleashed. This apparently innocent thesis in the history of science ran into a blizzard of criticism from the right-wing academics in the Bund Freiheit der Wissenschaft, which had been founded in 1970 in response to the excesses of the student movement of the late 1960s. They saw in the finalisation thesis a plot to subject 'mature' science to external political direction of the kind experienced in totalitarian regimes. A conference on science in danger ('Gefährdete Wissenschaft') was held in Munich in 1976, followed by a book, *Die politische Herausforderung der Wissenschaft. Gegen eine ideologisch geplante Forschung*, in the same year (Hübner et al. 1976). (See also the book resulting from a counter-conference, *Konsequenzen kritischer Wissenschaftstheorie* (Hubig and von Rahden (1977).) Partly because of this controversy, the Institute was closed after von Weizsäcker's retirement, and Habermas returned to Frankfurt in 1981 (see Leendertz 2010, 2013; also Wehrs 2008 and Drieschner 1996). Drieschner, though disenchanted with the organisation of the Institute of which he had been a member, notes that the finalisation theorists 'in the end only wanted to make a serious contribution to an ongoing discussion in the history of science'. As one of the Starnberg group, Gernot Böhme (1980: 12) wrote:

> The thesis of the finalisation of science was intended to serve the self-enlightenment of scientists. It entered the discourse of *sociologists* of knowledge and the circus of ideology construction in the mass media.

Frank R. Pfetsch (1979) also outlines the intellectual and political context, noting that there was also criticism from the left that the Starnberg group were insufficiently materialist in their account of science (see Hieber 1975). Fuller (2000: 254) develops a similar critique:

> Because finalization remained fixated on the trajectory of the natural sciences as dictating the pace at which social policy could occur, their account of scientific change ended up shadowing the 'linear model' of technical development so common to both welfare state and Marxist economic policy thinking in the Cold War period: a scaled-up research base in the natural sciences corresponded to the formation of a capital base that was deemed necessary for reliable and efficient wealth production, while social scientific research later provided insight into how the surplus from that production may be most equitably distributed or, as the case may be, redistributed.

The bizarre if short-lived political controversy over finalisation is of course just one example of the controversies over the politics of science which bemuse those of us with an interest, however distanced, in the history and sociology of scientific knowledge, as well as those like Steve Fuller centrally concerned with these issues.[25] That the Science Policy Research Unit at Sussex University, founded only a little earlier (1966) and working on many of the same topics, survived more or less intact was largely due to the diplomatic approach and personal charisma of its long-standing and founding director, Chris Freeman (see Campos 2016). SPRU's sister unit, History and Social Studies of Science, was however poorly supported by the University and eventually absorbed into SPRU.[26] The Science Studies Unit at Edinburgh, also established in the 1960s, recently celebrated its 50th anniversary and its activities survive in a new structure.[27]

This chapter has traced the multifarious aspects of the relation between Kuhn's work on natural science and the activities of social scientists. Kuhn's knowledge of social science was uneven, and his own practice of it left him unsatisfied, but his model of paradigms and scientific change greatly strengthened post-positivist trends in the philosophy of science, including that of social science. The holistic approach which it shared with French traditions shaped structural(ist) Marxism in the English and German language areas. The term paradigm has become used in increasingly metaphorical ways, notably in the so-called 'paradigm wars' around social science methodology.[28] More generally in the social sciences, however, the relations between opposed *theoretical* 'paradigms', or whatever we choose to call them, have been largely pacified, as the rest of the world becomes more warlike and dangerous.

References

Barkin, J. Samuel. 2010. *Realist Constructionism. Rethinking International Relations Theory*. New York: Cambridge University Press.

Bayertz, Kurt, ed. 1981. *Wissenschaftsgeschichte und wissenschaftliche Revolution*. Pahl-Rugenstein.

Benjafield, John G. 2019. Major Paradigms and Approaches in Psychology. In *The Cambridge Handbook of the Intellectual History of Psychology*, ed. Robert J. Sternberg and Wade E. Pickren, 4–28. New York: Cambridge University Press.

Benton, Ted. 1977. *The Philosophical Foundations of the Three Sociologies*. London: Routledge.

Benton, Ted. 1984. *The Rise and Fall of Structural Marxism: Louis Althusser and His Influence*. New York: Macmillan.

Blackmore, John. 1999. Boltzmann and Epistemology. *Synthese* 119:157–189.

Blaug, Mark. 1975. Kuhn Versus Lakatos, or Paradigms Versus Research Programmes in the History of Economics. *History of Political Economy* 7 (4): 399–433. Reprinted in Gutting (1980); page reference is to the reprint.

Bloor, David. 1973. Wittgenstein and Mannheim on the sociology of mathematics. *Studies in History and Philosophy of Science* 4 (2): 173–191.

[25] See also Weingart (1983, 1997) and Leeming (1997).

[26] For a fuller discussion of science policy at Sussex, see Outhwaite (2017).

[27] https://www.ed.ac.uk/news/2016/celebrating-50-years-of-science-studies

[28] See for example Bryman (2008), Knappertsbusch (2023).

Böhme, Gernot. 1980. *Alternativen der Wissenschaft*. Frankfurt: Suhrkamp.

Böhme, Gernot, Wolfgang van den Daele and Wolfgang Krohn 1972, 'Alternativen in der Wissenschaft', *Zeitschrift fur Soziologie* 1 (4, October), 302–316.

Böhme, Gernot, Wolfgang van den Daele and Wolfgang Krohn 1973, 'Die Finalisierung der Wissenschaft', *Zeitschrift fur Soziologie* 2 (2, April), 128–144.

Bottomore, T B and Maximilien Rubel (eds), *Karl Marx. Selected Writings in Sociology and Social Philosophy*. London: Watts. Republished by Pelican 1963.

Bottomore, Tom. 1975a. *Sociology as Social Criticism*. London: George Allen and Unwin. Republished (Routledge Revivals) in 2010.

Bottomore, Tom. 1975b. Competing Paradigms in Macrosociology. In *Annual Review of Sociology*, ed. Alex Inkeles, James Coleman, and Neil Smelser, 191–202. Palo Alto: Annual Reviews.

Bourdieu, Pierre. 2004. *Science of Science and Reflexivity*, Translated by Richard Nice. Cambridge: Polity Press and Chicago: University of Chicago Press. First published 2001.

Breysig, Kurt. 1939. *Die Völker ewiger Urzeit. Nordländer, Nordwestamerikaner, Nordostamerikaner*. Berlin: de Gruyter.

Bruner, J. S., and Leo Postman. 1949. On the Perception of Incongruity A Paradigm. *Journal of Personality* XVIII:206–223.

Bryman, Alan. 2008. The end of the paradigm wars? In *The Sage Handbook of Social Research Methods*, ed. Pertti Alasuutari, Leonard Bickman, and Julia Brannen, 13–25. London: Sage.

Campos, Ângela. 2016. SPRU History project. A Report on its qualitative angle. http://www.sussex.ac.uk/spru/about/history

Carroll, Michael P. 1972. Considerations on the Analysis of Variance Paradigm in Sociology. *The Pacific Sociological Review* 15 (4): 443–459.

Cavell, Stanley. 2010. *Little Did I Know. Excerpts From Memory*. Stanford University Press.

Chartier, Roger. 2009. *Au bord de la falaise*. 2nd edn. Paris: Albin Michel.

Drieschner, M. Die Verantwortung der Wissenschaft. Ein Rückblick auf das Max-Planck-Institut zur Erforschung der Lebensbedingungen der wissenschaftlich technischen Welt (1970–1980). http://www.ruhr-uni-bochum.de/philosophy/staff/drieschner/beding.htm Accessed 13.1.16 Earlier version published in: T. Fischer and R. Seising (eds.): Wissenschaft und Öffentlichkeit. Frankfurt: Lang, 1996, 173–198.

Driver-Linn, Erin. 2003. Where Is Psychology Going?: Structural Fault Lines Revealed by Psychologists' Use of Kuhn. *American Psychologist* 58 (4): 269–278.

Duhem, Pierre. 1906. *La théorie physique: son objet, et sa structure*. Paris: Chevalier & Rivière.

Eckberg, Douglas Lee, and Lester Hill. 1979. The Paradigm Concept and Sociology: A Critical Review. *American Sociological Review* 44 (6): 925–937. (Reprinted in Gutting 1980).

Effrat, Andrew. 1972. Power to the Paradigms: An Editorial Introduction. *Sociological Inquiry* 42:3–34.

Elias, Norbert, Hermínio Martins, and Richard Whitley, eds. 1982. *Scientific Establishments and Hierarchies*. Dordrecht: Reidel.

Erlei, Mathias, and Justus Haucap, eds. 2019. Special Issue: Mainstream vs. heterodoxe Ökonomik: Forschungsprogramme im Vergleich. *List Forum* 44(4).

Fischer, Michael and Paul Hoyningen-Huene, eds. 1997. *Paradigmen*. Frankfurt: Peter Lang.

Fleck, Ludwik. 1935. *Entstehung und Entwicklung einer wissenschaftlichen Tatsache. Einführung in die Lehre vom Denkstil und Denkkollektiv*, Basel: B. Schwabe und Co., Verlagsbuchhandlung. English translation: *Genesis and Development of a Scientific Fact*, Ed. by T.J. Trenn and R.K. Merton, Chicago/London: Chicago University Press, 1979.

Fleck, Ludwik. 1983. *Erfahrung und Tatsache. Gesammelte Aufsätze. Mit einer Einleitung herausgegeben von Lothar Schäfer und Thomas Schnelle*. Frankfurt: Suhrkamp.

Forman, Paul 1970. 'Weimar Culture, Causality, and Quantum Theory, 1918-1927. Adaptation by Weimar Physicists to a Hostile Intellectual Environment', *Historical Studies in the Physical Sciences* 3, 1–115.

Friedrichs, Robert W. 1970. *A Sociology of Sociology*. New York: Free Press.

Freud, Sigmund. [1930] 1961. *Das Unbehagen in der Kultur*. Vienna: Internationaler Psychoanalytischer Verlag. Translated by James Strachey as, Civilization and its Discontents', Standard Edition, vol. XXI.

Fuller, Steve. 2000. *Thomas Kuhn. A Philosophical History for Our Times*. Chicago University Press.

Giegel, Mans-Joachim. 1977. Gesellschaftliche Bedingungen des Theorienpluralismus in der Gesellschaftswissenschaft. In Hubig und von Rahden, 216–233.

Gieryn, Thomas F. 2010. Paradigm for the Sociology of Science. In *Robert K. Merton: Sociology of Science and Sociology as Science*, ed. Craig Calhoun, 113–139. Columbia University Press. http://www.jstor.org/stable/10.7312/calh15112.10.

Golinsky, Jan. 2011. Thomas Kuhn and the Interdisciplinary Conversation: Why Historians and Philosophers of Science Stopped Talking to One Another. In *Integrating History and Philosophy of Science. Problems and Prospects*, ed. Seymour Mauskopf and Tad Schmaltz, 13–28. Springer BV.

Gouldner, Alvin. 1965. *Enter Plato. Classical Greece and the Origins of Social Theory*. New York: Basic Books.

Gouldner, Alvin. 1970. *The Coming Crisis of Western Sociology*. New York: Basic Books.

Greshoff, Rainer. 2010. Die Theorievergleichsdebatte in der deutschsprachigen Soziologie. In *Soziologische Kontroversen*, ed. Georg Kneer and Stephan Moebius, 182–216. Berlin: Suhrkamp.

Greshoff, Rainer, Gesa Lindemann and Uwe Schimank. 2007. Theorienvergleich und Theorienintegration – disziplingeschichtliche und methodische Überlegungen zur Entwicklung eines paradigmenvermittelnden "conceptual framework" für die Soziologie. Oldenburg: (Diskussionspapiere/Arbeitsgruppe Soziologische Theorie 1–2007. https://www.uni-oldenburg.de/fileadmin/user_upload/sowi/ag/ast/download/dp/ast-dp-1-07.pdf

Gunnell, John. 2014. *Social Inquiry after Wittgenstein and Kuhn. Leaving Everything as it is*. New York and Chichester: Columbia U.P.

Gutting, Gary, ed. 1980. *Paradigms and Revolutions. Appraisals and Applications of Thomas Kuhn's Philosophy of Science*. Notre Dame: University of Notre Dame Press.

Habermas, Jürgen. 1968. *Erkenntnis und Interesse*. Frankfurt: Suhrkamp. Translated as *Knowledge and Human Interests*. London: Heinemann.

Halfpenny, Peter. 1982. *Positivism and Sociology. Explaining Social Life*. London: Allen and Unwin.

Heise, Arne. 2020. Ideology and Pluralism in Economics: A German View. *International Journal of Pluralism and Economics Education* 11 (2): 114–129.

Hieber, Lutz. 1975. Zum Konzept der Wissenschaft. *Leviathan* 3:449–472.

Hubig, Christoph, and Wolfert von Rahden, eds. 1977. *Konsequenzen kritischer Wissenschaftstheorie*. Berlin: De Gruyter.

Huebner, D. R. 2013. Wilhelm Jerusalem's Sociology of Knowledge in the Dialogue of Ideas. *Journal of Classical Sociology* 13 (4): 430–459.

Hübner, Kurt, Nikolaus Lobkowicz, Hermann Lübbe, and Gerard Radnitzky, eds. 1976. *Die politische Herausforderung der Wissenschaft, gegen eine ideologisch verplante Forschung*. Hamburg: Hoffmann und Campe.

Kaiser, David. 2016. Thomas Kuhn and the Psychology of Scientific Revolutions. In *Kuhn's Structure of Scientific Revolutions at Fifty. Reflections on a Science Classic*, ed. Robert J. Richards and Lorraine Daston, 70–95. Chicago University Press.

Kelvin, William Thomson. 1904. *Baltimore Lectures on Molecular Dynamics and The Wave Theory of Light*. London: C. J. Clay and sons; Baltimore: Publication agency of the Johns Hopkins University.

Knappertsbusch, Felix. 2023. From Paradigm Wars to Peaceful Coexistence? A Sociological Perspective on the Qualitative-Quantitative-Divide and Future Directions for Mixed Methods Methodology. *Forum Qualitative Social Research* 24, 1, Art. 2: http://qualitative-research.net

Kuhn, Thomas S. 1970. Reflections on my Critics. In *Criticism and the Growth of Knowledge*, ed. Imre Lakatos and Alan Musgrave, 231–278. Cambridge: Cambridge University Press.

Kuhn, T. S. 1992. *The Trouble with the Historical Philosophy of Science*. Robert and Maurine Rothschild Distinguished Memorial Lecture, 19 November 1991. Department of the History of Science of Harvard University, Cambridge, MA.

Kuhn, Thomas S. 2000. Rationality and Theory Choice. In *The Road Since Structure. Philosophical Essays 1970–1993*, ed. Kuhn. Chicago: University of Chicago Press.

Leeming, William. 1997. Revisiting Finalization. *Social Science Information* 36 (3): 387–410.

Leendertz, Ariane. 2010. *Die pragmatische Wende. Die Max-Planck-Gesellschaft und die Sozialwissenschaften 1975–1985*. Göttingen: Vandenhoeck & Ruprecht.

Leendertz, Ariane. 2013. "Finalisierung der Wissenschaft". Wissenschaftstheorie in den politischen Deutungskämpfen der Bonner Republik. *Mittelweg 36* 22 (4): 93–121.

Lehman, T., and R. T. Young. 1974. From Conflict Theory to Conflict Methodology: An Emerging Paradigm for Sociology. *Sociological Inquiry* 44:15–28.

Martins, Hermínio. 1972. The Kuhnian "Revolution" and Its Implications for Sociology. In *Imagination and Precision in the Social Sciences—Essays in Memory of Peter Nettl*, ed. Stein Rokkan, A. H. Hanson, and Tom Nossiter, 13–58. London: Faber and Faber.

Merton, Robert K. 1945. Paradigm for the Sociology of Knowledge (reprinted in *The Sociology of Science: Theoretical and Empirical Investigations* (University of Chicago Press, 1973), 7–40).

Mullins, Nicholas C. 1973. *Theories and Theory Groups in Contemporary American Sociology*. New York: Harper and Row.

O'Donohue, William. 1993. The Spell of Kuhn on Psychology: An Exegetical elixir. *Philosophical Psychology* 6 (3): 267–287.

Offe, Claus. 1977. Praxisbezüge der Sozialwissenschaft als Krisenwissenschaft. In, ed. Hubig and von Rahden, 216–251.

Outhwaite, William. 1983. *Concept Formation in Social Science*. London Routledge.

Outhwaite, William. 1987. Laws and Explanations in Sociology. In *Classic Disputes in Sociology*, ed. R. J. Anderson, J. A. Hughes, and W. W. Sharrock, 81–98. London: Allen & Unwin.

Outhwaite, William. 2017. *Zagadnienia Naukoznawstwa* 2 (212): 157–183.

Outhwaite, William. 2018. Kuhn and Social Science. In *Time, Science and the Critique of Technological Reason. Essays in Honour of Hermínio Martins*, ed. José Esteban Castro, Bridget Fowler, and Luís Gomes, 81–98. Basingstoke: Palgrave.

Pfetsch, Frank R. 1979. The 'Finalization' Debate in Germany: Some Comments and Explanations. *Social Studies of Science* 9(1 European Issue): 115–124.

Popper, Karl. 1970. 'Normal Science and its Dangers', in I. Lakatos and A. Musgrave (eds), Criticism and the Growth of Knowledge. Cambridge: Cambridge University Press.

Príncipe, João. ed. 2015. *Évora Studies in the Philosophy and History of Science. In Memoriam Hermínio Martins*. https://dspace.uevora.pt/rdpc/handle/10174/16647

Rasmussen, Stig Alstrup. 1982. Finalization and Completed Theories. *Zeitschrift Für Allgemeine Wissenschaftstheorie* XIII (2): 359–369.

Reingold, Nathan 1991. *Science, American Style*. Rutgers University Press

Schäfer, Wolf. 1983. *Finalization in Science: The Social Orientation of Scientific Progress*. Dordrecht: Reidel.

Schnelle, Thomas. 1982. *Ludvik Fleck: Leben und Denken*. Universität Hamburg: Philosophisches Seminar.

Schroyer, Trent. 1984. On Finalization in Science. *Theory and Society* 13:715–723.

Shapin, Steven. 1995. Here and Everywhere: Sociology of Scientific Knowledge. *Annual Review of Sociology* 21:289–232.

Sharrock, Wes, and Rupert Read. 2002. *Kuhn*. Cambridge: Polity.

Scheler, Max. 1926. *Die Wissenformen und die Gesellschaft*. Leipzig: Der Neue Geist Verlag.

Sherman, L. W. 1974. Uses of the masters. *American Sociologist* 9:176–181.

Shils, Edward. 1970. Tradition, ecology and institution in the history of sociology. *Daedalus* 99 (Fall): 760–825.

Smith, Mark J. 1998. *Social Science in Question*. London: Sage.

Sternberg, Robert J., and Wade E. Pickren, eds. 2019. *The Cambridge Handbook of the Intellectual History of Psychology*. New York: Cambridge University Press.

te Heesen, Anke. 2022. *Revolutionäre im Interview. Thomas Kuhn, Quantenphysik und Oral History*. Berlin: Wagenbach. [The publisher's website includes the original English quotations.]

Therborn, Göran. 1976. *Science, Class and Society*. London: New Left Books.

Trompf, Garry. 2011. The Classification of the Sciences and the Quest for Interdisciplinarity: A Brief History of Ideas from Ancient Philosophy to Contemporary Environmental Science. *Environmental Conservation* 38 (2): 113–126.

van den Daele, Wolfgang and Peter Weingart 1975. 'Resistenz und Rezeptivitat der Wissenschaft', *Zeitschrift fur Soziologie* 4 (2, April), 146–164.

van den Daele, Wolfgang, Wolfgang Krohn and Peter Weingart 1979. (sic) Geplante Forschung. Vergleichende Studien iiber den Einfluss politischer Programme auf die Wissenschaftsentwicklung (Suhrkamp, Frankfurt a/Main, 1979).

Walsh, David. 1972. Sociology and the social world. In *New Directions in Sociological Theory*, ed. Paul Filmer, Michael Philipson, David Silverman, and David Walsh, 15–36. Cambridge, Mass.: M.I.T. Press.

Wehrs, Nicolai. 2008. "Tendenzwende" und Bildungspolitik. Der "Bund Freiheit der Wissenschaft" (BFW) in den 1970er Jahren. *Potsdamer Bulletin Für Zeithistorische Studien* 42:7–17.

Weingart, Peter. 1983. Verwissenschaftlichung der Gesellschaft - Politisierung der Wissenschaft. *Zeitschrift Für Soziologie* 12:225–241.

Weingart, Peter. 1996. *Wissensproduktion und soziale Struktur.* Frankfurt: Suhrkamp.

Weingart, Peter. 1997. From "Finalization" to "Mode 2": old wine in new bottles? *Social Science Information* 36 (4): 591–613.

Part II
Origins, Development, and Methodologies of the Social Sciences

Chapter 9
From Objective Methodology to Situated Research: The Changing Place of the Researcher in French Sociology

Pauline Launay and Cherry Schrecker

Abstract This chapter retraces and analyses the evolution of the relationship between researcher and research object in French sociology from its academic and institutional origins until the present day. Our story, as do the majority of histories of French sociology, begins with Durkheim and his quest for an objective methodology. Over time, this approach weakens and is challenged by new ideas and empirical methods, inspired, for many, by American sociology or by French sociologists returning to France after the Second World War. Rather that considering the researcher as divorced from the social situation she or he is studying, these new approaches encourage them to consider the effects of their person on the situation observed and on the manner in which they interpret it.

Keywords Sociology · History · Methodology · France · Objectivity · Situated knowledge · Empirical research

9.1 Introduction

In French sociology, the relationship between the researcher and his or her research object has varied over time and in keeping with the theoretical approach which guides the research. Very schematically, the ideal research position has varied between a researcher whose ideal was to disengage themselves from the reality studied and one which stipulates that such a stance is impossible, the researcher being part of the reality studied. The first attitude was that developed by Émile Durkheim in his *Règles de la méthode sociologique* (1894). The second was

P. Launay
Caen University Hospital, Caen, France
e-mail: pauline.launay@unicaen.fr

C. Schrecker (✉)
Université Grenoble Alpes, Saint Martin d'Hères, France
e-mail: Cherry.Schrecker@univ-grenoble-alpes.fr

Y. Shan (ed.), *History and Philosophy of the Social Sciences*,
https://doi.org/10.1007/978-3-032-08167-4_9

developed in sociologie of action, in ethnomethodology and in anthropology. Early and progressive formulations of this standpoint can be found in the works of Max Weber, who carried out fieldwork (Weber 1892; Aldenhoff-Hübinger and Bruhns 2018) and gave some suggestions on ways in which the relation to the field could be envisaged theoretically (Weber 1968 [1922]). It is very clearly formulated by the philosopher and sociologist Alfred Schütz, who bases his reflections, among others, on Weber and the American pragmatists. Schütz suggests that the researcher, as co-participant of the situation they are observing, should put aside their personal biography (Schütz 1954) so as to understand the point of view of the other participants. Whilst these latter have personal objectives linked to the situation, the researcher is present for purely scientific reasons. In turn Schütz is cited by Erving Goffman (1974), Harold Garfinkel (1967) and Aaron Cicourel 1935 [1972]. The integration of this standpoint into French sociology was, as we shall see, a much later development. With reference to the concept of axiological neutrality (Weber 1917 [1965]) it is suggested that a way of attaining this is that researchers' social characteristics should be described and analysed in order to clarify her or his point of view and possible sources of bias. Thus, it could be argued that the researchers' social characteristics have become an important part of their research material.

This chapter will retrace and analyse the evolution of the relationship between the researcher and the research object in French sociology from its academic and institutional origins until the present day.

9.2 The Quest for an Objective Methodology: French Sociology and the Durkheimian School

We will begin with an evocation of the origins of sociology in France centred around the hegemonic figure of Émile Durkheim and his quest for an autonomous scientific sociology. Not until after his death did empirical currents other than statistical emerge, as will be shown in our presentation of the contribution of his nephew, Marcel Mauss. Other initiatives, which included the foundation of the *"College de Sociologie"*, were cut short by the *Vichy régime*, during which a large number of sociologists emigrated. Some of those who remained behind continued their research in government approved institutes, others were ostracised or died in captivity.

9.2.1 Sociology as an Independent Discipline: Methodological Rules and Academic Institutions

Durkheim's ambitions were both disciplinary and institutional. He describes in detail the theoretical basis of his sociology and, more importantly, the methodology by which sociology was to become a science in the *Règles de la méthode sociologique* (Durkheim 1894). Researchers were invited to shed their personal preconceptions and value judgments in order to treat social facts as observable from afar as if unknown to the researcher: “The first and most fundamental rule is: consider social facts as things.” (Durkheim 1894, 32) This involved adapting the methodology of the natural sciences whose methods were “positive”, that is to say based on empirical data. In keeping with the idea of the scientific object, it was necessary to observe society from a distance. In this conception of sociology as a scientific discipline the researchers’ viewpoint is considered as disembodied, neutral and objective (and thus neither situated nor judgemental). In Durkheim’s view, data was objective in so much as it was not altered by human interpretation. As we all know, statistical data was considered to have this quality (even though, more recently, the way in which the statistics referred to by Durkheim were compiled and interpreted has come under scrutiny and provoked much discussion (Besnard 1973). In the preface of the second edition of *Les Règles de la méthode sociologique*, Durkheim affirms that he does not seek to deny that human activity is important in social life but, as he states later in his study of suicide (Durkheim 1897), he believes that individual motivations, in this case for killing oneself, are inaccessible even to the person concerned. If the person were able to analyse their motivations there was also the danger that they might lie about these. As such, personal reports emanating from individuals cannot be observed empirically and are too unreliable to serve as a base for the definition of a research object (PUF 9th edition 1930, p. 4). To a great extent also, society is seen as determining human behaviour. Thus, there is no place for the individual and her or his point of view in Durkheim’s methodology, be it as a subject of the research or as a researcher. Only at this price was it possible to see sociology as scientific and separate from neighbouring disciplines notably psychologie (based on the individual) and philosophy (purely speculative as devoid of an empirical methodology).

As far as institutionalisation is concerned, Durkheim’s ambition was to create sociology as an independent academic discipline centred around himself and his disciples. The first university post in which sociology was taught, entitled Social Science and Education, was created at Bordeaux university in 1887 and occupied by Durkheim who in fact taught relatively few sociology courses (Lukes 1973). Becoming hegemonic also involved eliminating potential disciplinary and personal competitors, refusing to acknowledge competing literature and to publish in scientific journals coordinated by potential competitors. Among these latter were Frédéric Leplay, Gabriel Tarde and René Worms whose research concentrated on the individual or on smaller scale situations. It was qualified as unscientific by the criteria established by Durkheim (Karady 1979; Mosbah-Natanson 2008; Masson and Schrecker 2016, 2017). Two of these authors, Leplay and Tarde, carried out smaller scale empirical research, in which the researcher was physically present, this, as we have seen, was not the case of the Durkheimian researcher. Though their research is better known to contemporary sociologists it was little referred to in the mainstream sociology of the time in France.

Political differences can also be observed, the Durkheimian point of view has been qualified as republican (Lubek and Apfelbaum 1981), such a political stance involved attention toward political equality rather towards individual achievement as would be the case in more Liberal politics which give greater importance to individual motivation and initiatives. In the republican ideology, the individual, be they researcher or researched, takes second place to larger political or cultural entities.

After the death of several members of the Durkheimian school during World War I, including Durkheim's son and Durkheim himself (in 1917), the remaining members continued to promote the Dukheimian tradition. According to Johan Heilbron (1985, 2015) they were divided into two groups, the members of one, which included Maurice Halbwachs, Marcel Mauss and François Simiand, concentrated on research. The other group, among whom Celestin Bouglé, Paul Fauconnet, Georges Davy and Paul Lapie, spent most of their time on teaching activities. In the early 1920s only four chairs existed in sociology and most of the teaching was in "sociology and morality", a certificate which was part of a degree in philosophy. Thus sociology was not very visible in larger academic circles. What is more, disputes arose as to the intellectual quality of the teaching group, the researchers refusing not only to help them set up sociology courses at university level (Marcel 2001) but also to be involved in any teaching or administrative activities. The researchers insisted on the scientific nature of Durkheim's, and, by extension, their own, production and put great emphasis on the need for scientific proof. A large part of their publications consisted of economic and philosophical developments of Durkheim's works often linked to the concept of collective representations. They were theoretical, highly specialised and hardly accessible to the general public. Only Mauss, who was not a great communicator and whose ideas attracted little attention during his lifetime, developed an interest for empirical research.

9.2.2 Marcel Mauss and the Contribution of British Anthropology

It is important to emphasise that the founding figures of French sociology, with the possible exception of Alexis de Tocqueville and Frédéric Le Play, neither conducted field research in the true sense of the term nor formulated the fundamental principles of the then emerging sociology in connection with empirical methodology. As we have seen, Durkheim, influenced by Auguste Comte, whose use of the historical method involved the consultation of numerous documents, neither engaged in small-scale fieldwork, nor contemplated theorising this approach. Like Comte and Durkheim, a large majority of French sociologists at the beginning of the twentieth century were trained in philosophy, and situated empirical studies were far from their preoccupations. To a great extent then, small-scale empirical methodology in French sociology today stems from the adoption of methods developed in foreign sociology, be it American or German, or in anthropology, stimulated by the works of Marcel Mauss.

Whilst he did not himself carry out fieldwork, Mauss advocated an ethnographic approach which owed much to British anthropology, thus creating intellectual links between this latter and Durkheimism. In his *Manual of Ethnographie* (Mauss 1947) Mauss accorded great importance to field studies. His objective was to describe and develop a 'total social phenomenon' with important theoretical implications. It is interesting to note that Mauss characterised the ethnographic approach as relatively independent of a theoretical framework. The objects studied are not approached in a neutral manner but are examined through the prism of implicit knowledge. This includes that of the relationships maintained in the field with other dimensions of social reality and the accumulated knowledge available to the researcher, both within and outside the discipline. The interpretation of meaning is not conducted solely via the understanding of the immediate circumstances, it refers to a set of propositions necessary for the comprehension of the behaviour observed. In order to achieve the distance necessary for an objective understanding of the situation, the use of an interpretative language is required. It is constituted by a collection of general propositions not directly derived from the field. Thus, the interpretation of a field involves a dual process of comparison: that between wider social contexts and the specific situation studied. Once the social situation to be studied is characterised and identified, the wider social situations are defined via comparisons made between various known social behaviours, drawing on their similarities and differences. This background of comparisons generates a tacit theoretical knowledge that is mobilised in the interpretation of the field.

In this manner the base for empirical investigation was established, but Mauss greatly suffered during World War II, during which his health and living conditions declined. In order to avoid dismissal by the Nazi regime, he resigned from his post at the *Collège de France.* From this point, his intellectual production was greatly reduced and the anthropological techniques he had developed were never applied to concrete circumstances. On the 26th of July 1944 Halbachs was arrested and taken to Buchenwald where he died in March 1945 (Mucchielli and Pluet-Despatin 1999). Mauss, tired and discouraged, died in 1950.

Among the initiatives observed during the interwar period, a short-lived adventure in the form of the *Collège de Sociologie* deserves our attention in that it seeks to venture beyond the beaten tracks of positivist sociology in order to forge connections with literature and to move beyond the strictly rational aspects of social activity.

9.2.3 The Collège de Sociologie*: The Alliance of Knowledge and Action*

In 1937, a group of intellectuals, which included well-known figures such as Georges Bataille (1897–1962), Roger Caillois (1913–1978), Michel Leiris (1901–1990), and Jules Monnerot (1909–1995), signed the inaugural Declaration of the College of Sociology, published in the journal *Acéphale*. The College was conceived of both as an experience and as a means of experimentation and those who participated considered themselves as both experimenters and subjects of observation.

Their relationship with the Durkheimian school was ambivalent. The Collège was named in the wake of Marcel Mauss and in reference to Durkheim, though it sharply departed from the Durkheimian paradigm. For example, in 1946 Jules Monnerot authored a work titled *Les faits sociaux ne sont pas des choses* (Monnerot 1946). Two aspects highlight deviations from Durkheimian principles (Bridet 2007). The College of Sociology positions itself as an institute dedicated to "the study of social existence in all its manifestations where the active presence of the sacred is evident" (Caillois 1938). While the choice of focus is clearly within the Durkheimian tradition, the College does not confine itself to the consideration of social facts. Indeed, it seeks to identify "points of coincidence between the fundamental obsessive tendencies of individual psychology and the guiding structures that preside over social organisation and command its revolutions" (*Ibid.*). This acknowledgment of psychological determinations based on the biological reality of humans represents a significant departure from the French sociological tradition of the time.

It seems pertinent to add, as a second transgression, that the objectives set by the Collège were not solely scientific. They also encompassed moral and political dimensions. Had the College merely claimed to understand contemporary society in order to inform political choices, it would have been closely aligned with the perspectives of Mauss and Durkheim, who believed that sociological knowledge must ensure the consolidation of republican institutions. However, members of the College of Sociology went further by engaging in politics within the framework of their sociological activity. The activist nature of sociology is not the result of an arbitrary decision; it is grounded in specific objects of study with reciprocal influence: the sacred, power dynamics, myths, impulsivity, affectivity, and eroticism. The College asserts the necessity of a theory that is directly linked to political practice, thereby challenging the separation between knowledge and action. The researchers' position is seen here as engaged rather than distant or neutral as had previously been the case in French sociology.

The activities of the Collège were abruptly interrupted in July 1939 due to the outbreak of World War II, which put an end to the exchanges that had begun with members of the Frankfurt School. Nevertheless, its impact persisted after the War, especially in the fields of literature, phenomenological sociology, and clinical sociology (*cf.* Hollier 1979; Poutot 2013 for more detail on the *Collège de Sociologie*).

Other initiatives during the second half of the 1930s involved younger researchers, of whom Jean Stoetzel. Stoetzel, who had been on a study visit to America in 1938, began to explore empirical methodologies, mainly via the development of statistical methods founded on the Gallup polls. According to him French sociology in general was on the upturn at the end of the 1930s. As was the case for the *Collège de Sociologie*, this developpement was interrupted by the Nazi occupation of France and the onset of World War II (Stoetzel 1957).

9.2.4 *Sociology During the* Vichy Régime

The Vichy regime severed the institutional bases of sociology in France, deeming it subversive due to the involvement of some of its figures in the "socialist lobby" (Ferrette 2012). As from 1940, sociology professorships were gradually eliminated, and the *Centre de documentation social* (CDS) of the *École Normale Supérieure* (ENS) was disbanded. Research programs were cancelled or redirected towards institutions like the *École des cadres d'Uriage* (Uriage Executive School), established in 1940, or the *Alexis Carrel Fondation*, established in 1941. This latter included sociologists who were to become prominent during the postwar era, such as Paul-Henry Chombart de Lauwe, Alain Girard, and Jean Stoetzel who was director of social sciences. The topics of study at these institutions were influenced by the political agenda of the Pétain régime, focusing on subjects like birth rates, eugenics, and ruralism.

French sociologists who were not prepared to integrate these government-propelled institutions or who, for other reasons, could not hold academic posts during the War, experienced diverse fates. Max Bonnafous and Marcel Déat joined the government, Georges Gurvitch emigrated to New York where he integrated the *Ecole libre des hautes études* (ELHE) at the New School for Social Research in the company of, among others, Jacques Maritain Alexandre Koyré and Claude Lévi-Strauss. Georges Friedmann became a member of the Résistance before going into hiding in 1943, and Raymond Aron joined General De Gaulle in London. Only Maurice Halbwachs continued his research in France until he was arrested and deported.

With the death of Maurice Halbwachs in 1945, following earlier deaths such as those of Simiand in 1935, Fauconnet in 1938, and Bouglé, who committed suicide when the Germans entered Paris, in 1940, the "golden era" of Durkheimian sociology, already weakened by the end of the 1930s, came to a close, at least temporarily as, as we will see, Durkhiemian influence, after a period of latence, has been much discussed and continues to this day to be considered important by sociologists.

9.3 The Aftermath of World War II: A Diversity of Theoretical Approaches and Empirical Inquiries

French sociology, greatly weakened by the end of World War II, found a new dynamic via the contribution of sociologists returning from the United States or from other places of exile. Many of them were looking for new approaches to replace Durkheimianism, often seen as out-moded. Particularly important for our purpose is that renewed empirical approaches were part of their baggage. As we will see, German sociology was also to open new horizons.

On an institutional level, French sociology benefited from the foundation of two new institutions: the National Institute for Demographic Studies (INED) in 1945 and the National Institute for Statistics and Economic Studies (INSEE) in 1946. Other significant developments included the reopening of the *Centre d'Etudes Sociologiques* (CES) probably initiated by Gurvitch on his return from the US (Heilbron 2015). The CES was the first public research laboratory of the post-war period under the direction of the CNRS. The creation of the sixth section of the *Ecole Pratique des Hautes Etudes* (EPHE) in 1946. Though this latter was at first mainly a venue for scholarly lectures, it became, as from 1948–49, a place where empirical research was encouraged and carried out. Nevertheless, French sociology still faced challenges due to a lack of organisational structure within a broader political context of reconstruction, resulting in a multitude of individual scholarly pursuits.

From the 1950s French sociology became ever more diversified and institutionalised. The 1960s and 1970s marked a tumultuous period, characterised by the polarisation of various intellectual currents. The debates and intellectual conflicts of this era had lasting repercussions on the discipline in France. Several factors contributed to this polarisation. Firstly, the rapid expansion of sociology in France following World War II, with the establishment of new sociology departments in universities, attracted a substantial influx of students and researchers. This growth nurtured a diverse array of sociological perspectives and ideas. Moreover, the 1960s and 1970s were marked by significant social and political upheaval in France. Student protests, labour strikes, and profound cultural shifts fueled sociological discussions. Sociologists felt compelled to engage with the pressing issues of the time, thus contributing to the polarisation. During the 1960s and 1970s, sociology faced essential questions regarding its ability to identify laws, produce objective discourse, or be considered a science. Simultaneously, the criteria for scientificity itself came under scrutiny.

9.3.1 Sociology from Overseas and the Development of an Empirical Methodology

Many of the major developments in empirical methodology were to come from the other side of the Atlantic Ocean, brought back by sociologists returning from America where they had been exiled during World War II.

The exchange of sociological ideas across the Atlantic and the consequent crossed influence of European and American sociology and anthropology are not new (*cf.* Schrecker 2010). As from the end of the nineteenth century and during the first decades of the 20th study visits in Europe were of great importance for American sociologists seeking to enrich the discipline in their home country. Major imports were the works of Émile Durkheim, Max Weber and Georg Simmel which were to inspire, among others, the journalistic approach of Robert Park and the functionalism of Talcott Parsons, two perspectives which, each in their own way, were to contribute to the development of the sociologies of action. Later, many

European sociologists visited America. Some, such as Paul Larzarsfeld, remained, others brought back ideas which were to renew methodological approaches. This was the case of Stoetzel, for example, after his study visit in America who, inspired by the Gallup polls and Theodor Adornos's attitude scales, was instrumental in the development of statistical methods in France after the war. Another returnee from the United-States at the end of the War was Georges Gurvitch. Before the war Gurvitch had held a post in the university of Strasbourg and, prior to this, had trained in Germany. On his return to France Gurvitch attempted to integrate American and French sociology, proning the ideas of liberty and social commitment and attempting to combine empirical and theoretical elements. Heilbron (2015, p. 133) argues that the resulting collection of typologies failed to give clear guidelines on either. For this reason and for others linked to his difficult personality and his manner with students, Gurvitch never gained key positions in French sociology. Others did: in 1955 Stoetzel was appointed as a professor at the Sorbonne. Along with Claude Lévi-Strauss and Raymond Aron he came to represent the new generation of sociologists (Heilbron 2015).

Claude Lévi-Strauss, a major figure in post-war structuralism, carried out fieldwork on a long term basis particularly in Brazil. His meeting with Roman Jakobson during his stay at the New School in New York is legendary and often cited as at the origin of structuralism (Durand and Albert 2011). Among other things, and as far as we are concerned in this chapter, Lévi-Strauss contemplated the question of subjectivity and writing in ethnological fieldwork (1955). Mondher Kilani (1990), notes that there seems to be a contradiction in his epistemological stance which combines Durkheimian postures and a growing reflexivity. Lévi-Strauss conceives of the field as an inductive collection of data, as, according to him, "all facts must be exactly observed and described, without allowing theoretical prejudices to alter their nature and importance" (Lévi-Strauss 1958, p. 307). When LéviStrauss also asserts that "facts must be studied in themselves," he posits a conception of reality external to the observer. However, his viewpoint appears closer to constructivism when he later declares in his *Introduction à l'œuvre de Marcel Mauss* (1980, p. XXVII–XXVIII) that, in a science where observation is "of the same nature as its object, the observer is himself a part of his observation." He pursues: "To properly understand a social fact, it must be apprehended in its entirety, that is to say, from the outside as an object, but as an object of which the subjective (conscious and unconscious) apprehension that we would have of it, inevitably being humans, is an integral part, living the fact as indigenous rather than observing it as an ethnographer." (our translation). We will see below some of the ways in which Structuralism was developed by the next generation.

Another returnee, this time from England, was Raymond Aron (1905–1983). Aron played a crucial role in the evolution of French sociology, bridging the gap between Durkheim's followers and the new generation. He emphasised the need for empirical research in the transition from philosophy to sociology indeed, most French sociologists during the immediate post-war years and even later were trained in philosophy. This was the case of Aron whose academic career had begun at the *Ecole Normale Supérieure* (ENS), where, in 1928, he passed his aggregation in philosophy. He went on to teach in Germany from 1930 to 1933. On his return to France, he became secretary at the Centre for Social Documentation from 1934 to 1939. During World War II he worked in General de Gaulle's staff in London and founded the journal *La France libre*. On his return from London after the War, Aron became Professor at the *Institut d'Etudes Politiques de Paris, from 1945 to 1954*, followed by a sociology chair at the Faculty of Letters in Paris from 1955.

Aron was instrumental in introducing German sociology in France. In his book *Contemporary German Sociology* (Aron 1935) he aims to reconcile the French and German sociological traditions.

His doctoral thesis in 1938 had investigated Historical knowledge, which held a major place in German methodological debates. Aron distanced himself from holistic and deterministic sociology, advocating a combination of explanation and understanding. His impact extended beyond theory, influencing intellectuals such as Pierre Bourdieu, his student and later his collaborator. In 1970, Aron took on the chair of Sociology of modern civilization at the *Collège de France*. While Aron's theoretical contributions were significant, he did not develop practical applications for his sociology. As we will see, this gap was to be filled by Pierre Bourdieu. Though Bourdieu was trained as a philosopher, empirical observation was to become an important element of his sociological practice.

9.3.2 Structuralist Sociology and Marxist Sociology: Behind the Structures, the Eclipse of the Subject?

The influence of structuralism, in the wake of Claude Lévi-Strauss, and of Marxist sociology, on the social sciences was significant. While these two perspectives do not entirely overlap, they converge by their emphasis on the weight of structures and social determinism. They diverge, however, on the issue of societal dynamics, their historicity, and the question of dialectics (Lefebvre 1963). Critics have accused them of distancing themselves from concrete realities and individuals, whilst emphasising an abstract analysis of society.

Throughout the decades of the 1960s and 1970s, marxists like Henri Lefebvre contemplated the issue of alienation and the subjectivity of actors (Le Roulley 2021). Clinical sociology, which emerged in the 1960s, was directly influenced by Lefebre and Cornelius Castoriadis and by psychoanalysis, which was in vogue at that time. Castoriadis, alongside Claude Lefort, co-founded the group and journal *Socialisme ou barbarie*, which was of Marxist and anti-Stalinist orientation. In his work *The Imaginary Institution of Society* (1975), Castoriadis emphasised the dialectical, symbolic, and imaginary dimensions of the social, defining the institution as a dialectical relationship between the instituting and the instituted. Among the leading figures of clinical sociology is René Lourau, who developed the concept of "institutional analysis" (1975), an approach aimed at unveiling power dynamics and issues within social institutions. He employed an engaged methodology, developing intervention within organisations. Additionally, Georges Lapassade, a proponent of socio-analysis, played a pivotal role by introducing participatory research methods and underscoring the importance of researchers' involvement in the field. Clinical sociology encourages researchers to adopt a reflexive stance and acknowledge their own positioning in their research. It often favours qualitative methods such as participant observation and in-depth interviews to capture the richness of the social contexts under study. This school of thought argues that it is inadequate to analyse an institution without considering the lived experiences of those who shape it or are subjected to it. Clinical sociology proposed an epistemological rejuvenation, advocating an interdisciplinary approach, especially with psychoanalysis and anthropology. It has also made theoretical contributions, particularly in the comprehension of institutions and their unconscious facets, and instigated a methodological shift toward active immersion within organisations.

A generation later structuralism was developed by Pierre Bourdieu. He advocated a methodological approach to the analysis of social and cultural structures, emphasising the pivotal role of these structures in individual lives. He introduced concepts like cultural capital and the social field (Sapiro 2020) in order to explain how social inequalities were shaped by structural forces. Bourdieu's conception of the researcher's and his place in the field varied throughout his career. In his early works, notably *Les Héritiers* (1964), he adopted a detached approach to the field, emphasising an objective observation of social structures and *habitus*. Over time, his perspective on the researcher's role was transformed. Bourdieu developed the process of objectivation, emphasising the importance of a more direct involvement of the researcher in the context studied. He affirmed that a deep understanding of social realities required active participation and engagement in the field, whilst remaining aware of the pitfalls of subjectivity. His evolution over time can also be observed in the development of socio-analysis, an approach that has gained prominence and extends beyond the field of sociology today. Socio-analysis encourages a thorough examination of social mechanisms by integrating the researcher's personal involvement and social inheritance. This perspective was innovative in that it combined the idea of scientific objectivity with a recognition of subjective dynamics and power relations present in the field. Bourdieu's intellectual legacy, marked by his shift toward more situated research, continues today to influence disciplinary and interdisciplinary approaches.

9.3.3 The Researcher's Involvement: Alain Touraine, Raymond Boudon and Edgar Morin

The academic landscape became polarised around the figures of Pierre Bourdieu and Raymond Boudon, while Alain Touraine's sociology aimed to amalgamate actionism and sociological interventionism, involving sociologists actively in the observed social movements. Clinical sociology, which integrated the influence of psychoanalysis, also challenged conventional approaches to investigation and the significance of participants' prior experiences.

Other perspectives were thus developed at the same time as those of Bourdieu, creating intense debate during the 1960s and 1970s. Questions regarding the researcher's neutrality came to the fore, accompanied by question of their role in sociological research. The dispute between Bourdieu and Raymond Boudon gave rise to a polarisation in the sociological field, divided between proponents of holism and those advocating methodological individualism. Holism, as seen by Bourdieu, was to expose social injustice, thus contributing to social change. Boudon argued in favour of strict neutrality on the part of the researcher in order to maintain objectivity and detachment from any form of engagement. This current emphasised the importance of methodological rigour in sociological research. Methodological individualism, as its name suggests, is centred on individuals and their rational choices, downplaying the role of social structures in the explanation of human behaviour.

Alain Touraine also made a significant contribution, particularly in the domain of the sociology of work and organisations, by introducing the concept of sociological interventionism. He advocated that sociologists play an active role in impulsing social transformation by using their research to inform social movements and public policy. Touraine supported the idea of sociologists' engagement in addressing social issues.

These contrasting perspectives gave rise to a variety of debates and positions encompassing holism and individualism, determinism and autonomy, constraint and freedom, neutrality and engagement, agent and social actor. Researchers were thus obliged to consider the part played by individuals and social structures in their analyses of society and to reflect on the way of integrating (or not) these elements in their research. They were encouraged to examine their own position, values and influences, and to consider that sociological research cannot be entirely devoid of their own implication.

Edgar Morin remained outside this triumvirate. As a sociologist and philosopher, he is primarily known for his manner of approaching reality in all its richness and complexity. Although he is not specifically associated with particular research methodologies, he has written about his conception of research, the field, and the researcher's involvement. Firstly, via a series of works titled *La Méthode* (1977–2008) in which he explores the complexity of reality. He emphasises the importance of a holistic approach and of considering the multidimensionality of phenomena under study. He suggests that the researcher should be open to a global and contextual understanding of the field. Secondly, Morin's encouragement of the researcher's involvement, not only implies active participation in the field but also an awareness of their own subjectivity and its influence on their results. He affirms that the researcher cannot be entirely neutral and must recognize their role in knowledge construction. Thirdly, he insists on the necessity of transcending reductionist and linear approaches. According to him, reality is complex, consisting of relationships and interactions between a large number of elements. Thus, the researcher should be willing to use methods that capture this complexity, freeing themselves from overly rigid frameworks. Morin also advocates a transdisciplinary approach that integrates various perspectives so as to address the complexity of the phenomena studied, in order to attain this end various methods can be mobilised. Though Morin did not develop a specific research methodology, his ideas have been influential in that they encourage researchers to reconsider their methodological approaches in order to restitute the complexity of the field and to reflect on how to approach reality in a more comprehensive and nuanced manner.

9.4 Methodological Evolution and Epistemological Reflection

The 1980s and 1990s were a pivotal period for French sociology, marked by a significant evolution in the conception of fieldwork and the researcher's involvement. This era witnessed a decline in holistic perspectives, particularly structuralism and Marxism, giving way to methodological and epistemological pluralism influenced, in part, by ongoing reflections in the field of anthropology.

9.4.1 Qualitative Methods and Endotic Anthropology

The emergence of endotic anthropology, which, we suggest, is situated in the wake of American and British community studies, influenced the way in which sociologists approached methodology and their connection to the field. In France, anthropologists such as Michel Agier and Marc Augé promoted this approach, emphasising the study of local realities, daily practices, and cultural meanings intrinsic to specific communities. This perspective revitalised field investigations, encouraging sociologists to immerse themselves in the context of their study and to consider local perspectives.

Exotic anthropology faced a crisis, notably linked to movements against colonialism and the rise of independence movements. As a result, the CNRS established *Les missions du Patrimoine* and reconnected with the tradition of European folklore. Consequently, many ethnologists, in the latter part of their careers, practised a Proximate Ethnography (endotic rather than exotic) on French territory. Influential figures such as Pierre Bourdieu, Jeanne Favret-Saada, Gérard Althabe, Hélène Clastres, Solange Pinton, and Yvonne Verdier emerged in this context, significantly shaping the restructuring of qualitative research in anthropology and sociology. The methods of ethnology and qualitative sociology became very close, adopting an inductive approach in which the researcher knows in advance what they are looking for, but are also interested in what the research subjects reveal during the research. What is observed depends on the role played in the environment studied, sometimes involuntarily, as illustrated by Jeanne Favret-Saada's work. In *Les Mots, la mort, les sorts. La sorcellerie dans le bocage* (Favret-Saada 1977). Immersion in the field is taken to its extreme and the author affirms that it is illusory to consider the possibility of objective understanding. In her study of witchcraft in the Bocage

Favret-Saada abandons her own norms, beliefs, customs and everyday practise in order to adopt those of the groups studied. The mere presence of the researcher constitutes an intervention in the lives of their subjects speech is considered as performative, thus the researcher interprets what is said not as truth or falsehood but as an act aimed at producing an effect. Favret-Saada's work renews the theoretical scope of qualitative inquiry, and concludes when the researcher understands their role in specific situations during the investigation. Rather than becoming captive to the interpretations of the subjects, researchers can incorporate them and compare them with their own observations.

In this context, the issue of the researcher's involvement was to become crucial in sociology as well. Sociologists, including Pierre Bourdieu and philosophers such as Michel Foucault were major figures contributing to this discussion. They acknowledged the importance of active engagement in the field, challenging the distanciation of previous decades. The idea that total neutrality was impossible gained ground, and researchers sought to integrate their own positions, subjectivities, and experiences into the research process.

Simultaneously, debates on the political engagement of the researcher gained prominence. Some researchers, influenced by movements like activism and critical sociology, advocated for more direct involvement in social issues. Others, drawing inspiration from the Durkheimian tradition, emphasised the need to maintain a critical distance to preserve scientific objectivity. These discussions contributed to shaping a new generation of French sociologists characterised by an increased sensitivity to the nuances of the field and a thoughtful reflection on the ethical issues related to the researcher's involvement.

9.4.2 Action Theories and the Place of the Researcher

During the same period we observe a renewed interest in qualitative methodologies. Currents like symbolic interactionism, along with authors such as Erving Goffman, influenced qualitative research by emphasising the importance of social interactions and symbolic meanings. Sociologists explored more flexible and contextual approaches, adapting to the complexity of the social phenomena studied. The significance accorded to a nuanced understanding of social realities led to increased interest in methods such as participant observation and in-depth interviews which necessitated long-term involvement in the field.

This movement was fed by the translation of several authors working in North America in various institutions, among which the University of Chicago and the New School for Social Research. Translations into French of American research involving theories of action resulted from an interest in this approach by a restricted number of French researchers, these works thus became generally available to French sociologists. A majority of these translations were published during the 1970s (Goffman, Cicourel) and 1980s (a collection of works of Alfred Schutz (1987), and Howard Becker's *Outsiders* in 1985). Surprisingly Howard Garfinkel's work, which grapples with questions linked to the perception of social norms and the research methods appropriate to their understanding, was translated some 20 years later, in 2007.

In relation to these translations and many others, was a growing interest in the Chicago School of Social Sciences which led to the publication of large numbers of articles, and later several books, on the history of the Chicago school and the methodology employed (for example, Chapoulie 2001; Guth 2004, 2008). Knowledge of the works of Robert Park and Ernest Burgess enlarged the possibilities for sociological fieldwork. Though Schütz had explored the means of scientific observation (the researcher should put aside their personal biography in order to understand the motivations and viewpoint of the persons involved in the situation), the observers in the works of Becker, Goffman and Garfinkel often seem invisible or devoid of social characteristics. The development of qualitative research and the immersion in the field gave rise to questions as to the influence of the researcher on the research situation and the way in which the material collected and analysed is influenced by their personal characteristics.

Several methodological stances were inspired by action theories, among them Methodological individualism which gained importance as an approach focusing on understanding social phenomena through the prism of individual actions and decisions. Influenced by Boudon, this approach emphasised the importance of considering individual rationality and choice in sociological analysis. Actionalism emerged in an attempt to integrate dynamics between social structures and individual actions. Ethnomethodology, influenced by Harold Garfinkel, focused on studying the methods individuals use to make sense of their social world, involving detailed examination of interactions and daily practices, challenging established norms and assumptions about social reality, highlighting the importance of understanding how individuals construct and interpret meaning in their social interactions. Actor-Network Theory (ANT), developed by Michel Callon and Bruno Latour, emphasised the role of non-human entities, such as technologies or institutions, in constructing social realities, contributing to a more comprehensive understanding of the complex interaction between human and non-human elements in sociological inquiry. These latter pioneered ethnographic description in the laboratory, challenging the Popperian paradigm of objective scientific knowledge.

9.4.3 *Relativism and Epistemological Reflection*

The shift in epistemological reflection led, in some camps, to a more relativistic perspective, represented, for exemple, by Michel Callon and Bruno Latour. Though it remained fairly marginal in sociology, Postmodernism, involving key figures like Jean-François Lyotard, questioned grand narratives, emphasising the fragmented and contingent nature of knowledge. It encouraged a critical reflection on power dynamics embedded in language, representation, and discourse, challenging traditional sociological frameworks. In this intellectual landscape, epistemological reflection went beyond methodological innovations, engaging in the translation or reexamination of past or neglected authors. Researchers critically reassessed research methods, promoting a deeper understanding of the inherent complexities of the research process. Concurrently, works dedicated to epistemology, such as those by Passeron (1991), Schapper (1998) and Berthelot (2001) emerged, questioning the specificities of sociological thought and writing.

The translation or reexamination of past or marginalised authors became a notable feature of epistemological discourse which involved revisiting the works of earlier sociologists, shedding light on overlooked perspectives, and contributing to a more comprehensive understanding of the historical foundations of the discipline. Researchers engaged in a critical dialogue with the sociological tradition, acknowledging the contingent nature of knowledge production and the importance of diverse intellectual contributions.

At the same time, critical reflection on research methods gained importance. Sociologists questioned the underlying assumptions of established methodologies, paving the way for more reflective and context-sensitive approaches. This critical stance toward methods was part of a broader effort to dismantle rigid frameworks and encourage a more dynamic and adaptable understanding of sociological inquiry. Works explicitly dedicated to epistemology played a significant role in shaping the intellectual landscape. The authors explored the intricacies of sociological thought and writing, contributing to a reassessment of underlying assumptions and inherent biases in sociological practice. This stimulated an increased awareness of the challenges and complexities inherent in sociological knowledge production.

9.4.4 Implication, Reflexivity, and Commitment: New Critical Theories of Implicit Normativities

During the twenty-first century in French language sociology, gender studies and post-colonial theories, such as Standpoint Theory and Black Thought, have given rise to criticisms further challenging the notion of a disinterested and universally objective scientific standpoint. These theories assert that scholarly production is socially rooted and entangled in power relations. Rather than distancing themselves from these roots, many sociologists now prefer acknowledging them, contending that this acknowledgment leads to the production of more reliable knowledge.

Andrew Abbott emphasises contemporary shifts in the relationships between empirical horizons (pertaining to truth or falsity) and normative horizons (concerning good or evil). Abbott advocates moving beyond empirical studies to develop a new normative ontology that embraces an understanding of societies beyond the Western context. This new ontology should facilitate the assessment of social inequalities and gender relations from perspectives divergent from the ontological framework of contractual liberalism, predominant in Western societies, but not universally accepted. This critique aligns with post-colonial theories, Standpoint Theory, and Black Thought.

Post-colonial theories scrutinise power dynamics between the Global North and South, including ethnocentric Western epistemology. Figures like Mahmood Mamdani from Uganda (1996) and Achille Mbembé from Cameroon (2005) analyse current relations between formerly colonised and colonising countries, as well as the colonial past's impact on colonised nations. They explore power dynamics within decolonised regimes in a context of globalised neoliberalism affecting all societies and seen as weakening the oppressed.

Standpoint Theory, also known as "positionalism" or "situated knowledge," originates from feminist and Marxist works. It challenges the idea of disinterested and universally objective scientific objectivity, proposing that scholarly production is socially rooted and embedded in power relations. Rather than distancing oneself from this rootedness, the approach is to acknowledge and describe it as a pathway to more reliable knowledge. This theory aligns with Black Thought or African American thought. Bell Hooks, also known as Gloria Jean Watkins, is a key figure in this context. In 1981, she published *Ain't I a Woman? Black Women and Feminism*, (Hooks, 1982) examining the connections between race, class, and gender in various social spheres. These perspectives have significantly influenced certain currents of French sociology since the year 2000, advocating a fundamental reassessment of the existing social order. Drawing inspiration from Marxist traditions, the critical theory of the Frankfurt School, and activist sources linked to social movements, these currents question multiple fronts of emancipation, encompassing not only social class but also feminism, anti-colonialism, ecology and discrimination based on age or disabilities.

In their work *Les Politiques de l'enquête* (2008), Alban Bensa and Didier Fassin emphasise the researcher's significance in their field and the necessity of reflexive work on the socio-historical context of the interaction between the investigator and research subjects, and seek to ascertain the effects of asymmetrical power dynamics within the investigative relationship. Positioned within the social space and immersed in the field of inquiry, sociologists are also products of a socialisation process that shapes political and ideological inclinations as from birth. This affects the research process from the construction of the object to the presentation of findings. Their involvement renders imperative a reflexive process, questioning the conditions of researchers' production.

Reflexivity, as perceived at present, involves an auto-socio-analysis undertaken by the sociologist in order to objectify her or his own unconscious bias, and is considered to be a fundamental condition for ensuring the scientific rigor of sociological practice. Contesting the principle of researchers' non-neutrality (Naudier and Simonet 2001) has become challenging, as advocating for a supposedly neutral stance ultimately implies adhesion to the established social order. Aligned with the reflexivity developed by Bourdieu's followers, among others, since the 1990s. Failure to acknowledge one's social position and trajectory would amount to the "illusion of making an illusion" (Mauger 1991, 128). In the absence of a reflexive posture, the historical and social determinants, conscious or unconscious, weighing on the researcher become embedded in the research process, potentially leading to biased investigative practice.

It is recommended that researchers employ their research tools for self-analysis—critically examining their engagement in the sociological discipline to enhance methodology and approach. Works such as Didier Eribon's *Retour à Reims* (2009) and Marie-Rose Lagrave's *Se ressaisir* (2021) serve as examples. Many articles and books published recently devote whole sections to these aspects of the research, which has become a *passage obligé* in theses based on qualitative empirical methods. The social effects of researchers in the field and the effects the field has on them is also the central concern of a large number of articles, many published recently (Bizuel 1998; Lavigne 2007; Lefrançois 2017; Richard-Frève 2017 are some among many examples). This procedure is a means of recontextualising the entire research framework within their institutional and social trajectory, and attempts to show how this affects the relationship with the people studied, the possible bias in the material collected as well as in analysis. It also takes into account the possible emotional effects on the researcher. As such, this reflection doesn't merely involve self-disclosure but attempts to ensure scientific rigour and a respect for the ethical dimensions of research (Morelle and Ripoll 2009). Within this perspective, various reflections emerge around investigations in "challenging" (Dionigi 2001), "sensitive" (Bouillon et al. 2005), or "difficult" environments (Boumaza and Campana 2007). A recent collective work attests to the diversity of researchers' encounters with destabilising or even dangerous investigative situations (Le Roulley and Uhel 2020).

As inevitably engaged, many researchers also tend to take part or position themselves regarding the social processes under investigation, and the stance of activists and researchers can have reciprocal influence (Dunezat 2011) thus linking with the sociology of intervention and institutional analysis. This is evident, for example in the "experimental" approaches of Irène Pereira (2016) or Pascal Nicolas-Le Strat (2008).

9.5 The End of the Grand Divide. De-anthropocentrism in the Humanities?

Contemporary sociology faces challenges as it incorporates the perspective of non-humans, such as objects and animals, into its field of analysis. This orientation expands the traditional sphere of sociology, focused on human interactions, paving the way for new ethical, methodological, and ontological considerations. Object-oriented sociology, notably popularised by Bruno Latour, has emerged as a major reference in this approach. In works like *Laboratory Life* or *We Have Never Been Modern*, Latour advocates considering objects as full-fledged social actors actively participating in the construction of social realities. As a result, researchers increasingly integrate objects into their analyses, reconsidering materiality and non-human elements as constitutive elements of social dynamics.

The inclusion of animals in the sociological field, known as animal studies, also emerges as an innovative domain. Authors such as Donna Haraway, with her concept of "cyborg kinship," and John Law, exploring cohabitation between humans and animals in "assemblages," provide key perspectives. These approaches challenge the traditional boundary between humans and nonhumans, prompting a reevaluation of the complex relationships created in these circumstances. By opening up to non-humans, sociologists face methodological challenges, such as the need to adapt the tools of observation and interaction for use with non-human actors. This redefinition of the field also requires rethinking analytical methods, by considering the relationships between humans and non-humans as complex networks of interactions.

What is more, this orientation raises ethical questions, particularly concerning the treatment of nonhumans as study subjects. Researchers need to demonstrate increased sensitivity to the consequences of their presence and interactions in environments involving non-humans, raising additional challenges to the traditional position of the distant observer. The integration of nonhumans in contemporary sociology opens up new perspectives, redefining our understanding of social dynamics. The works of Latour, Haraway, Law, and others have played a crucial role in transforming the discipline, emphasising the importance of considering non-humans as legitimate social actors while prompting deep reflections on the methodology and ethics of sociological research.

The works of Philippe Descola, a French anthropologist, make a significant contribution to the discourse on nature and culture, providing a particular insight into the contemporary challenges of sociology that incorporate non-humans. In his *Beyond Nature and Culture* (2005), Descola proposes a thorough analysis of the various ways in which human societies conceive relationships between humans and non-humans. He develops a typology based on four distinct ontologies: totemism, animism, naturalism, and analogism, challenging the traditional dichotomy between nature and culture, highlighting the diversity of relationships between humans, objects, and animals through the lens of culturally specific thought systems. By incorporating Descola's perspectives, contemporary sociology has enriched its understanding of how different cultures conceive and interact with the non-human world, offering valuable insights for thinking about social dynamics beyond traditional boundaries between nature and culture.

9.6 Conclusion

In contrast to the ideal research position advocated by Durkheim, that of a researcher whose objectivity involves seeing the situation from afar and avoiding personal implication with the research object and procedures, many French researchers seek to ensure axiological neutrality (Weber 1919). The much-debated question of

Weber's intentions has been resolved in a large number of studies by the principal of total transparency as to the researchers' personal implication and the conditions under which the research was carried out. This involves the presentation of the researcher's self and situation and a projection of the ways in which his or her situated self may have an effect on the collection and analysis of material. Taken to an extreme this gives the researcher a central position in her or his research. Paradoxically then, the quest for objectivity which, at the origins of French sociology, involved the distance and quasi-disappearance of the researcher may now bring sociologists to become the central object of their own research.

It is important of course to situate our discussion by emphasising that many researchers today use statistical methods and reflect on the way in which the choice of questions may play on the material gathered, without recourse to the depth of self-analysis developed in qualitative research as described in our chapter.

References

Albera, Dionigi. 2001. « Terrains minés », *Ethnologie française*, vol. 31, no 1, p. 5–13.

Aldenhoff-Hübinger, Rita, and Hinnerk Bruhns. 2018. Enquêtes rurales et industrielles: le chemin de Max Weber vers la sociologie. *Les Études sociales* 167–168 (1): 333–362.

Aron, Raymond.1935. *La Sociologie allemande contemporaine*, Paris: Alcan.

Becker, Howard. 1985 [1963]. *Outsiders. Etudes de sociologie de la déviance.* Paris: Métailié.

Bell Hooks. 1982. *Ain't I a Woman? Black Women and Feminism.* Boston: South End Press.

Berthelot, Jean-Michel (dir). 2001 *Epistémologie des sciences sociales.* Paris: Presses Universitaires de France.

Besnard, Philippe. 1973. Durkheim et les femmes ou le Suicide inachevé. *Revue française de sociologie* 14 (1): 27–61.

Bizeul, Daniel. 1998. Le récit des conditions d'enquête: exploiter l'information en connaissance de cause. *Revue française de sociologie* 39 (4): 751–787.

Bridet, Guillaume. 2007. Roger Caillois dans les impasses du Collège de Sociologie. *Littérature* 2 (146): 90–103.

Bouillon Forence, Fresia Marion and Tallio Virginie. 2005. (dir.), *Terrains sensibles. Expériences actuelles de l'anthropologie*, Paris: CEA-EHESS, 208 p.

Caillois, Roger. 1938. Introduction [*NRF*, n° 298], repris *in* Denis Hollier (1995 [1979]), *Le Collège de Sociologie (1937–1939).* Paris: Gallimard.

Chapoulie, Jean-Michel. 2001. *La Tradition sociologique de Chicago (1892-1961).* Paris: Seuil.

Cicourel, Aaron. 1973 [1972]. *Cognitive sociology.* Harmondsworth: Penguin.

Dunezat, Xavier. 2011. « Travail militant et/ou travail sociologique? », in Delphine Naudier et Maud Simonet (dirs.), *Des sociologues sans qualités* ?, Paris: La Découverte, p. 80–97.

Durand, Jacques, and Jean-Pierre Albert. 2011. Roman Jakobson et Claude Lévi-Strauss: linguistique et anthropologie structurales. *Caravelle* [En ligne] 96.

Durkheim, Émile. 1894. *Les Règles de la méthode sociologique.* Paris: Presses Universitaires de France.

Durkheim, Émile. 1897. *Le Suicide.* Paris: Presses Universitaires de France.

Favret-Saada, Jeanne. 1977. *Les Mots, la mort, les sorts. La sorcellerie dans le bocage.* Paris: Gallimard.

Ferrette, Jean (dir.). 2012. *Les Sociologues sous Vichy*, Collection Anamnèse. L'Harmattan.

Garfinkel, Harold.1967. Studies in Ethnomethodology, Englewoods Cliffs: Prentice-Hall.

Garfinkel, Harold. 2007 [1967]. *Recherches en ethnométhodologie*, trans. by Barthélemy Michel, Dupret Baudouin, de Queiroz Jean-Manuel, Quéré Louis. Paris: PUF.

Goffman, Erving. 1974. *Frame Analysis: An Essay on the Organization of Experience*. New York: Harper Row.

Guth, Suzie. 2004. *Chicago 1920. Aux origines de la sociologie qualitative*. Paris: Téraèdre.

Guth, Suzie. 2008. *Modernité de Robert Ezra Park. Les concepts de l'Ecole de Chicago*. Paris: L'Harmattan.

Heilbron, Johan. 1985. Les métamorphoses du durkheimisme, 1920–1940. *Revue française de sociologie* 26 (2): 203–237. ("La Sociologie Française dans L'Entre-Deux-Guerres").

Heilbron, Johan. 2015. *French Sociology*. Ithaca and London: Cornell University Press.

Hollier, Denis. 1979. *Le Collège de Sociologie*. Paris: Gallimard.

Karady, Viktor. 1979. Stratégies de réussite et modes de faire-valoir de la sociologie chez les durkheimiens. *Revue française de sociologie* 20 (1): 49–82.

Kilani, Mondher. 1990. Les anthropologues et leur savoir: du terrain au texte. In *Le Discours anthropologique*, ed. J. M. Adam, M. J. Borel, C. Calame, and M. Kilani. Paris: Méridiens Klincksieck.

Lavigne, Chantal. 2007. À qui appartient l'objet de recherche ? Penser l'implication du chercheur dans son objet: le handicap (surdité). *Nouvelle revue de psychosociologie* 2 (4): 23–39.

Lefebvre, Henri. 1963. Réflexions sur le structuralisme et l'histoire. *Cahiers internationaux de Sociologie. Nouvelle série* 35:3–24.

Lefrançois, Claire. 2017. Enquêter sur l' "emploi des seniors" quand on est "une petite jeune de 25 ans": un terrain dans un rapport social d'âge. *SociologieS* [Online].

Lévi-Strauss, Claude. 1958. *Anthropologie structurale*. Paris: Plon.

Lévi-Strauss, Claude. 1980. Introduction à l'œuvre de Marcel Mauss. In *Sociologie et anthropologie*, ed. Marcel Mauss. Paris: PUF.

Le Roulley, Simon and Uhel, Mathieu. 2020. Chercheur.e.s critiques en terrains critiques, Bordeaux: Le bord de l'eau.

Lubek, Ian and Erika Apfelbaum. 1981. Histoire de psychologies sociales perdues. Le cas de Gabriel Tarde. *Revue française de sociologie*, 22 (3). Sociologies françaises au tournant du siècle. Les concurrents du groupe durkheimien. Etudes réunies par Philippe. Besnard, p. 361–395.

Lukes, Steven. 1973. *Émile Durkheim*. Harmondsworth: Penguin.

Mamdani, Mahmood. 1996. *Citizen and Subject. Contemporary Africa and the Legacy of Late Colonialism*. Princeton: Princeton University Press.

Marcel, Jean-Christophe. 2001. *Le Durkheimisme dans l'entre-deux-guerres*. Paris: PUF.

Masson, Philippe, and Cherry Schrecker. 2016. *Sociology in France After 1945*. London: Palgrave Pivot.

Masson, Philippe and Schrecker Cherry. 2017. "How Does the Individual Find a Place in French Sociology?", *American Sociologist*: Springer Verlag, p. 1–18.

Mauger, Gérard. 1991. « Enquêter en milieu populaire », *Genèses. Sciences sociales et histoire*, vol. 6, no 1, pp. 125–143.

Mauss, Marcel. 1947. Manuel d'ethnographie, Paris: Payot.

Mbembe, Achille. 2005. *De la postcolonie. Essai sur l'imagination politique dans l'Afrique contemporaine*. Paris: Karthala.

Monnerot, Jules. 1946. *Les Faits sociaux ne sont pas des choses*. Paris: Gallimard.

Morelle, Marie et Ripoll, Fabrice. 2009. « Les chercheur-es face aux injustices : l'enquête de terrain comme épreuve éthique », *Annales de geographie*, n° 665-666, no 1, p. 157–168.

Mosbah-Natanson, Sebastien. 2008. Internationalisme et tradition nationale: le cas de la constitution de la sociologie française autour de 1900. *Revue d'Histoire des Sciences Humaines* 18:35–62.

Mucchielli, Laurent, and J. Pluet-Despatin. 1999. Halbwachs au Collège de France. *Revue d'Histoire des Sciences Humaines* 1:179–188.

Naudier, Delphine and Simonet, Maud (dirs.). 2001. *Des sociologues sans qualités ? Pratiques de recherche et engagements*, Paris: La Découverte.

Nicolas-Le Strat, Pascal. 2008. Quand la sociologie entre dans l'action: la recherche en situation d'expérimentation sociale, artistique ou politique, Rennes: Editions du Commun.
Pereira, Irène. 2016. Travailler et lutter: essais d'auto-ethnobiographie, Paris: L'Harmattan
Passeron, Jean-Claude. 1991. *Le Raisonnement sociologique. L'espace non-popperien du raisonnement naturel*. Paris: Nathan.
Poutot, Clément (dir.). 2013. *Le Collège de sociologie*, Collection Anamnèse. Paris: L'Harmattan.
Richard-Frève, Émilie. 2017. Quand une bergère se fait anthropologue dans le milieu pastoral en Provence: comment conserver une double identité sur son terrain d'étude et son terrain d'appartenance? *SociologieS* [Online].
Sapiro, Gisèle. 2020. *Dictionnaire international Bourdieu*. Paris: CNRS Editions.
Schapper, Dominique. 1998. *La Relation à l'Autre. Au cœur de la pensée sociologique*. Paris: Gallimard.
Schrecker, Cherry, ed. 2010. *Transatlantic Voyages and Sociology, the Migration and Development of Ideas*. Aldershot: Ashgate.
Schütz, Alfred. 1954. Concept and theory formation in the social sciences. *The Journal of Philosophy* 51 (9): 257–274.
Schütz, Alfred. 1987. *Le chercheur et le quotidien*, trans. by Nochis-Gilliéron Anne. Paris: Méridiens Klincksieck.
Simon, Le Roulley. 2021. *Introduction à la sociologie d'Henri Lefebvre. Anamnèse*: Bordeaux: Le bord de l'eau.
Stoetzel, Jean. 1957. Sociology in France: an Empiricist View. In *Modern Sociological Theory in Continuity and Change*, ed. H. Becker and A. Boskoff. New York: Dryden Press.
Weber, Max. 1892 [1894]. *Die Lage der Landarbeiter im ostelbischen Deutschland*, ed. by Martin Riesebrodt. Tübingen: Mohr.
Weber, Max. 1917 [1965]. *Essais sur la théorie de la science*. Paris: Plon.
Weber, Max. 1921. Gesammelte Politische Schriften, Munich: Drei Masken Verlag.
Weber, Max. 1968 [1922]. *Economy and Society*. New York: Bedminster.

Chapter 10
American Sociology: Between Science and Reform

Stephen Turner

Abstract American sociology is associated with quantification, as well as an alternative tradition of qualitative research. The development and relations of these traditions were always accompanied with issues over objectivity, the role of subjectivity, causality, statistical methods, and the relation between the aspiration of sociologists to be scientific and their relation to the reform movements that provided many of their topics. These basic issues were already established in the precursor to academic sociology, the surveys and statistics of the labor statistics movement, which was a collection of state and federal research bureaus. This research went beyond labor questions to deal with many social problems. The "scientific" aspirations of sociology were already present in the writings of these researchers but were transformed by Franklin Giddings into a science based on the philosophy of Mach and Pearson, and on a new model for measuring social facts. This was contested by critics who preferred an organic view of social causality and a higher idea of objectivity. The issues were focused in the thirties and forties by the rise of attitude surveys, which led to the postwar synthesis of Columbia sociology, which sought scientific status for casual claims through generalizability. This project was influential but ultimately disappointing, and was replaced in the sixties by causal modelling, which provided its own model of causality. This, together with an acceptance of the inevitability of irreducible perspectives, largely ended the aspiration to "science" in the theoretical sense.

Keywords Causality · Objectivity · Labor statistics · Subjectivity · Measurement

---

S. Turner (✉)
University of South Florida, Tampa, FL, USA
e-mail: turner@usf.edu

Y. Shan (ed.), *History and Philosophy of the Social Sciences*,
https://doi.org/10.1007/978-3-032-08167-4_10

10.1 Introduction

It is normal in histories of American sociology to focus on its early academic institutionalization, which preceded that of German and French sociology. But all these sociologies arose from the reform movements of the nineteenth century that preceded them and set the terms for them by identifying the issues that sociologists needed to address. Intellectually, the great puzzle was accommodating socialism to Darwinism. Politically, it was the labor question. But there were many ancillary movements and problems, such as the status of women, divorce, crime, suicide, the relation of cities to rural life, and issues of race and ethnicity, to name only a few. Social change, especially the rapid urbanization across the world of the 1880s was an important driver for these issues, which no longer could be understood in terms of traditional morality, and gave rise to novel conflicts.

There were common earlier roots in the statistics movement of the mid-nineteenth century and there was a great deal of intellectual exchange over labor statistics in the later part of the nineteenth century. But in contrast to Europe, where institutionalization occurred mostly later and in a different form, separate from statistics, in the United States this work led more or less directly to institutional sociology. This meant that the practices of the labor statistics movement set the standards of data collection for American sociology, but not in Europe, despite having many similar kinds of investigations during the reform period.

The story of data collection has two basic sides, quantitative and qualitative, and this divide continues to characterize American sociology (Akresh 2017). The core methodological practices of American sociology, historically and to a large extent in the present, are survey research and the secondary analysis of quantitative data collected for other, normally administrative reasons. Qualitative research, which goes by different names and has many forms (from participant observation, ethnography, field research, to autoethnography), nevertheless plays a large and currently numerically dominant role. The historical relation between the two sides is complex. One reason for the complexity has to do with issues of data analysis, which have been the subject of changing methodological ideologies. In what follows I will trace the major phases of development on each side and their interactions.

These basic methodological patterns of American sociology were, as with much else, already set in the period of social reform of the late nineteenth century. The basic problems motivating reform research were shared with Europe, and there was considerable overlap. But the specific "methodological" strategies arose in a different administrative and intellectual context and developed in a different way. With a few important exceptions, notably the influence of Karl Pearson and the vague and indirect but powerful influence of Logical Positivism, primarily in the 1950s and 60s, the tradition developed without explicit philosophical sources. It developed in response to opportunities and practical exigencies that were closely related to the development of an extensive body of labor legislation, the short hours movement, and other issues, such as protections for women workers.

Labor statistics still survive as an important form of bureaucratic social research, especially in the US, where the Bureau of Labor Statistics is a well-funded and important bureaucratic player, with the largest survey research program in the world, whose reports on employment, compensation, and productivity drive stock markets and public policy. But the focus of this bureaucracy narrowed to these economic issues at the turn of the century, after the passage and standardization of massive amounts of labor legislation in the late nineteenth century. The kind of social inquiry which was a small but nevertheless significant part of its earlier mission, however, continued, and, indeed, with both academic and philanthropic support in the twentieth century, expanded with the "survey movement."

10.2 The Methodological Legacy of the Bureaus of Labor Statistics

Leaving aside the census, which was itself a form of social research, and used in college instruction, the distinctive form of social research that led to American academic sociology was originally established as a part of labor statistics. It was a self-consciously American invention, related to but distinct from official statistics.[1] One of the problems that motivated the establishment of the first bureau, in Massachusetts, was the question of whether the returning veterans of the civil war would displace the workers in the factories, notably the shoe factories, who had been employed for the war effort. The concern was shared but opinions were divided. The governing board was itself divided between labor and businessmen, a pattern that was continued in later inquiries, where "objectivity" and general acceptance of findings was the goal of the legislative bodies that funded the research. Research was to act as a referee in questions of public concern.

Objectivity, understood in a particular way that came to be better defined, was central to this role. The first commissioner of the Massachusetts bureau was removed for violating what the legislators who funded the bureau regarded as a pro-labor bias, and indulging in speculation in his official writings for the bureau. But what was objectivity, and what was the role of research and reporting for the bureaus? The question is bound up with practical questions of how to produce facts, and a distinctive answer emerged. The initial authorizing discussions for the various state bureaus, which characteristically in American constitutional practice preceded and were the model for the later Federal bureau, envisioned a factual inquiry which took into account differences of opinion.

What the legislators had in mind was a quasi-judicial inquiry akin to any other commissioned legislative inquiry, in which sides would be heard, facts revealed, and guidance for legislation emerge. How did surveys and ordinary social research evolve out of this original impulse? The sticking point in several legislatures, and in

[1] There were, however, previous wage studies, for a survey see Charles J. Bullock (1899).

the courts, was whether the bureaus had the power to compel responses. The Massachusetts bureau became stuck on this issue: the officers of the bureau were not officers of a court. Other states went back and forth: there was a reluctance to grant the power to compel testimony, but some granted it. In practice, the bureaus did not have enough staff—often they were the director and an assistant—to pursue compulsion. Instead, they hit on the idea of large numbers of blanks which could be sent by mail and returned by mail, tabulated, and made into tabular presentations of the results. Like the census, these were counts, not rates or anything that was further statistically analyzed. The solicitation of opinions was another matter. This was like legislative testimony, as we will see. The 1885 New York report commented on the aims, after listing the representative cites in which they held sessions:

> In these manufacturing centers it was possible to cover a large number of different industries, and to secure the more intelligent and skilled mechanics as witnesses. Setting out with a fixed determination to confine the testimony to representative manufacturers and mechanics, and to exclude the vaporings of theorists and professional laboring men-which, by the way, are the curse of honest organized labor. (New York Bureau of Statistics of Labor 1886: 10)

The initial steps establishing these bureaus thus reflected the legislative origins of the inquiry. The bureaus were small commissions, with small budgets, instructed to produce information relevant to legislation.

The category labor statistics was nevertheless defined very broadly, to include virtually every topic on the reform agenda:

> Among the most important topics which have claimed the attention of the State bureaus, and under which original investigations have been made, there may be mentioned: the employment of child labor in manufactories; the education of such children; the condition of tenement houses or homes of low-grade laborers in cities; the hours of labor; wages and earnings; strikes; cost of living; relation of savings banks to the people; condition of operatives; moral, industrial, and sanitary condition of workingwomen; the truck system of payment of wages; accidents in manufacturing establishments; co-operation; conciliation and arbitration; comparative wages and prices in different countries; pauperism and crime; the unemployed; convict labor; drunkenness and liquor-selling; crime; divorces; the sanitary condition of working people in their homes and employments; effects of certain forms of employment on female health; the influence of temperance upon crime; profits and earnings; liability of employers for personal injuries to their employees; industrial education; the working of mines; Sunday labor; health statistics of female college graduates; profit-sharing; food consumption; farm mortgages; and many other topics of more or less importance. (State of Rhode Island and Providence Plantations 1890: 12)

What this implied, methodologically, was initially quite unclear. The model of judicial inquiry, meaning, in the common law tradition a contest between two sides, was never far removed from the thinking of the legislators who funded the bureau. But voluntary co-operation, which took the form of "blanks" with questions which were sent to large numbers of people and compiled statistically became the de facto norm for all the bureaus.

The large scale collections of blanks which were meant to be filled out by respondents, and interview schedules filled out by enumerators, as in the census, especially on labor related data—wages, household expenses, women's work, and so forth—generated huge amounts of material that needed to be tabulated. The initial questionnaires were unusual from a present point of view in that they did not merely ask for opinions understood as attitudes, but instead asked the respondent to explain or analyze the problem that was the subject of the study. This kind of questioning did not last long as part of the surveys, because the opinions were too various and impossible to summarize and analyze. But the impulse to collect all the relevant opinions remained.

Among the many innovations and variations that occurred during the labor statistics era was the change to individual rather than household responses, the use of distributions and ranges rather than averages in tabular presentations, and the introduction of index numbers to represent inflation. But the most consequential innovation was mechanization. The Massachusetts bureau developed a mechanical device that depended on color chips that was fast, but required human intervention to use the colored chips, and was beaten in a speed competition by the new Hollerith machines, which used punch cards and did not require the same degree of human intervention. The Hollerith technology was exported to Europe, but was not used commercially until the New York Central took it up for billing purposes, after which it became standard business practice (Austrian 1982).

From a present point of view the strategy of data collection with the schedules or blanks was odd, but nevertheless purposive and with an aim of representation relative to the problem being addressed. Sampling in the sense of sampling theory did not exist. But there were three basic strategies. The first was to send out mail blanks to targeted individuals, with the goal of collecting large numbers of responses. The second was to ask people with some kind of authority, postmasters or business proprietors, to report on their customers or employees. This allowed them to generate very large numbers for the tables. The third was to select a representative and named locale, and to produce as complete a census of that locale as possible with respect to a large set of questions. The first two methods had the greatest sampling problems, and it was known and acknowledged that the aggregate numbers would be incomplete. But for topics like household income and expenses and average and highest and lowest wages, the method was adequate, because these economic phenomena tended to equilibrate within a locale.

The methods, as they became conventional in the various bureaus by 1890, saw the initial model of testimony yielding to a basic pattern: surveys followed by "personal calls" to informants.

> The methods now open to them are the same, of course, that are open to all statistical offices devoted in any way, or to any extent, to original investigations. There are three such methods: first, the method of securing information by the use of uniform schedules or blanks, sent to parties from whom facts are expected; second, through public hearings; third, through the efforts of special agents, using prescribed forms of inquiry for the purpose of securing uniform information and for facilitating tabulation, etc. The first method named has been proven to be of little use in the past, except under the most favorable conditions, although now with a keener interest in statistical accuracy this method is becoming more efficient, some of the bureaus testifying that it works well. The second is that adopted by legislative committees, and, as a rule, simply results in bringing together a mass of incongruous statements not easily classified, and, in many instances, utterly incapable of classi-

> fication. This is the reason why the investigations made by legislative committees, relative to the labor question, have resulted in voluminous reports of testimony, unaccompanied by crystallized and classified results by the committees conducting the investigation. The best practical and, therefore most useful method of securing information is the employment of especial agents or experts to make personal calls upon parties from whom information is desired. The experience of nineteen years proves this to be the most trustworthy method open to the bureaus: when it is desired to collect official statements, and from official sources; a special letter or blank is useful, and usually accomplishes the desired end. (Connecticut 1890: 39)

In practice, what this meant was studies of numbers of workers in particular occupational categories: data on wages, hours worked, and so forth, provided most of the purely statistical content of the bulletins, which consisted of many pages of tables. But this was complemented by qualitative interviews which enabled interpretation.

Although the most impressive quantitative achievements of the surveys were thus in counting workers in different categories and giving a picture of working conditions in particular occupations, the bulletins of the state bureaus also included a significant amount of research and commentary on other topics of interest to reformers, as well as theoretical writing on "the Labor Question" in its various aspects, including reports on legislation, on European and British practices, and on alternative models of ownership and compensation. Indeed, the idea of co-operativism, in all its three dimensions—labor, consumption, and credit—was one of the central concerns of this literature, and a frequent subject of the papers included in the annual bulletins of the bureaus. In addition, there were studies of co-operative and utopian communities and their histories and ideologies.

Beyond this, there were community studies that prefigured and set the standard for future studies. The Federal bureau had even more flexibility, and in the later stages of the movement's existence funded such things as the Hull House community surveys and surveys done by W.E.B. Dubois, including the famous Philadelphia Negro study (Dubois 1967 [1899]). So the bureaus represented a nascent research oriented discipline. The difficulties of survey research were all apparent during the period. Non-response, refusal to respond, suspicions about what would be done with the data—these were endemic issues, and they led the bureaus to emphasize visitors who could report on situations, which is to say qualitative inquiry.

The surveys, and the intellectual leader of the movement Carroll Wright, were not without their critics, e.g., H. F. Bliss (1897a, b, c, 1898, 1900a, b) and A. L. Bowley (1895). The criticisms involved generic statistical issues. But they were presented informally, prior to the development of the modern statistical and methodological nomenclature. Moreover, the problems pointed out by the critics have not gone away: the issue of representativeness, the problems of non-response, sampling, the assumptions behind indexes, the problems inherent in understanding the respondent's interpretation of questions, all are there today. These were acknowledged, and Wright and his contemporaries were if anything more skeptical of survey methods than later enthusiasts, as we will see.

10.3 Troubles over Causation and Objectivity

One criticism that was particularly telling, having to do with causation, came from the emerging sector of academic sociology, from Albion Small (1902). To understand the issues here it is necessary to understand the practical notion of causality that informed most of this research. A paradigm case of this can be found in the blanks of the Charity Organization society, which provided services to the poor, visited the poor to ascertain circumstances and eligibility support, and also to find out why they were in need. The blanks they filled out typically listed causes—among them injuries that prevented work, illness, death of the provider, and particularly alcohol and its effects. These reports, and in the case of surveys in which the respondents themselves provided similar information, were taken as evidence of cause. Thus, one might show that a higher percentage of foreign-born criminal offenders were affected by alcohol. In this case the "cause" was something an enumerator elicited from the subject.

The basic statistical "method" here was the comparison of percentages or rates, and the ongoing concern was the appropriateness of the comparisons: whether the methods of collection were sufficiently similar, or whether the categorizations were comparable. The results of this were problematic, in part because of the inadequacy of the causal theories the methods were designed to vindicate. The reformers ignored confounding, assumed the causal efficacy of the facts they collected, and came to catastrophically bad policy conclusions. The Bureaus of Labor Statistics were not innocent of this either. The obvious case is the role of alcohol and the campaign for temperance, which was heavily buttressed on all sides by statistics that associated drink with any number of bad outcomes—insanity, crime, poverty, divorce, and so forth. It was then concluded that prohibition would have the effect of eliminating these outcomes, or diminishing them. The cause was of course confounded with many other causes, the result rather than the cause, or multiple redundant causes which produced the same result.

Ordinarily there was just one stage to the comparison of percentages, and the causal reasoning was presented in the commentary on the differences. Yet there were efforts to go beyond this by assessing net effects. A typical example of the reasoning is found in this passage from a study of female college graduates motivated by claims to the effect that higher education was deleterious to women's health:

> That during college life there was nearly two and one-half per cent less deterioration in health as compared with the deterioration in health reported during working time by the working girls of Boston. (Massachusetts Department of *Labor* and Industries 1885: 530)

> That the present health of graduates seems to have been affected according as their parents have enjoyed either good or poor health, the figures showing three per cent increase in health for those whose parents were both in good health, and a decline in health of over 17 per cent for those whose parents were both in poor health. (Massachusetts Department of *Labor* and Industries 1885: 529)

> a certain deterioration in health on the part of some of the graduates. On the other hand, an almost identical improvement in health for a like number was reported, showing very plainly that we must look elsewhere for the causes of the greater part of this decline in health during college life. If we attempt to trace the cause, we find that this deterioration is largely due, not to the requirements of college life particularly, but to predisposing causes natural to the graduates themselves, born in them, as it were, and for which college life or study should not be made responsible. (Massachusetts Department of *Labor* and Industries 1885: 531)

This was a sophisticated example of net effects thinking (Turner 1997). Rather than taking the totals for reports of ill-health or the overall percentages as evidence, the cases were divided into causal groups, in this case categories of parental health, so that the percentage differences within groups could be used as an estimate of the effect of college education. As with many such exercises of what was later called partialling, the supposed effects disappeared.

The point of this study was not to invent a causal account, but to test the various causal claims made by others. This was the norm for this body of work. The goal was to adjudicate existing causal claims statistically, using self-reports or visitors' reports on causes rather than to propose novel causal hypotheses, or to measure them as variables. One such supposed effect was "nervousness," which was carefully traced through self-reports of before, during, and after college experiences, and found to be non-existent. The results provided confirmation, or support, usually carefully expressed, or lack of support, as in this case.

The sheer existence of the practices of data presentation of the bureaus, the tables and numbers, were a considerable constraint on the researchers. The formats were not particularly novel: they resembled the census. But the topics were different and adapting the basic formats to the presentation of such things as average wages, ranges of wages, marital status and many other such facts, and putting these in numerical form, represented a considerable achievement. And the desire for comparability was strong, and served to enforce uniformity among the state bureaus, despite the differences in the industries and topics that were their primary concern. This fit nicely with the conception of the bureaus as providers of facts to the legislature and as quasi-judicial fora, in which issues with identifiable sides, or problems with multiple solutions, were to be openly and dispassionately addressed. But the effect was broader: the tables of numbers and percentages, and the range of tables, set a standard of what was to be expected of a statistical analysis of a topic or place.

This set of practices developed under a cloud of suspicion: the movement was allied closely to the labor movement. The practices were a response, and a successful one from the point of view of the bureaus, to the challenge of demonstrating impartiality in the face of bitter conflict. Wright himself was chosen for the position as head of the newly created federal bureau over other nominees precisely because of his apparent political neutrality (Goldberg and Moye 1985: 2, 5–6). The governance structures of state bureaus often involved a balanced board of supervisors—a practice repeated in later research projects on controversial topics. But the bureaus were self-consciously composed of practical men, not theorists, and did not develop a principled defense of the objectivity of their methods or of its "scientific" or "sociological" character. Yet they used the terms, along with "social science," to describe their contribution. This was to come later. But the issues were never far from the surface: reform movements and philanthropies also purported to be "scientific." Yet they had a clear conception of their special role.

> [The] bureaus are engaged in a grand mission. Their integrity in this mission is unassailable, and the results they are bringing out constitute a most valuable series of contributions to social science. The popular education of the masses in the elementary facts of political and economic sciences is the greatest educational end of the day. The bureaus of statistics of labor are in the line of facilitating this grand work by their faithful investigations into all the conditions where facts should be known, and into all causes of bad conditions of whatever nature, and by their fearless promulgation of the results such investigations. To attempt to turn such a sphere of labor to base purposes is a crime not easily punishable by law, but which can be punished by an unwritten law which reaches the violator through a decree more to be dreaded than any merely judicial order or sentence—the sentence public opinion passes upon the man who prostitutes the cause of humanity. (Connecticut 1890: 32)

This sounds hyperbolic, but it reflects two features of the American situation that bear emphasis and cast a long shadow on the future discussion of the problem of objectivity in American sociology. In the first place, although the collection of labor statistics seems uninspiring and intellectually banal, the labor conflicts of the time were sufficiently violent and impressive that this task and related tasks placed the bureaus in the heart of the major social issue of the day. But, in contrast to the situation in other countries (Reick 2023) the labor movement had democratic power—thus public opinion mattered, and both educating it and avoiding being judged as a partisan also mattered. Wright had the role of fact finder in one of the most brutal conflicts over the mining of anthracite (Anthracite Board of Conciliation 1908), something that he could not have credibly done if he was seen as biased.

"Objectivity" as an ideal in this context has an instrumental character: it is a means of dealing with volatile conflict without being drawn into it. But it has a complicated relation to the movements that motivated reform in the first place, in this case the labor movement. The bureaus were supported because they were seen to be useful for the cause. But the relation was a delicate one. To be seen as an advocacy organization would lead to the bureau's abolition. They had a defined role:

> Legislatures are using them to carry on special investigations, to make distinctive inquiries on matters coming up for legislation; and so long as politics do not enter into the administration of such offices, so long as governors will look to the interests of education and not of politics, the *personnel* of the bureaus will be kept outside of political ranks. (Connecticut 1890: 31)

Objectivity was thus a shield. But it took a specific form: they looked at the statistical evidence and asked what it did or did not establish. The model was judicial, and was especially apparent in the cases where labor issues were actually adjudicated, such as the case of the anthracite strike in Pennsylvania in 1902, which was billed as the greatest and most damaging strike ever. The commission looking into to it for President Theodore Roosevelt included Wright as Chief Recorder and Charles P. Neill as assistant (Anthracite Board of Conciliation 1908; Goldberg and Moye 1985: 46).

But this was not as limiting as it might seem. The bureau heads met routinely to discuss not only practical matters but the great questions that motivated the creation of the bureaus. They were in this sense not merely practical men, but theorists. They understood that "The supreme menace is the tendency of the concentration of wealth. That unequal distribution of wealth and its aggregation by the few has always been a disintegrating force is sufficiently well known" (Missouri Division of Industrial Inspection 1893: 6). They quoted Spencer at length to each other, both for his formulations of the conflict of collectivism and individualism and for his views on the origins of property, which were not far from Proudhon or Henry George.

> We discuss matters here when the ideas underlying them are in embryo. We throw out suggestions of lines of work, and we discuss propositions which it is the privilege of the bureaus to crystallize into well-directed investigations. With all this, statistical science is making rapid progress in this country. (State of Rhode Island and Providence Plantations 1890: 15)

These meetings were thus a prototype of the meetings of the disciplinary academic associations that were beginning to be established in the same period.

The educative function they assigned to themselves was to raise the level of discussion of "evils" to a higher level. It is not easy to draw a line between them and the moralistic reformers on whom they relied to identify issues and to provide a side to serve in framing a "judicial" role for either statistics or analysis. It is nevertheless easier to cite exemplary works and deviations. The concern with the maldistribution of wealth as a danger to society was overtly stated, but at a higher level of analysis, citing the effects of the concentration of wealth on ancient civilizations on their decline. The empirical discussion of profit sharing, written by Giddings (1887), who also co-authored a key theoretical article in economics with John Bates Clark on the topic (Clark and Giddings 1888), but published, as was the practice of the time, under the name of Wright (e.g., Wright 1887), was not only judicious but analytic and sympathetic. It explained the rational basis of the choices of workers, why they preferred to invest in enterprises other than their employer, and their preferences for a stable paycheck (Giddings 1887: 374–6.) This was a paradigm of balanced analysis, free of moralism. Balance and a desire to let the facts decide distinguished them from the reformers.

The now-famous writings of W.E.B. Dubois, written under "the supervision of the bureau and employing its established methods, explicitly including, in the case of the *Philadelphia Negro*, reference to the Hull House maps (an earlier Wright sponsored project), were less balanced and lacked the balanced supervisory structure of the bureaus. They faithfully followed the Bureau's practices of data collection and presentation, but described, for example, 40% of the Black population of Farmville as "idlers," and, in the *Philadelphia Negro* (1967 [1899]), included highly negative evaluations of preaching and levels of civilization. And, as noted above, the bureaus were not free from moralism either. Temperance, which had become an obsessive "scientific" crusade (Okrent 2010), was treated unambiguously as a good and rum as an evil whose statistics were to be investigated rather than sympathetically understood.

10.4 The Reformers

In the labor statistics period before the academicization of sociology, there was an active body of reform movements (Bliss 1908) which worked in a similar way. The reform movements, including labor movements, temperance movements, and the American Social Science Association, a confederation of topical reform movements, considered the labor statistics bureau its "ally"; all used statistics, and sometimes produced them. Their focus, however, was on education in the sense of propaganda. They recommended books, published pamphlets, and were especially interested in academic authorities that supported their causes. The issue of property was a favorite: Denis Fustel De Coulanges, who influenced Durkheim (see Jones 1993) was a favorite source (Fustel de Coulanges 2000 [1889]). This was an intellectually rich space of dialogue and contestation. Yet in connection with actual research, they operated for the most part with pre-established causal theories about the domain in which they sought reform and sought to provide grounds in support of them. This body of research paralleled the Booth survey in London (Bales 1991). They were attempts to call attention to evils which they thought they had the answers to.

The differences in audience or educational intent made for differences in presentation. Visual representations of statistical facts were important for presenting to the public: pages of tables were not as effective. The reform movements developed excellent charts, with colored figures, and used such methods as presenting symbols representing facts. These methods had a long subsequent history, and like other parts of statistical practice became part of normal public discourse and rhetorical practice. The Massachusetts Bureau stalwart Charles Pidgin, writing in a volume edited by George McNeill, who had been removed from the Massachusetts Bureau for his favoritism to labor, explained their appeal.

> The historian who uses figures has an advantage over the one who relies on words alone. His readers can grasp in a one-page statistical table what ten pages of words would fail to express as plainly. Now, the artist can come to the aid of the statistical historian with colored maps, charts, and diagrams….in statistical presentation the table or diagram supplies the statement and the argument, while words are used but for description and explanation. (Pidgin 1886: 628)

This emphasis was greatly enhanced during what came to be called the Survey Movement, with the support of the Russell Sage Foundation.

The larger world of reformism was less interested in research, or novel views of causality. Their common view, realized in different ways, was that the research was done, the facts were now known, but that the message of how to live and organize policy had not been presented to the public in a way that was effective. As one of the Sage foundation's key figures, with a background as a journalist, put it, the research effort consisted of "A great reservoir, a tiny spigot" (Kellogg and Deardorff 1928: 828). The aim of the survey movement and of Sage philanthropy was to enlarge the spigot, to simultaneously enlarge the public communication of the social knowledge relevant to reform, and to tailor research to local needs and audiences. The major triumphs of this movement were the Pittsburgh Survey, a multivolume presentation of social ills using journalistic techniques, such as life-histories of victims and photographs. The volumes were followed by a temporary museum which included displays representing the educational message, which was toured by citizens and schoolchildren. These were strikingly innovative and visually stunning: the Pittsburgh one had a frieze near the ceiling representing cholera deaths, for example. Here and in the Springfield survey, maps were used to show the distribution of ill-health in the city (Greenwald and Anderson 1996). The same methods were adapted to rural settings by C. Josiah Galpin, who used schoolchildren to collect information used in maps presented at a county fair, illustrating the benefits of "socialization" or interacting with the town, for isolated "poor hoe farmers" who were regarded as the social problem of rural life (Turner 2016: 13; see Galpin 1918). A related impulse guided the Rand School of Social Science, which trained reform organizers in the urban setting of New York City.

These methods used surveys, at term which had many meanings, not to add to the statistical work of the Labor Statistics Bureaus so much as to publicize issues to promote local reforms. Many of the "surveys" consisted in reports that resulted from the co-operation of leading charities in defining and publicizing local social problems, and "surveyed" only local charity leaders. This effort overlapped with a movement for municipal "efficiency," and indeed many of the "surveys" included in the bibliography of the movement related to such topics as efficient heating of public schools. What these surveys lacked was a concern for the subjective experience and point of view of the subjects: they assumed a standard of correct behavior and of the aims of their subjects without investigating them and without much concern for them. Indeed, their aim was to change attitudes, beliefs, and practices, not to understand them. In this respect the judicial model of the labor statistics movement was different, in that it sought out the opinions of the participants on the opposite sides of issues, even though the work of understanding was done through open ended questions which were rarely analyzed, and the non-statistical reports of the visiting inquirers who spoke to the subjects.

The labor statisticians and reformers prior to 1900 used terms like sociology, social science, and sometimes Christian sociology and socialism interchangeably, and did not make sharp distinctions between reformism and "science." With the creation of the Sage foundation, the organization of the social sciences into disciplines, and the retirement of Wright from the labor statistics bureaucracy to become an academic sociologist and university president, and the gradual narrowing of the concerns of the labor statisticians, this changed. The reform elements focused on action, notably publicity or public education. The university sociologists were concerned with defining the new discipline, creating a canon, and training students. The mutual relation between sociology and social reform movements was a defining feature of the American discipline throughout its history. But the terms of the relationship changed repeatedly, as did the reform movements and the discipline itself. The academic side of sociology never completely abandoned these reformist concerns, and indeed the topics of dissertations of sociologists into the 1920s closely tracked the social topics of the Labor Statistics Bureaus, especially the investigation of local workers and community institutions and co-operatives.

The separation from reformism was slow and incomplete, but it operated in this way: students were recruited from the reform-oriented public and converted into "scientific" sociologists. What this meant varied from university to university. The two main rivals were Columbia and Chicago, which had different visions. The field quickly developed its own distinctive conflicts over methodological issues: this is where the division between qualitative and quantitative sociology begins to be defined in the rivalry between Chicagoan Albion Small and Columbian Franklin Giddings. The key philosophical issues that motivated the academic field—objectivity, causality, the relation between qualitative and quantitative knowledge, the idea of theory and its relation to ideology, and the question of the nature of science itself—all came to be defined through this rivalry.

10.5 Enter Giddings and the Problem of Cause

Franklin Giddings was a friend of Wright, and they influenced one another. As noted, Wright published a study by Giddings (1886) under his name (Wright 1887), and a lengthy paper on profit-sharing in the Annual Report, which was reprinted as a separate publication (Wright 1886). Giddings was a quantitative innovator. He had experimented with index numbers to gauge inflation. The idea was used by Wright in his own estimate of growth in income between 1860 and 1890. But Giddings turned from economics to sociology and carried these quantitative concerns with him.

In 1901, once he had established himself as the leader of Columbia sociology, Giddings published a puzzling book: *Inductive Sociology*. The book became a bible for his students over the next decade, students who were to become leading figures in the leading departments of sociology in the interwar years. It was a radical extension of the measurement thinking that he had himself begun with the problem of measuring inflation. Giddings proposed quantifying a vast array of sociological and labor related concepts that figured in either public discourse or academic commentary for the purpose of subjecting them to statistical test. This was a two part novelty. The idea that one could quantify theoretical concepts in sociology turned the past uses of statistics upside down. Writers like Émile Durkheim (cf. Turner 1996) or Enrico Morselli (1882 [1879]) interpreted tabular statistics collected for other reasons in terms of their preferred theoretical concepts. Thus religious confession could be made to stand in for an unmeasured theoretical variable, such as integration into the community. In contrast, Giddings started with the concepts, and devised quantitative measures to fit them.

His most developed foray into this territory in the 1901 book was incomplete, but indicative of his aims. He theorized that the most progressive states of the union would be characterized by an intermediate degree of heterogeneity or social sympathy, and overly homogenous or overly heterogenous states would be less progressive. This was based on a theory of subjective values: the progressives are "those who have so long enjoyed varied experiences and have cultivated manifold interests that their subjective values make up totals that are highly complex, and yet, at the same time, harmonious" (1901: 178–81). This, he claimed, applied at the population level as well:

> *A population that has only a few interests, which, however, are harmoniously combined, is conservative in its choices. A population that has varied interests, which are as yet inharmoniously combined, is radical in its choices. Only the population that has many, varied, and harmoniously combined interests is consistently progressive in its choices.* (Giddings 1901: 181; italics in original)

He then constructed an index of social heterogeneity or degree of social sympathy, based on proportions of persons in divergent social groups, and compared it to three categories of states based on the index numbers, and observed that the states in the middle category were indeed the progressive ones.

This was a crude quantitative hypothesis test. But it was also novel: the alternative would be to classify states as progressive or not, and comment on the features that might explain why they were. Moreover, with Giddings, the "theory" came first, the test second. Indeed, in this volume, Giddings elaborates a general categorization of levels of social groupings, focused on the idea of "concerted volition" and "like-mindedness" as a source of "social values" (1901: 159), and this was the source of the hypothesis. The reasoning depended on his more general account of social sympathy. The links in this argument were perhaps tenuous and disputable, but the goal was clear: to go from a highly general account of social life to testable claims.

The second element of Giddings' novel approach involved correlation. Under the influence of Ernst Mach and Karl Pearson, each of whom was aware of the argument of August Comte that "cause" was a part of the pre-positive metaphysical stage of scientific development and would be replaced by prediction. The claim of Mach and Pearson was that there was nothing to cause other than description of a relationship. For Pearson, this meant that there was nothing to cause but correlation (Giddings 1922: 127–43; Porter 2004). Giddings makes the same point:

> The notion "causation" like the notion "explanation" has had a past of which the less said the better. It has kept company with metaphysics. But that is no reason why it should not usefully be employed in science, under a watchful eye. If we talk about "explanation," we must talk about "causation" because, as will appear, explanation is a discovery of causation.
>
> Loosely we speak of a dominant factor (outranging guns, over-whelming numbers), or of a last contributed factor (the lighted match applied to the fuel), or to a factor to which responsibility attaches (a casting vote) as a cause, or as *the* cause of something. This is rhetoric only, but it is harmless and convenient. (Giddings 1922: 130; emphasis in original)

Giddings turned to correlation as developed by Pearson and Francis Y. Edgeworth which in Edgeworth's hands had involved multiple variables, (Edgeworth 1892a, b), to replace the traditional percentage-based analyses of the labor statistics era.

This had problematic consequences: clearly there was a demand for causal understanding in the face of social problems. Merely identifying correlations, of which there were an infinite number, did not seem like a sufficient answer, either on the practical level of intervention or the still obscurely defined "scientific" level. To go beyond correlations to identifying the "real causes" of a social phenomenon required something more than the model of adjudicating between sides. Ideally, one would be able to point to something like sociological laws. Alternatively, one could move the problem of cause to the side of practice, and intervention. Allyn Young in a later review of the essays of Edgeworth related to economics, observed that

> practical considerations, and especially our interest in knowing at just what point we ought to lay hold of the economic mechanism for purposes of control, will suggest that, for some concrete purpose, some one variable factor will play the part of cause better than the others. (Young 1925: 724)

This solved the problem in a convenient way: what one took to be the cause—or to play the part of cause—depended on one's concrete purposes.

Giddings struggled with the problem: giving up cause as part of the theoretical structure of sociology was a step too far in the direction of theoretical austerity. His solution was different. One could distinguish correlations in terms of the extent to which they persisted in the face of variation in other variables. If a correlation was partialled (Yule 1907), that is to say the data points it summarized were divided by a potentially related variable, and the correlation persisted, it was in some sense vindicated as genuine. If the correlation disappeared it was shown to be not genuine. Giddings analogized these relations to constants in physical theory, and suggested that sociology pursue the "constants," or rather their probabilistic analogue, in the social world (Giddings 1922: 132–3). Needless to say, this was not a successful strategy: certain large correlations were relatively stable; not enough were ever found to become the basis of a science or to be regarded as a surrogate for sociological laws.

In short, this was a "solution" on the philosophical level that had little to do with emerging practice, which more closely resembled Allyn Young's formulation: local correlations intelligible as causes in relation to a particular practical purpose (1925). Theory in the more abstract sense of Giddings was thus far removed from the rapidly expanding enterprise of correlational analysis of the early interwar period. What analysts did then, and do today, is to identify plausible causal variables in the specific context in question and test their relations to one another through correlational analysis, preferably with multiple correlations (Woofter 1920) and partialling (Ross 1924).

There was, however, an alternative approach to causality. Throughout the early period of US sociology, Giddings had been opposed by Albion Small on methodological grounds. The issue was objectivity. Giddings was heir to the austere notion of statistical evidence inherited from the labor statistics movement and the austere version of descriptivist positivism of Mach. Those closer to the reform tradition rejected this austerity, on the grounds of their rejection of its notion of objectivity and defense of an alternative notion. The economist F.H. Knight, in a review (1925) of *Origins of Sociology* by Albion W. Small (1924), quotes the views Small advanced in multiple publications during this period about what sociology added to past discourse about society.

> In the preface the movement is described as "a drive from relatively irresponsible discursiveness toward … 'objectivity'," and in the final chapter it is again defined as including "*everything that followed the expression of a deliberate and avowed purpose to work for a scientific interpretation of cause and effect in human society at large*" (page 337; italics Professor Small's). (Knight 1925: 725)

"At large" was the operative concept here. This was a vision of sociology motivated by the idea of society as an organism that Small took from Albert Schäffle. It reflected his opposition to what he claimed was Giddings' individualism and psychological reductivism. Objectivity was to be found elsewhere, namely in a holistic concept of community or society.

Small's reasoning, which reappears in various forms throughout the dominant forms of "theory" in American sociology, was that true cause and effect was to be found in the interconnected processes of society at large: this conflict did not go away, but it was only a small part of the expanded methodological soul-searching and conflicts of the 1930s. It revived with the dominance of "functionalist" theory of the late 1930s and 40s and "systems" thinking of the 1950s and 60s. (Parsons 1954 [1945]). The details, transformations, and ebbs and flows of these ideas are too complex to deal with here, but there is an important core explanatory structure that also relates to the next twist in the development of empirical sociology, so it needs to be explained here.

One can discern a pattern in responding to the complexity of the subject matter of social life which can be summed up as the analogy-ontology-causality sequence. A theoretical concept is constructed by analogy: society is like an organism. It is then ontologized into "society is an organism." Then causal consequences and arguments, including functional arguments, are derived based on the existence of this reality. The effect of this is to avoid the need to construct casual accounts based on individuals and their actions. Instead, these actions are subsumed by the constructs, which replace and purport to explain them. This kind of reasoning does not apply only to "social systems." The concept of attitude is a product of the same kind of reasoning. An attitude is a theoretical entity, based on an analogy with the attitude or orientation of a rigid body in physics, but turned into a measurable object with explanatory causal power. It is worth noting that even Spencer warned against this kind of reasoning, which he called "remote analogical reasoning," where "we find much unlikeness between the predicates, as well as the subjects" (Spencer 1872: 76, §299). But he added that despite its weakness as a form of inference, its conclusions could be independently validated by inductive confirmation (Spencer 1876: 613).

10.6 The Thirties: Subjectivity and Science

This was only part of the background to the decade of intense methodological discussion that followed in the 1930s? There were both internal and external reasons for its renewed intensity. The external ones were significant: Prohibition had proven to be a failure, which discredited the reform movements that had regarded it as a panacea. This had methodological implications: the reformers, who called themselves "social workers," claimed a special kind of social work knowledge that supported prohibition, based on their professional experiences (Bruere 1927). Foundations, disturbed by the disaster of prohibition, supported the idea of a more rigorous social science. This meant that the problem of *vera causa* was not merely a philosophical one: the practice of attributing social problems to one's preferred evils had to be replaced.

Academic sociology had been allied with the "profession" of social work, which was essentially a term for reformism, up to the early 1920s. But was relentlessly separating itself from it from the early 20s on, and social work itself was re-professionalizing around the psychologically inflected "case method." In the early 1930s, dues for the American Sociological Society were changed to rid the society of the reform-oriented non-academics who comprised a significant part of its membership. This was part of a larger movement to better define what was scientific about sociology. The idea of being scientific was also promoted by the newly established and foundation funded Social Science Research Council, leading to similar discussions in neighboring disciplines. Part of the discussion focused on measurement. The new political science of Robert Merriam, who had been taught by Giddings, asked what the discipline's unit of measurement was, and answered "the vote." Harvard had a "social ethics" program but no formal sociology program, appointed a committee in the mid-1920s to investigate the scientific status of sociology, and replaced the social ethics program with a sociology program in the 1930s. The other social science disciplines each had a movement to be more scientific and objective; in two cases, psychology and sociology, creating organizations within the discipline to move it a scientific direction. This was a reflection of the internal desire to be scientific, rooted in the early history of the discipline. But it was an ambiguous and contested aspiration: did it require theories and laws, or more rigorous empiricism? And would that empiricism include subjective reality? And how would it relate to the scientific claims of adjacent disciplines?

The point at which the concept of objectivity became hotly contested and associated with the qualitative quantitative distinction is difficult to identify (Danziger 1990: 66; cf. Bannister 1987), but a key moment in the late 1920s involved the Institute for Social and Religious Research, which itself was an outgrowth of an ambitious attempt to survey the adherents of the world's religions, which failed. In form its studies resembled the statistical presentations of the labor statistics movement, though it began to experiment with advanced, causally oriented, correlational analysis. The Institute had hired a social reformist minister, Robert S. Lynd, to study a community in Indiana, particularly with an eye to the role of its religious institutions. This study led to a manuscript with few statistical tables, but with a great deal of ethnographic observation, and an ideological message: against American individualism, which was portrayed as the source of the community's failure to address social problems. The institute refused to publish it. Lynd went to an anthropologist friend and it was published anyway; it became one of the most influential sociological books of the period: *Middletown* (Lynd and Lynd 1929).

What was compelling about the book was its descriptions of community life. This was similar to the kinds of studies of communities, occupations, and social settings that came to be the Chicago empirical tradition, which had a somewhat different orientation to its subjects. Robert Park provided a particular model. He was famous for displacing reformist censoriousness, of which Lynd was a late example, with an understanding of subjects in their specific social settings, or as he put it "Investigation of the moral and personal social world in which the individual lives and to which he is responsive" (quoted in Rauschenbusch 1979: 148). This depended on more than statistics. But it did not exclude them. The methodological practice was exemplified in a never published study also funded by the ISRR led by the leader of 1920s Chicago sociology, Robert Park, on race relations in the Pacific west. It involved a two-step process: first to get the pro- and anti-Orientalists to agree on the statistical facts; second "to try and get white people interested in the Orientals as human beings" (Rauschenbusch 1979: 114). Park had an academic concern here, but also a meliorative one. But for Park, the result was not to be condemnation but understanding.

Where was objectivity here? Park had a moral aim: of improving on the defective understandings of the reformers and communicating these improvements to the academic world and the public. Although he never articulated it himself, he conveyed to his students the need to understand people on their own terms. This was itself a kind of moral imperative. When the concept of value-free science was imported to American sociology, his follower and successor Louis Wirth was vocally against it. But there was another contradiction in their strategy: the moral and personal social world that did the explaining for them was local and subjective; but they also sought generalizations about group relations. It was difficult to see how this was possible, and indeed it was not (see Wirth 1951).

The role of interpretation, and the contrast between interpretation and measurement, which was assimilated to the distinction between the subjective and the objective, was a question that consumed much discussion during the 1930s. A defining moment came during Giddings' protégé William Fielding Ogburn's "Presidential Address" to the American Statistical Association in 1930, where he treated the scattergram, which is to say the unanalyzed set of data points, as fact, and compared its interpretation to the interpretation of an editorial cartoon in a newspaper (Ogburn 1930 [1929]). This was an expression of the kind of austerity that rejected the Lynds and defined a pole of the dispute over objectivity. Another was produced by Ogburn when he caused a truncated version of a saying of Lord Kelvin to be engraved on the new social science building at the University of Chicago: "When you cannot measure your knowledge your knowledge is meager and unsatisfactory" (Bulmer 1986: 182; cf. Danziger 1990).

But the idea of measurement was no panacea either. The measurements were indirect: indices. Insiders were skeptical about indices. F.A. Ross, who pioneered the use of partialling in causal analysis to produce accounts akin to later causal modelling, wrote

> Personally, I prefer to call all but a very few "indicia." The series seldom measure the exact thing they are supposed to measure. They are not true indices. Ordinarily they give but a vague indication, rather than constitute accurate measurement. (Ross 1934: 463)

Ross, however, had in mind the kinds of indices Giddings had promoted: objective but indirect measures. Ogburn's *Social Trends* (1933) was full of these, such as attempts to assess cultural change by the counting magazine articles and the changes in numbers (Hart 1933: 382–442). But the discussion of measurement was already being transformed in a different and confusing direction by another change: the importation from psychological social psychology of a quantitative concept of attitude, and methods of measuring them.

The practitioners knew that the indexes or measures were selected from a wide array of possibilities: they were simply useful some of the time. They also knew also that the correlations were a product of the practical choices of what variables to measure, the indexes to select, assumptions about sampling and heterogeneity to make, and what correlations to calculate, partial, and refine. It is a more or less arbitrary selection out of an infinite universe of possibilities. As Ogburn student and future Harvard sociologist Sam Stouffer wrote in an official publication of The American Sociological Society on method:

> It is repeating the obvious to say that, even if the sociologist has successfully estimated the probability of an observed relationship arising from a suitably defined population of possibilities, his task usually will not be ended. He seeks to supply a reasonable explanation of the relationship found, unless, as is seldom the case, the explanation is self-evident from the manner in which the problem was set up. This search for an explanation may lead him to the use of historical methods, case methods, or any other methods, including more statistics. The vision of a mathematical regression equation, with its range of chance error known, as the goal of one's research, may help thinking out one's problem, but often is not a very useful conception to follow to the literal end. The constants in the population of possibilities underlying a good many of our problems are probably shifting too fast in time or varying too much from place to place to justify spending money on the intricate kind of study which can be made profitably in physics or even biology. And, too often, numerical indexes of crucial factors are of unknown reliability or non-existent. In such circumstances, sole reliance on a few statistics expressed with pretentious mathematics may lead to a specious sense of security which inhibits a real analysis. (Stouffer 1934: 485–6)

This was the view from the quantitative side in the 1930s, and it was skeptical: it knew that the numbers were based on abstractions and assumptions. It was entirely consistent with Ogburn's more pungently expressed views on the elimination of statistics, and with the critiques of misuse of statistics by Wilson cited by Stouffer (Wilson 1926; Stouffer 1934: 487).

10.7 The Conflict Defined: Subjectivity Quantified

One can think of the 1930s as the decade of controversy over subjectivity. It began with objectivity measures being used as surrogates for direct measure of attitude, but with social psychology committing to the attitude concept, and its psychological theory to the effect that attitudes were the causes of action. This was challenged by Richard LaPierre in a famous study showing that in a real world context what was taken as an expression of attitude had no effect on action, and that other parts of the context did (LaPierre 1934).

One of the defining documents of the era was Herbert Blumer's critique of the famous work of decades earlier, *The Polish Peasant in Europe and America* (Thomas and Znaniecki 1918), which was important for introducing the concept of attitude to sociology.

> My own feeling is that this work of Thomas and Znaniecki (1918) presents a dilemma as far as social research goes … in the following form: on the one hand, an inescapable need of including the subjective element of human experience, but, on the other hand, an enormous, and, so far, un-surmounted, difficulty in securing devices that will catch this element of human experience in the way that is customary for usable data in ordinary scientific procedure in other fields (Blumer 1939: 111). (Hammersley 2010: 72)

Thomas conceded the point: "it is evident that statistical studies of the behavior of populations will have a limited meaning so long as the statistical data are not supplemented by individual case histories" (Thomas 1939: 87). But he held out the hope that "these inferences in turn must be continually subjected to further statistical analysis as it becomes possible to transmute more factors into quantitative form. Blumer radicalized his critique by pointing to the intrinsic mismatch between attitude theory and any plausible understanding of human action. But at the same time he rejected understanding, what his early mentor Charles Ellwood called sympathetic introspection, as unscientific.

This left a gap that Blumer himself could not resolve. The phrase "customary for usable data in ordinary scientific procedure in other fields" goes to the heart of the problem. It set the standard for science too high for any available empirical methods. It represented, *in nuce*, the parting of the ways: either one gave up on the idea that the subjective element could be reduced to numbers, and therefore gave up on sociology as a "science," or one committed to refining the measures and their statistical analysis and calling this science, and defending the idea that this approximated in some way "ordinary scientific procedure." The quantitative route became the continuous process of verification and of hypothesis formation (Thomas 1939: 87), in which qualitative knowledge had a role, but only a preliminary one. This was the outcome of the many discussions of methods at Columbia in the 1930s, best captured in Theodore Abel's "The Operation called *Vertstehen*" (1948). For Abel, *Verstehen* was useful only as a source of hypotheses, which could then be refined. But his own major study, of the life histories of Nazis, belied this. So the problem was more basic, for the reasons LaPierre had established.

The resolution to these issues—always partial and rejected by Blumer's followers and many others—was a result of a *deus ex machina.* Entirely apart from sociology, there was a wartime demand for studies of morale. This was on the same level as the use of correlation: it was "useful" rather than "scientific." And the utility had already been established in public opinion research, where it had become transmitted back to the public and become part of the culture itself (Igo 2007). This merged with academic social science in the classic study *Studies in Social Psychology in*

World War II: The *American Soldier* (Stouffer et al. 1949). At this point, during the war, an often vitriolic three-way epistolary discussion between Stouffer, P.F. Lazarsfeld, and Louis Guttman took place. Lazarsfeld attempted to provide a mathematical basis for treating scales as genuine measurement, an effort which took place against the backdrop of Stanley S. Stevens attempt to rethink measurement theory in a way that justified calling psychophysical "measurement" genuine (Mitchell 2007).

Wartime interdisciplinary research on morale, the effects of bombing, and a renewed interest in racial attitudes, led to a closer alliance with psychological social psychology, and a new post-war identity—as a part of the "behavioral sciences." This alliance, which shifted sociology from its traditional relation with economics, also shifted its statistical methods: a change from correlation to non-parametric statistics like χ^2, and to tests against the null hypothesis. This change was not really argued for.

The differences were these. In the older model of correlational analysis the distinction between cause and correlation was key. Giddings' "constants" analogy did not work, and in practical contexts something else did: the correlations were more or less validated by the kinds of interpretative material mentioned by Stouffer, and by common sense. Correlations were incomplete explanations on their own. Providing quotations and life histories made a correlation into an explanation, but one subject to further interpretation and further partialling. And these were not "theoretical" explanations, but practical ones.

The new model, adapted from psychological experiment, was to simply demonstrate an effect by providing a hypothesis that divided subjects or other kinds of data into categories, quantitative or not, and overcoming the null hypothesis—the no-difference hypothesis. The preferred statistic for this was χ^2, though there were many variants (Chapin 1955 [1947]; McCormick 1941).

> It should final be observed that a four-fold correlation table includes some of the basic elements of experimental design. Thus, in Table 61, we have an independent factor or treatment, Agriculture; a dependent factor, Imprisonment for Crime; an experimental group, the Prison population; and a control group, the Non-Prison population. On the other hand, dichotomies are used instead of classes based on measurement. In Table 61, sex and (roughly) age have been held constant, and there is nothing in the method that precludes as rigorous factor control as seems worthwhile. Even the broad 2 × 2 table may, therefore, be a valuable analytic device. (McCormick 1941: 217)

This was a problematic method in its home territory of randomized experiments. The confirmation offered by a demonstration that the null hypothesis was probably false was weak.

And psychologists routinely advanced the experimental tradition by adding conditions that further divided categories into new casual links.

The transfer into sociology was by analogy: observational studies were interpreted as experiment-like if there was some assurance that other factors did not produce the relevant effects (see Danziger 1990: 152–4). But the same kind of weak confirmation of defeating the null hypothesis could produce results. Simply generating new experiments with new conditions was not practical. But the results could be validated in a new way, by showing that they generalized to new settings. The poster child, and clearest example of this came from Stouffer's *American Soldier* study (1949): the concept of relative deprivation. The idea was that soldiers did not think they were being treated unfairly in the process of discharging them from service because they compared themselves not to the general process of the Army, but to their reference group: the soldiers around them. This was simultaneously a sociological idea, of reference groups, and a psychological one that did not depend on interpretation or the articulated opinions, justifications, and concepts of the subjects. This was an important change from the past approach to subjectivity.

10.8 The Behavioral Science Revolution: 1945–1965

The change did not yet fit into a broader account of sociological knowledge, in the sense of developing sociology as a theoretical science. Prominent sociologists other than Blumer complained about this. Louis Wirth, for example, wrote to the Rockefeller officer Joseph Willets noting the weak intellectual and theoretical yield from this research, and outlined an alternative approach to attitudes that understood them in their local settings, as a result of a genealogical process, and in terms of their relation to action. But at the time he was doing this funding for morale research took off, interdisciplinary teams were created in the military, quick and dirty methods, such as the Likert Scale, were devised and widely used, and the moment passed.

The methodological expression of this enthusiasm came in the form of an articulation of the approach at Columbia under the influence of Robert Merton and Paul Lazarsfeld and a wide group of associates in two volumes entitled *The Language of Social Research* (Lazarsfeld and Rosenberg 1955). The general thought behind the volumes was that smaller studies, typically involving scales of attitudes, could have their results validated not by reference to the sayings of the subjects or participants but by being successfully generalized.

What the new form of research promised was generalizability, the use of sociological or social psychological concepts, and results. Because of the scales, it was "scientific," not subjective. And because the practice of testing against a null hypothesis and testing for significance at the 0.05 level lowered the standards of acceptability, and non-parametric statistics methods were developed to take advantage of this lowering, it became easy to generate results. Moreover they were often results of a particular kind: they associated attitudes or other variables with standard demographic variables, and thus almost always showed that there were differences, and therefore "social" causes for the differences. This was a powerful rhetorical device, which promised eventual policy relevance, on the basis of the widespread assumption that what was social was also open to change by manipulating the conditions that led to the differences. Thus, if lack of schooling was a variable that predicted a disfavored outcome, more schooling was therefore a solution to the differences that were attributed to schooling. Needless to add, this was a promise that was rarely fulfilled, but the practice of identifying social differences in outcomes motivated and still motivates a great deal of social and health care policy. But the failures of these policies also motivated more research of the same kind, with more statistical refinement.

From the point of view of the scientific ambitions and credentials of sociology, the issue was establishing generalizations with some "theoretical" content. Columbia sociology under Merton and Lazarsfeld, and the large team that contributed to *The Language of Social Research*, produced a methodological manifesto and model that standardized the practices, though older forms of statistical analysis persisted unchanged. The difference was in the idea that statistical studies could be turned into generalizations. The strategy worked by taking the explanations and terms of the initial study and raising them to a higher level of abstraction, stating them in more general and more generally applicable contexts, and seeing if the same results could be obtained in other contexts.

This strategy had undeniable appeal, in the sense that it made great promises. One of the students of the era, Maurice Stein, captured the period nicely in later comments:

> S: Merton believed that we were all pre-Newtonian, that we were making the building blocks out of which some synthetic genius would one day make the great theory, the grand theory or the grand synthesis. I mean, it's possible (mutual laughter).
>
> D&E: Merton and Lazarsfeld generated excitement about imminent new discoveries?
>
> S: Tomorrow! Tomorrow! Just on the verge of everything! Lazarsfeld is going to develop the perfect scale and Merton was going to develop the perfect theory about that scale. And sociology was onto a new plane! They were very smart men, no question. And when Parsons' *The Social System* (1951) arrived, they thought he was Newton. (Stein et al. 2011: 135)

This was indeed, albeit briefly, the sensibility. By the mid-fifties the tone had changed: the big achievements failed to occur.

There were two explications of the Columbia strategy in terms recognizable to the history and philosophy of science. The first was done by a student of Chapin who had attached himself to Columbia, Hans Zetterberg, an enterprising Swede who self-published *On Theory and Verification in Sociology* (1963 [1954]), which went through several editions. It was the sole explication that attempted to align these practices with the contemporary language of logical positivism. It was limited, as all later writings on "theory construction" by statistical sociologists were, by the fact that the theoretical generalizations were not "general" but were based on either correlations or disproofs of null hypotheses. Therefore, there was a logical gap between the formalism required for deductive theory and the methods: a point made to Lazarsfeld by the young Patrick Suppes, who commented that his methods did not lead to a theory "in the accepted sense of the word," to no effect (International Sociological Association 1961: 350–51). In the late 1960s and early 70s there was a spate of books on theory construction that foundered on the problem (Blalock 1972; Glaser and Strauss 1999 [1967]; Greer 1989 [1968]; Hage 1972). These had the same problem: they were attempts to make correlations into deductive theory.

This strategy was taken up by Barney Glaser, who was exposed to it at Columbia, and applied to qualitative research: the explanations that were given in a qualitative case were simply restated in more general terms to see if they could be applied elsewhere. This was the basis for the influential book with Anselm Strauss, *The Discovery of Grounded Theory* (Glaser and Strauss 1999 [1967]). In both cases there was a certain circularity: to decide whether the new context was sufficiently similar to count as an application depended on interpretation, and the interpretation of the context depended on the theory. So these were "apply where they apply" theories rather that genuine universals. And the same held for the strategy in general.

The theoretical yield of the strategy was minimal. As with Giddings' constants, there were few higher-level generalizations to be found beyond the big correlations of race, class, gender, and such things as political outlook and status-attainment, and even these were correlations. But the idea of balance was also lost: if the categories were psychological and not a matter of the articulated views of people, the psychological or attitudinal categories supplanted them and overrode them. That indeed was the point of such scales as the F-Scale in the Authoritarian Personality studies: it was a measure of depth psychology, not of mere justifications and rationalizations.

The rise and fall of influence of Logical Positivism in the 1950s and 60s had a marked impact on American sociology, but an odd one. The main effect was to associate "positivism" with quantification, and to promote a conflict between quantifiers and non-quantifiers, which became elevated into a wide-ranging controversy that eventually included all of the issues of sociology and philosophy, ideology, and politics. This was a strange misunderstanding. The basic ideas of Logical Positivism were inapplicable to sociology; there were no laws, and no theories with a meaningful deductive structure. But sociologists nevertheless tried to mimic such theories by converting correlations into laws and by loosening the notion of deduction to make it appear that there were deductive theories (Collins 1975), or by misapplying texts in the philosophy of science. A favorite was R.B. Braithwaite, who had provided an analysis of functional explanation which was taken to legitimate the uses in sociology (1953). But the gap between the models of theory described in these texts and the sociological versions was large. The term "positivism" nevertheless was a red flag for critics, who equated it with quantification, and developed elaborate critiques.

The standard model of research nevertheless remained hegemonic. The questionnaire method with its perfect scales was fruitful largely through creating an illusion of quantitative solidity. Stein commented on his experience of actually working with this method under Merton-Lazarsfeld protégé Alvin Gouldner (1954a, b). As Stein later recalled the experience:

> During the Gyp Study, we threw out the questionnaire we had administered because we knew that the data was meaningless. We learned that you had to drink with the miners. And you had to go down into the mine with them and work alongside them. (Stein et al. 2011: 142)

By the middle of the 1960s, this sense of the artificiality of the quantified phenomenon had developed into a critique, which emerged in a systematic way under the heading of "ethnomethodology." A subtitle of one of its texts captures its point: "reality in everyday and sociological discourse" (Pollner 1987). The critics distinguished the "methods" which ordinary people used in classifying social events and phenomena to the methods used by sociologists, to make the point that sociologists relied on "data" that was in fact constructed according to routine practices, "mundane reason," as Pollner put it, that undermined its apparent solidity. Official records, for example, were the result of informal understandings and bureaucratic short cuts, rather than representing raw facts.

10.9 After 1965: Causal Models and Beyond

The 1960s were a transformative period in American sociology, as elsewhere. From the point of view of the problem of subjectivity and quantification, its many conflicts, fueled by the student movement, but also by a reaction against the model provided by Parsons, Merton, and Lazarsfeld (Mills 1959; Stein and Vidich 1963) served to harden a division between quantitative and non-qualitative forms of sociology that has persisted to the present (Akresh 2017: 551). But the issues were overlain by other issues, many of which, like the political conflicts, were imported from other sources. From Kuhn's influential *The Structure of Scientific Revolutions* (1996 [1962]) came the idea of paradigms, which was reinterpreted to apply to sociology in the form of the idea that the field contained multiple incommensurable paradigms, which included methodological paradigms. Viewpoints proliferated. Merton himself, during a sojourn at CASB (Center for the Advanced Study of Behavioral Sciences), wrote up a defense of his own work as an approach, which he called structural, to differentiate from structural functionalism, rather than as a unifying, overarching scheme for sociology (Merton 1975). The idea of a sociology as a unified science was effectively dead.

Path analysis provided what appeared to be an almost magical resolution to these issues. It purported to give numerical weights to the causal contributions of various causal variables. In a sense, this was no different from Yule's use of partialling, from which it derived. But Yule had been conservative in his use of the method. The casual relations he dealt with, and indeed which Sewall Wright's original use of path analysis in population genetics dealt with, were known if imperfectly understood causal relations, with few potential confounders or risks of spuriousness. In the new uses it was being put to it was unclear what qualified something as a cause in the first place, and even what a causal relation in sociology was. This was cheerfully ignored and the methods were applied as though the quantified contributions within the model provided the answer. Causality was treated simply as an assumption like other assumptions that needed to be made to construct a model. The strategy was to minimize the intellectual content of these assumptions in the hope that the data analysis would eliminate the causally irrelevant variables.

The models they built were, especially in the beginning of this period, causal structural equation models of a kind shared with econometrics and other disciplines, but less focused on prediction than on casual analysis (Blau and Duncan 1967). Elaborate attempts at scaling largely disappeared: the constraints were given by the form of the available data, which typically corresponded to policy goals. In time, economists themselves took over many of the topics of quantitative sociology. As topics, such as status attainment, were exhausted, new areas were studied, often related to policy issues. Statistically, however, these efforts were problematic: one consequence of the development of statistical methods was to make them less transparent and more sensitive to the manipulation of assumptions.

From the point of view of the history and philosophy of science, this was effectively the end of the quest to make sociology a "science." The use of statistics to address policy questions survived. The elevation of policy issues, such as inequality, racial differences, gender differences, and similar topics into "theoretical" issues continued, but without the sense that a unified "science" of sociology would be the result. This is not to say that "theories" continue to be proposed, and viewpoints established, but the idea that these will constitute or contribute to a "science" or overarching scientific theory of society of the kind envisioned by the labor statisticians, Small, Giddings, and their successors, is no longer part of the project. Theoretical images of society as a whole were imported from Europe.

The aim of becoming a science like the others simply vanished. In its place was the idea that what sociologists did was itself "scientific" because it was statistical. The tools for research that the quest to make sociology a science remained and were deployed in research on these externally driven topics. This is what sociology became: something that the labor statisticians would have recognized as similar to their own work, and with similar problems over objectivity, presentation to the public, bias, and causality, and a similar dependence on the explanations given by ordinary people. But the kind of transparency that the labor statisticians aspired to was lost, and along with it the influence over public discussion that the labor statisticians had achieved.

References

Abel, Theodore. 1948. The Operation Called Verstehen. *American Journal of Sociology* 54 (3): 211–218.

Akresh, I. R. 2017. Departmental and Disciplinary Divisions in Sociology: Responses from Departmental Executive Officers. *American Sociologist* 48 (3–4): 541–560.

Anthracite Board of Conciliation. 1908. In *Anthracite Board of Conciliation*, ed. W. L. Connell, vol. 2. Wilkes-Barre, PA: Record Press.

Austrian, Geoffrey D. 1982. *Herman Hollerith, Forgotten Giant of Information Processing*. New York: Columbia University Press.

Bales, Kevin. 1991. Charles Booth's Survey of Life and Labour of the People in London 1889–1903. In *The Social Survey in Historical Perspective: 1880–1940*, ed. Martin Bulmer, Kevin Bales, and Kathryn Kish Sklar, 66–110. Cambridge: Cambridge University Press.

Bannister, Robert C. 1987. *Sociology and Scientism: The American Quest for Objectivity, 1880–1940*. Chapel Hill, NC: University Of North Carolina Press.

Blalock, Hubert. 1972. *Theory Construction: From Verbal to Mathematical*. Hoboken, NJ: Prentice Hall.

Blau, P. M., and O. D. Duncan. 1967. *The American Occupational Structure*. New York: Wiley.

Bliss, H. F. 1897a. Eccentric Official Statistics. *American Journal of Sociology* 2 (4): 515–531.
Bliss, H. F. 1897b. Eccentric Official Statistics II. *American Journal of Sociology* 3 (1): 79–99.
Bliss, H. F. 1897c. Eccentric Official Statistics III. *American Journal of Sociology* 3 (3): 355–377.
Bliss, H. F. 1898. Eccentric Official Statistics IV. *American Journal of Sociology* 4 (1): 79–93.
Bliss, H. F. 1900a. Eccentric Official Statistics V. *American Journal of Sociology* 6 (1): 105–113.
Bliss, H. F. 1900b. Plutocracy's Statistics, Statistical Liars, and Liars Official and Unofficial, Exposed. *Rabid Socialist*. Chicago: C. H. Kerr. https://rsdiscover07.hsp.org/Record/marc-27394. Historical Society of Pennsylvania.
Bliss, H. F. 1908. *The New Encyclopedia of Social Reform*. New York/London: Funk & Wagnells Company.
Blumer, Herbert. 1939. Critiques of Research in the Social Sciences. In *An Appraisal of Thomas and Znaniecki's 'The Polish Peasant in Europe and America'*, vol. I. New York: Social Science Research Council.
Bowley, A. L. 1895. Comparison of the Rates of Increase of Wages in the United States and in Great Britain, 1860-1891. *The Economic Journal* 5 (19): 369–383.
Braithwaite, Richard. 1953. *Scientific Explanation: A Study of the Function of Theory, Probability and Law in Science*. Cambridge: Cambridge University Press.
Bruere, Martha Bensley. 1927. *Does Prohibition Work? A Study of the Operation of the Eighteenth Amendment Made by the National Federation of Settlements, Assisted by Social Workers in Different Parts of the United States*. New York: Harper and Brothers.
Bullock, Chas J. 1899. Contributions to History of Wage Statistics. *American Statistical Association* 6:187–218.
Bulmer, Martin. 1986. *The Chicago School of Sociology: Institutionalization, Diversity, and the Rise of Sociological Research*. Chicago: The University of Chicago Press.
Chapin, F. Stuart. 1955 [1947]. *Experimental Designs in Sociological Research*, Rev. Edn. New York: Harper & Brothers.
Clark, John Bates, and Franklin H. Giddings. 1888. *The Modern Distributive Process. Studies of Competition and its Limits, of the Nature and Amount of Profits, and of the Determination of Wages, in the Industrial Society of To-Day*. Boston: Ginn & Company.
Collins, Randall. 1975. *Conflict Sociology: Toward an Explanatory Science*. Cambridge, MA: Academic Press.
Connecticut Bureau of Labor Statistics. 1890. *Fifth Annual Report of the Bureau of Labor Statistics of the State Of Connecticut, for the Year Ending November 30, 1889*. Hartford: Press of The Case, Lockwood and Brainard Company.
Coulanges, Fustel de. 2000 [1889]. *The Origin of Property in Land*, trans. Margaret Ashley. Kitchener, Ontario: Batoche Books.
Danziger, Kurt. 1990. *Constructing the Subject: Historical Origins of Psychological Research*. Cambridge: Cambridge University Press.
Dubois, W.E.B. 1967 [1899]. *The Philadelphia Negro: A Social Study*. New York: Benjamin Blom. https://archive.org/details/philadelphianegr001901mbp.
Edgeworth, F. Y. 1892a. Correlated Averages. *Philosophical Magazine (Fifth Series)* 34:190–204.
Edgeworth, F. Y. 1892b. The Law of Error and Correlated Averages." Philosophical Magazine (Fifth Series), 34: 429–438, 518-526.
Galpin, Charles J. 1918. *Rural Life*. New York: The Century Company. https://archive.org/details/rurallife00galpgoog (accessed 17 July 2023).
Giddings, Franklin H. 1886. Profit Sharing. In *The Seventeenth Annual Report of the Massachusetts Bureau of Labor Statistics*, 1–83. Boston: Wright & Potter Printing Company. This is published under Wright's name.
Giddings, Franklin H. 1887. The Theory of Profit-Sharing. *Quarterly Journal of Economics* 1 (1): 367–376.
Giddings, Franklin H. 1901. *Inductive Sociology: A Syllabus of Methods, Analyses and Classifications, and Provisionally Formulated Laws*. New York: Macmillan.

Giddings, Franklin H. 1922. *Studies in the Theory of Human Society*. New York: The Macmillan Company.

Glaser, Barney G., and A. L. Strauss. 1999 [1967]. *The Discovery of Grounded Theory: Strategies for Qualitative Research*. Chicago: Aldine Pub. Co.

Goldberg, Joseph P., and William T. Moye. 1985. *The First Hundred Years of the Bureau of Labor Statistics*. Washington, DC: U.S. Government Printing Office.

Gouldner, Alvin W. 1954a. *Patterns of Industrial Democracy*. Glencoe, IL: The Free Press.

Gouldner, Alvin. 1954b. *Wildcat Strike*. Yellow Springs, OH: Antioch Press.

Greenwald, Maurine W., and Margo Anderson, eds. 1996. *Pittsburgh Surveyed: Social Science and Social Reform in the Early Twentieth Century*. Pittsburgh: University of Pittsburgh Press.

Greer, Scott A. 1989 [1969]. *The Logic of Social Inquiry*. New Brunswick, NJ: Transaction Publishers.

Hage, Jerald. 1972. *Techniques and Problems of Theory Construction in Sociology*. New York: Wiley-Interscience.

Hammersley, Martyn. 2010. The Case of the Disappearing Dilemma: Herbert Blumer on Sociological Method. *History of the Human Sciences* 23 (5): 70–90.

Hart, Hornell. 1933. Changing Social Attitudes. In *Recent Social Trends in the United States: Report of the President's Committee on Social Trends*, ed. William Ogburn, 382–342. New York: McGraw-Hill, chap. VII.

Igo, Sarah E. 2007. *The Averaged American: Surveys, Citizens, and the Making of a Mass Public*. Cambridge, MA: Harvard University Press.

International Sociological Association. 1961. *Transactions of the Fourth World Congress of Sociology (1959)*. Louvain, Belgium: International Sociological Association.

Jones, Robert Alun. 1993. Durkheim and *La Cité Antique*: An Essay on the Origins of Durkheim's Sociology of Religion. In *Emile Durkheim: Sociologist and Moralist*, ed. Stephen Turner, 23–50. London: Routledge.

Kellogg, Paul, and Neva R. Deardorff. 1928. Social Research as Applied to Community Progress. In *Proceedings of the First International Conference on Social Work (9–13 July)*, 784–831. Paris: lmprimé Union.

Knight, F. H. 1925. Review of *Origins of Sociology* by Albion W. Small. *The American Economic Review* 15 (4): 724–727.

Kuhn, Thomas. 1996 [1962]. *The Structure of Scientific Revolutions*. 3rd edn. Chicago: The University of Chicago Press.

LaPierre, Richard. 1934. Attitudes vs. Actions. *Social Forces* 13 (2): 230–237.

Lazarsfeld, Paul F., and Morris Rosenberg, eds. 1955. *The Language of Social Research*. Glencoe, IL: The Free Press.

Lynd, Robert S., and Helen Merrell Lynd. 1929. *Middletown: A Study in American Culture*. New York: Harcourt Brace & Co.

Massachusetts Department of Labor and Industries. Division of Statistics. 1885. *Sixteenth Annual Report of the Bureau of Statistics of Labor*. Boston: Wright & Potter Printing Company State Printers.

McCormick, Thomas Carson. 1941. *Elementary Social Statistics*. New York: McGraw Hill.

Merton, Robert. 1975. Structural Analysis in Sociology. In *Approaches to the Study of Social Structure*, ed. Peter M. Blau, 21–52. New York: Free Press.

Mills, C. Wright. 1959. *The Sociological Imagination*. Oxford: Oxford University Press.

Missouri Division of Industrial Inspection. 1893. *Fifteenth Annual Report Bureau of Labor Statistics and Inspection*. Jefferson City, MO: Tribune Printing Company, State Printers and Binders.

Mitchell, Joel. 2007. Measurement. In *Philosophy of Anthropology and Sociology*, ed. Stephen Turner and Mark Risjord, 71–120. Amsterdam: Elsevier.

Morselli, Enrico. 1882 [1879]. *Suicide: An Essay on Comparative Moral Statistics*. New York: D. Appleton and Company.

New York Bureau of Statistics of Labor. 1886. Third Annual report of the Bureau of Labor Statistics of the State of New York for the Year 1885. In *Documents of the Senate of the State of New York Vol. 3, One Hundred and Ninth Session*, 2–11. Albany, NY: Weed, Parsons and Company.

Ogburn, William. 1930 [1929]. Presidential Address, "The Folkways of a Scientific Sociology." American Sociological Association, December 1929. *The Scientific Monthly* 30 (4): 300–306.

Ogburn, William, ed. 1933. *Recent Social Trends in the United States: Report of the President's Committee on Social Trends*. New York: McGraw-Hill.

Okrent, Daniel. 2010. Last Call: The Rise and Fall of Prohibition. New York: Simon and Schuster.

Parsons, Talcott. 1954 [1945]. The Problems and Prospects of Sociological Theory. In *Essays in Sociological Theory*, Rev. Edn, 212–237. New York: The Free Press.

Pidgin, Chas F. 1886. Statistical Diagrams. In *The Labor Movement: The Problem of Today*, ed. George E. McNeill, 628–638. Boston: A. M. Bridgman & Company.

Pollner, Melvin. 1987. *Mundane Reason: Reality in Everyday and Sociological Discourse*. Cambridge: Cambridge University Press.

Porter, T. M. 2004. *Karl Pearson: The Scientific Life in a Statistical Age*. Princeton, NJ: Princeton University Press.

Raushenbusch, Winifred. 1979. *Robert E. Park: Biography of a Sociologist*. Durham, NC: Duke University Press.

Reick, Philipp. 2023. 'A Phobia of Numbers?' The Labour Movement and Social Surveys in the German Empire. *Labor History*. https://doi.org/10.1080/0023656X.2023.2230905.

Ross, F. A. 1924. *School Attendance in the United States in 1920*. Dept of Commerce, Bureau of the Census. Washington, DC: Government Printing Office.

Ross, F. A. 1934. The Use of Statistical Data and Techniques in Sociology. In *The Fields and Methods of Sociology*, ed. L. L. Bernard, 458–475. New York: Ferrar & Reinhart, Inc.

Small, Albion. 1902. Reviewed Work(s): Inductive Sociology, a Syllabus of Methods, Analyses and Classifications, and Provisionally Formulated Laws by Franklin Henry Giddings. *Science* 15 (383): 700–706.

Small, Albion. 1924. *Origins of Sociology*. Chicago: The University of Chicago Press.

Spencer, Herbert. 1872. *The Principles of Psychology*. Vol. 2. 2nd ed, 76. London: Williams and Norgate. Ch. VII, §299.

Spencer, Herbert. 1876. *The Principles of Sociology*. London: Williams and Norgate.

State of Rhode Island and Providence Plantations. 1890. *Third Annual Report of the Commissioner of Industrial Statistics Made to the General Assembly at Its January Session 1890*. Providence: E. L. Freeman & Son State Printers.

Stein, Maurice, and Arthur Vidich, eds. 1963. *Sociology on Trial*. Englewood Cliffs, NJ: Prentice-Hall, Inc.

Stein, Maurice, Steven P. Dandaneau, and Elizabeth A. East. 2011. Listening to Sociological Elders: An Interview with Maurice R. Stein. *The American Sociologist* 42 (1): 129–144.

Stouffer, Samuel A. 1934. Sociology and Sampling. In *The Fields and Methods of Sociology*, ed. L. L. Barnard, 476–488. New York: Farrar and Rinehart Inc.. https://archive.org/details/dli.ministry.12832/mode/2up.

Stouffer, S. A., Edward A. Suchman, Leland C. DeVinney, Shirley A. Star, and Robin M. Williams Jr. 1949. *Studies in Social Psychology in World War II: The American Soldier*. Princeton: Princeton University Press.

Thomas, W. I. 1939. Comment by W. I. Thomas. *Critiques of Research in the Social Sciences: An Appraisal of Thomas and Znaniecki's The Polish Peasant in Europe and America* [Bulletin 44], 82–87. New York: Social Science Research Council. https://brocku.ca/MeadProject/Thomas/Thomas_1939a.html.

Thomas, W. I., and F. Znaniecki. 1918. *The Polish Peasant in Europe and America*, 5 vols. Chicago: The University of Chicago Press.

Turner, Stephen. 1996. Durkheim Among the Statisticians. *Journal for the History of the Behavioral Sciences* 32:354–378.

Turner, Stephen. 1997. Net Effects: A Short History. In *Causality in Crisis*, ed. Vaughn McKim and Stephen Turner, 23–45. Notre Dame, IN: University of Notre Dame Press.

Turner, Stephen. 2016. Rural Sociology: A Slightly Personal History. In *Rural Sociologists at Work: Candid Accounts of Theory, Methods and Practice*, ed. Hans Bakker, 9–33. London: Routledge.

Wilson, E. B. 1926. Statistical Inference. *Science* LXIII:289–296.

Wirth, Louis. 1951. "The Significance of Sociology," opening address, First World Congress of Sociology. *International Social Science Bulletin* III (2): 197–201.

Woofter, Thomas J. 1920. *Negro Migration: Changes in Rural Organization and Population of the Cotton Belt*. New York: W. D. Gray.

Wright, Carroll D. 1886. Profit Sharing. In *The Seventeenth Annual Report of the Massachusetts Bureau of Labor Statistics*, ed. Franklin H. Giddings, 1–83. Boston: Wright & Potter Printing Company.

Wright, Carroll D. 1887. *The Fisheries of Massachusetts*. Massachusetts Bureau of Statistics. Wright & Potter Printing Company; also by Franklin H. Giddings.

Young, Allyn A. 1925. Review of Papers Relating to Political Economy by F. Y. Edgeworth. *The American Economic Review* 15 (4): 721–724.

Yule, G. U. 1907. On the Theory of Correlation for any Number of Variables, Treated by a New System of Notation. *Philosophical Transactions of the Royal Society, London Series A* 79:182–193.

Zetterberg, Hans. 1963 [1954]. *On Theory and Verification in Sociology*, Rev. Edn. New York: The Bedminster Press.

Chapter 11
Origins of Sociology in Germany

Stephan Moebius

Abstract This article outlines the history of sociology in Germany from 1842 onwards. As in France and England, the decisive social and intellectual processes that contributed to the emergence of sociology in Germany took place in the middle of the nineteenth century. Society was now no longer understood in terms of contract theory, that is, as something that first had to be produced by individuals, as was the case in early bourgeois political philosophy (Thomas Hobbes, John Locke). Rather, society came to be conceived as something that already existed, characterised either dynamically as a social movement, or more statically, as a social fact, and thus had to be "discovered". The central pioneers of sociology in Germany were Lorenz von Stein and Robert von Mohl, as well as Karl Marx and Friedrich Engels. For them, society was a phenomenon governed by its own logic and therefore became an object of knowledge in its own right. Around 1850, the most important contexts of social thought were historicism, Hegel's philosophy of history and in particular the "sciences of the state". Later on, it was historicism as well as the philosophy of Dilthey, Neo-Kantianism, Nietzsche, Folk psychology, and Darwinism. The "social question", and, at the turn of the century, the "cultural question", which became increasingly pressing, especially among members of the educated middle classes, marked the beginning of the development of sociology in Germany. The article traces the development of sociology from the mid-nineteenth century throughout the German Reich to the interwar period, and concludes with a brief outlook on the post-war phase after 1945 in West and East Germany up to the 1990s.

Keywords Sociology in the German Reich · Sociology in the Weimar Republic and interwar period · Sociology in Germany after 1945

S. Moebius (✉)
Universität Graz, Graz, Austria
e-mail: stephan.moebius@uni-graz.at

Y. Shan (ed.), *History and Philosophy of the Social Sciences*,
https://doi.org/10.1007/978-3-032-08167-4_11

11.1 Introduction

Social science in Germany began with spying activities of a German agent on French socialists and communists. In the 1840s, he "discovered" "society" as an independent object of knowledge in the social movements and ideas of these groups. The counter-revolutionary hope of the Prussian agent, that a greater knowledge about society would increase protection from impending revolutions, was one of the driving forces that lead to the emergence of the social sciences in Germany. This agent was Lorenz von Stein (1815–1890).

The historical background of these developments towards a changed conceptual view of the social, the "idea" or "discovery of society", were economic, political, and social transformation processes, and social problems, which were responsible for the emergence of sociological thought not only in the German-speaking world, but also in France (through Saint-Simon (1760–1825) and Auguste Comte (1798–1857)) and in UK (through Herbert Spencer (1820–1903)). The American War of Independence and the French Revolution had led to a new "awareness of ability", nurtured ideas of feasibility, formability, and change, and sparked new ideas of purposeful social planning, control, and guidance. These processes within liberal-bourgeois society went in line with the spread of the division of labour, the emergence of the capitalist economic system, industrialisation, and urbanisation. The new dynamics and antagonisms associated with these processes were increasingly understood as "social" struggles and antagonisms in the mid-nineteenth century, and the incalculable processes of impoverishment associated with capitalist commodity production became known as the "social question" (König 1949: 25; Pankoke 1970). The coincidence of a "awareness of ability" on the one hand and the feeling of being increasingly at the mercy of economic processes that were taking on a life of their own on the other, called for efforts of rationalisation and scientification as well as demands for the conscious organisation of the newly "discovered" social dynamics through "social" politics (Pankoke 1984: 1008).

Some of the early German social scientists believed that the best protection against future revolutions was the organisation of society and social reforms. The ordering and organisation of the social required knowledge about the current state and desirable future organisation. Due to historical developments, contemporaries were no longer inclined to think of "society" as a freely chosen association of independent individuals; instead, they started to comprehend it as an essentially economically determined structural entity with various interrelated conflicts of interest and dependencies. "Society" advanced to become an object of knowledge in its own right, whose dynamics, regularities, and laws needed to be discover, understand, and describe in more detail.

As in France or UK, in the German-speaking world too, the interest in scientific analyses of society emerged from the perceptions, and discourses of social crises, historical transformations and the generally changed historical experiences of the "Sattelzeit" (Koselleck 1976/1989). Social analysis was (and is) a specific way and part of a creative problem-solving, problem-formulation and interpretation of economic, political, social, and cultural processes that were considered problematic or at least “significant”, i.e. particularly worthy of interpretation and processing. Since the beginning of the nineteenth century, in Germany these problems included the aftermath of the French Revolution, the turbulent political period of the so-called “Vormärz” (the period preceding the 1848 March Revolution), the industrialisation and its consequences, the revolutions of 1848, the increasingly pressing “social question” and later, around 1900, the “cultural question” (crisis of culture, relativism), which was considered central, especially by a large part of the educated middle classes (Bruch 1989: 14; Bollenbeck 1994: 256ff).

Sociology itself were a “fait social”, a “social fact”, the result of social processes. Since these processes took different courses depending on society and local circumstances, and the intellectual observers of these processes had different social, ideological, and cultural backgrounds and associated perceptions; the reflexive, interpretative and discursive processes that are today referred to as “sociological” were, accordingly, different in each case. In the 1830s in France, Auguste Comte proposed the term “sociology” for the scientific observation, interpretation and explanation of the social, which had then become a separate object of scientific analysis. However, since Comtes terminology was linked to his “positivist scientific programme, those who did not share this view, or did not share it in its entirety, but were nevertheless convinced of the urgency of a science that adequately analysed social developments and upheavals, preferred other names.” (Mozetič 2018: 38).[1]

In Germany, the new academic discipline that “discovered” the social as a relatively autonomous sphere, that “objectified” it as a phenomenon with its own logic, and specified it conceptually, was initially (around 1850) called “social science” (“Gesellschaftswissenschaft”). The intellectual contexts of the emerging social science were historicism, Hegel’s philosophy of history and political science (Mikl-Horke 2001: 40; Lindenfeld 1997). In addition to historicism, the separation between state and society already conceptualised by Johann Gottlieb Fichte (1762–1814) and Georg Wilhelm Friedrich Hegel (1770–1831), as well as the idea of human perfection beyond all state constraints, and the idealism of Immanuel Kant (1724–1804) were particularly conducive to the preliminary stages and subsequent emergence of social science in Germany.

[1] The German quotations were translated by the author, for their very valuable assictance with the translation I would like to thank Antonia Schirgi and Frithjof Nungesser. The discussion from the section on the German Reich onwards draws on my more comprehensive study *Sociology in Germany. A History* (2021), which examines the period from 1945 to the present in greater detail.

11.2 The Early Social Scientists: Lorenz von Stein, Robert von Mohl, Karl Marx and Friedrich Engels

The "social question" was at the centre of the discourses and debates of the early German social scientists. It was the starting point both for the materialist social theory of Karl Marx (1818–1883) and Friedrich Engels (1820–1895) as well as for the emergence of "social science" ("Gesellschaftswissenschaft"), as it was espoused by Lorenz von Stein and Robert von Mohl (1799–1875).

The social struggles of 1830 as well as those on the eve of the revolutions of 1848 brought about a new perspective on the social. It was increasingly perceived and "discovered" as an independent sphere, strictly distinguished from the state. This enabled the early German social scientists, largely inspired by the early French socialists, to conceive society as a relatively autonomous, self-dynamic and antagonistic sphere of the social—a social sphere beyond the state. The metaphor of "discovery", which was used by von Stein, showed that, unlike Hegel, he did not perceive society as something that was historically unique, but rather something that always existed, but which first had to be revealed in its specificity by him as the discoverer.

The "discoverers of the social" like Lorenz von Stein and Robert von Mohl were usually political scientists and members of the liberal bourgeoisie. The conceptual separation they made of the society, which was engaged in class struggles, from the state, made it possible for them to emphasise all the more clearly the necessity of the state as a pacifying third party that stood above the struggling society. Lorenz von Stein, for example, had been arguing for welfare state reforms since the 1840s. According to him, reforms were the only way to prevent a revolution. The aim of the social scientists was to "objectify" the "social question" with the help of a social-scientific analysis and thus to give the state a new, scientifically supported relevance. Subsequently, the state was a third instance that could help to overcome the social antagonisms and that stood above the individual interests of the social classes. Von Stein sought to "abolish" (in the Hegelian sense) the conflict between capital and labour in a third party, the state, which he also called the "kingship of social reform" (*soziales Königtum*).

Von Mohl distinguished between three approaches to the discovery of society: firstly, "socialist partisans" who strove for a "practical transformation of the existing society"; secondly, those who dealt with socialist writings and ideas; thirdly, those philosophical minds, to whom he himself belonged, who in the course of their "research into human and especially state conditions have come to the conviction that there is an organisation of people, a society of its own, which is different from the state [...]." (Mohl 1855: 78) According to von Mohl (1855: 78), "the latter provide the actual scientific work". In contrast to von Stein, for von Mohl the economic interests of the social movements were only one moment of socialisation; other moments were worldview, tradition, customs, ways of life and passions. It is insightful to note that von Mohl developped a concept of the "social circles of life" ("Lebenskreise") that individuals belong to. This concept was a precursor of a concept that became prominent a few decades later, namely Georg Simmel's concept of "social circles" and his description of the "crossing of the social circles", that he developed, based on Moritz Lazarus' (1824–1903) psychology of peoples (*Völkerpsychologie*).

The need for a separate social science were highly controversial in the academic field. They were, in particular, opposed by state scientists who feared a loss of influence. The fiercest critic was the historian Heinrich von Treitschke (1834–1896), who was at the end of the 1870s a central figure in the Berlin anti-Semitism controversy as well as author of the multi-volume German History (1879–1894), a book series that no educated bourgeois household at the time could do without. In his habilitation thesis *Die Gesellschaftswissenschaft. Ein kritischer Versuch* (1859/1927), he vehemently opposed an independent theory of society as advocated by von Stein and von Mohl. Treitschke criticised their supposedly all too narrow concept of the state. According to him, state science and historiography were perfectly capable of grasping the dynamics of society. The threat to social cohesion should not be inhibited by social reforms, but by the consistent implementation of national unification. Treitschke considered questions posed by social science to be legitimate, but best dealt with in traditional (political) historiography; he therefore considered an independent social science to be completely "superfluous" (Riedel 1963: 59), and even dangerous. Von Treitschke regarded it as being close to social-revolutionary utopias. His criticism gave rise to a narrative that was taken up again and again in later times, which led to the "associative linking of 'sociology' and 'socialism'" (Rehberg 2010: 219).

With his book *Der Socialismus und Communismus des heutigen Frankreichs* (1842; engl. *Socialism and Communism in Contemporary France*), Lorenz von Stein unintentionally initiated a transfer of ideas. The book contributed significantly to the first awareness of French socialist and communist thought in Germany. It was Moses Hess (1812-1875), in particular, who introduced Marx and Engels to Stein's bookand to these socialist ideas in the 1840s. From 1844 onwards, the two began to develop the foundations of historical materialism together. ,.

If the observation of society as a class society was similar among the representatives of German social science and the representatives of historical materialism, von Stein and Marx were miles apart on in political matters. In contrast to Marx (who was closer to Hegel than the latter), von Stein, for example, unequivocally maintained the importance of private property as the "expression of a free personality" (Blasius 1977: 40f.). Moreover, von Stein sought the solution to the social crisis and antagonisms in the third party of the state (*soziales Königtum; lit.: social kingship*) and in the unification of the German Empire. Marx and Engels, on the other hand, regarded the state as a committee for the administration of the interests and business of the property-owning classes. For them, the solution to the social question lay not in social reform or in the nation state, but in the internationally oriented revolution.

Unlike the evolutionary social democrats around 1900, Marx and Engels did not understand the transition to socialism or communism as a teleological and inevitable process that would automatically, and independently of concrete class struggles, lead to a "happy end".. Consistently, in a radical historical and self-reflective manner, they also regarded their own theory as a social-historically mediated product that could only have developped from a certain point in the aggravation of social antagonisms and at the same time could in turn have an effect on social practice. In contrast to "utopian socialists", Marx and Engels refused a precisely planned, systematic intellectual construction of a future society. "The liberation is a historical, not a mental act," Marx and Engels wrote unequivocally in *The German Ideology* (1845–1847/2017 MEGA^2 I/5: 16). They went on to state: "Communism is for us not a state of affairs which is to be established, an ideal to which reality [will] have to adjust itself" (Marx and Engels 1845–1847/2017 MEGA^2 I/5: 37), because it "is not enough for thought to strive for realisation, reality must itself strive towards thought" (Marx 1844/2009 MEGA^2 I/2: 178). Similar to representatives of Critical Theory a hundred years later, who also distanced themselves from contemporary social planning and social-technological endeavours, Marx and Engels instead began by a "critique of the existing". It is only in the course of this critique that an alternative society emerges.

A few days before the February and March Revolutions of 1848, the world-famous *Manifesto of the Communist Party* appeared. It is not a document of historical determinism but deals with historical contingency and the human capacity to take the course of history into one's own hands and to (re)shape it collectively. No other writing by nineteenth century intellectuals had a comparable social-theoretical and political impact. The thought of Marx and Engels is still influential in social sciences today, even if the links and references differ significantly.

Unlike von Stein and von Mohl, who could not detach themselves from the narrower subject of state science, Marx, took a more abstract stance and criticised the state machinery and negated the need for state science as an independent discipline, since its object of research had ultimately proved to be the instrument of the ruling classes. Marx's argument was more sociological than that of the "social scientists", because in his eyes, the state existed neither alongside society nor above it in the sense of a third instance mediating the social antagonisms, but was to be understood genuinely as an immanent part and result of the social relations.

Towards the end of their lives, the focus of the analysis of Marx and Engels shifted towards an economic analysis of capitalism, without, however, giving up their political commitment. In addition to his work on the economic manuscripts, Marx had been involved in the International Workers' Association (IWA), which was founded in 1864. It became known as the *First International*, for which Marx wrote, among other things, the inaugural address (1864). One result of this increasing turn towards political economy was Marx's magnum opus *Das Kapital*, whose first volume was published in 1867. The publication not only coincided with the economic upswing of the so-called "founder's boom" (*Gründer-Boom*) of 1866–1873, but also with the founding of workers' associations and national workers' parties. Thanks to this book, Marx became the true contemporary sociological theorist of the Industrial Revolution. This is not the place to reproduce and critically comment on the entirety of *Das Kapital*. A few references must therefore suffice. The book begins with the commodity, which, according to Marx, constitutes the core of capitalist, commodity-producing society and economy. In contrast to earlier agrarian societies who produced goods primarily for their own use, in capitalism, the production of goods for exchange is predominant. As a result, goods increasingly become commodities. In capitalist societies in principal, everything can be objectified, turned into a commodity and thus alienated from its actual purpose. For Marx, the specificity of commodity production under capitalism is that human labour power, which is essential for the reproduction of human beings, also becomes a commodity that can be exchanged. By emphasising the factor of labour, Marx' argument differs from neoclassical economics, for which only the market price and the "marginal utility" of goods is the decisive economic criterion instead of labour. It is also characteristic of capitalist society that capitalists strive for a permanent increase in profit.

Even if the concept of society of the "social scientists" on the one hand and of Marx and Engels on the other differed, they had in common that the intellectual basis of their concepts emerged from their engagement with the philosophy of Hegel and the intellectual movements of thought that he had initiated. While von Stein insisted on the abolition of the dialectical process in a third party (the state), Marx and Engels insisted on the dialectical process as such. Hegel as a philosophical starting point separated them from other contemporary pioneers and representatives of sociology in Europe, such as Auguste Comte and Herbert Spencer. The theoretical concepts and political positions of the protagonists of the social discourse that began in Germany in 1840 were mainly motivated by their conviction that the "social question" urgently required a solution. Both intellectual camps, the social scientists Lorenz von Stein and Robert von Mohl, and the socialist theorists Karl Marx and Friedrich Engels, had a common starting point and perspective: for them, the social conditions were characterised by serious social inequalities and the resulting class struggles no longer coincided directly with the actions and functions of the state. However, the interpretations of social conditions as well as the answers to the "social question" differed according to the social, political, and ideological backgrounds of the protagonists.. Whereas, for von Mohl the state and society were at an equal level, von Stein hoped that the state was able to regulate the growing social conflicts; the conception of Marx and Engels' was diametrically opposed and took a historically new direction: Society no longer represented an area within or alongside the state, but the latter was ultimately nothing more than the political expression, the "managing committee", of society's ruling class. These two positions that constituted the field of German social analysis in the second half of the nineteenth century, were subjected to fierce criticism. In the case of Marx and Engels, the critics were a wide range of political and academic opponents; in the case of the social sciences of Stein and Mohl, the criticism came in particular from the political and historical sciences that dominated the academic field. Especially the historian Heinrich von Treitschke (1834–1896) was a harsh critic of the social sciences. He rejected a distinct area of knowledge of the social sciences as unnecessary (Treitschke 1859/1927). Society and the state could never be analysed separately. For him, the nation state represented the greater whole of all social relations. Accordingly, social cohesion, which was at risk, should not be achieved with the help of social reforms, but through the consistent realisation of national unification.

11.3 Sociology During the German Reich

National unification was realised in 1871. The foundation of the German Reich resulted from three wars, the Second Schleswig War (1864), the Austro-Prussian War (1866) and the Franco-Prussian War (1870/71). According to some researchers, the foundation of the Reich was an achievement of the then Reich Chancellor, Otto von Bismarck's, and his "revolution from above" (Wehler 1977: 235). The central characteristics of this time were social transformation, modernisation, high industrialisation, and the rise of the German Reich to the second most powerful industrial nation in the world, the expansion of the monetary economy, massive military armament, and an imperialist foreign policy. Social change was caused by an increasing population growth, an increasing division of labour, and rapid urbanisation, which was accompanied by housing and poverty problems. The labour movement as well as the women's movement grew rapidly. New scientific discoveries, and inventions such as electricity, the automobile or the zeppelin as well as the beginning of mass culture reinforced the impression of a rapidly changing time. At the same time, social criticism and cultural counter-currents like the youth movement and the so-called "life reform movement" (*Lebensreformbewegung*) emerged. Between 1871 and 1918, many new avant-garde styles and forms emerged in art, music, literature, and architecture. The economic, political, social, and cultural developments of the German Reich proved to be very favourable for the emergence, establishment, and institutionalisation of sociology in Germany. The rapid and condensed processes of industrialisation, modernisation, social differentiation, and cultural transformation required a broader perspective of analysis and interpretation of these economic, political, social, and cultural changes, a "sociological perspective" so to speak.

In the 1870s, under the influence of Comte, Charles Darwin (1809–1882), Spencer, modern biology , and other natural sciences, sociological concepts that understood society as analogous to an organism emerged and became predominant in Germany—for example those developed by Paul von Lilienfeld (1829–1903), Albert Schäffle (1831–1903), and, in a broader sense, Ludwig Gumplowicz (1838–1909). This tendency was, however, only temporary. From the late1880s onwards, it began to change: on the one hand, positivsm in the sense of Comte and Spencer , was increasingly criticised and condemned as a "Western" mode of thought; on the other hand, the growing popularity of other disciplines, philosophical orientations und new views on social and cultural problems gained influence. Yet many of these organismic approaches remained deeply rooted in idealism. Schäffle, for instance, despite his organicist metaphors, conceived the social primarily as a spiritual bond. By contrast, it was above all Marx and Engels who, having left idealism behind in their critique of Max Stirner (1806-1856) in *Die Deutsche Ideologie* (*The German Ideology,* 1845-1847) and drawing on their studies in the natural sciences, conceptualised society as a historically specific, dialectical, and mediating nexus between human nature and non-human nature. In doing so, they articulated critiques of the ecological destruction inherent in a capitalist economy oriented towards permanent accumulation—critiques that remain strikingly relevant today.

Following the period in which the state sciences, Marxism, Darwinism, and the natural sciences had provided the main intellectual framework, other academic disciplines and traditions in Germany gradually assumed a central role in the formation of sociology as an autonomous discipline. ,. Among them were historicism, the historical school of economics, as it was represented by Gustav Schmoller (1838–1917), or Neo-Kantianism that became one of the most influential philosophical movements through its skilful university policy (Köhnke 1986). Equally significant were the philosophical currents of Friedrich Nietzsche (1844–1900), Wilhelm Dilthey (1833–1911)with his foundation of the humanities (*Geisteswissenschaften*), and other various philosophies of life, above all Arthur Schopenhauer (1788–1860), and Henri Bergson (1859–1941). One particular feature of German sociology is its deep entanglement with philosophy. Important influences also came from the historical sciences, from Moritz Lazarus' (1824–1903) folk psychology (*Völkerpsychologie*), which was central to Georg Simmel (1858–1918), as well as from the legal sciences and phenomenology..

As can be seen, in particular, from works of Georg Simmel and Max Weber (1864–1920), the focus changed from the primordial central "social question" to the "cultural question". This shift of attention was due to the specific awareness of crisis among the educated middle classes in the German Reich. The early sociologists, now considered as classics, experienced the levelling of their ideals of culture and education ("*Kultur und Bildung*") as well as in the dissolution of traditional world views and cultural values, which was perceived and debated by many contemporaries of the educated middle class, as particularly problematic. The functional differentiation and increasing relative autonomy of social subsystems (economy, politics, art), secularisation and the increasing of divergent world views and individual cultural phenomena intensified the perception and awareness of a problematic social and cultural disintegration. It was the perception, diagnosis, and reflection of the prevailing situation as a "cultural crisis" that led to the "genesis" and "basic concepts of the social sciences" around 1900 (Acham 1996). It also fostered the emergence of sociology as *cultural sociology* in Germany at the turn of the century (Lichtblau 1996).

Unlike von Stein or von Mohl, the early sociological classics were no longer primarily concerned with the independence of "society" from the state, but with social and cultural cohesion. Not redistribution, surplus value production, the state or revolution, but ethics, lifestyle, world view, personality, individuality, and morality were the normatively charged key concepts of sociology around 1900. The essential terminological opposition in sociological discourse was no longer state versus society but community-society or society-individual and to the normatively highly charged differentiation between German "culture" and Western "civilisation" (Elias 1939/1977).

Especially historicism and the philosophy of Nietzsche had a great influence on the sociological classics' perception of the crisis. Both fuelled relativism. The problem of relativism raised by historicism and intensified by Nietzsche continue to preoccupy the social sciences and humanities to this day. It provoked a variety of reactions, such as scientific endeavours to find firm foundations in anthropology, ahistorically valid values, positivist science or in great syntheses (see Moebius 2025b). The normative revaluation of culture was also a reaction to this precarious relativism.

Besides the works of Marx, the first work that is today considered a sociological classic in Germany is Ferdinand Tönnies' *Gemeinschaft und Gesellschaft* (1887, Community and Society). It marks the transition from organistic sociologies to concepts of sociology based on cultural sciences and is hence in a tense relation to the natural sciences, Marxism, Hobbes and the humanities. Ferdinand Tönnies (1855–1936), an internationally renowned Hobbes expert, interpreted the transformation processes of his times as a shift from community to society. He lamented the decline of community but was also aware that there could be no turning back. When it first appeared, hardly anyone took notice of Tönnies' work. It was not until a few decades later, during the interwar period, that it was widely received. However, it was often misinterpreted and read as a romantic glorification of community or even as a proto-fascist propagation of a *Volksgemeinschaft* (people's community). Tönnies himself was close to social democracy and saw his work more as a basis for social reform to improve the situation of the working class. According to Tönnies communal aspects like cooperatives could contribute to a new social order in order to reduce the problems of capitalist society.

Georg Simmel was an ingenious analyst of the cultural transformation processes of his days. He was deeply influenced by the folk psychology of Moritz Lazarus (Köhnke 1996), from 1900 onwards also by Nietzsche and Bergson. His observations of modern culture from the perspective of cultural sociology range from analysing modern money culture, everyday phenomena, the opposition of socialism and individualism in society, politics, styles of thought and aesthetics to his theorem of the "tragedy of culture". For him, the "tragedy of culture" consisted in the discrepancy between "objective" and "subjective culture". By "objective culture" (a term he took from Lazarus) Simmel understood the totality of all material and spiritual "things" created by man (for example technology, science, but also art). The "subjective culture", on the other hand, is the need and the willingness of individuals to absorb the components of the objective culture created by subjective mind. With the progressive development of modern society, however, the products of objective culture multiplied to such an extent that individuals are ever less able to subjectively appropriate them and creatively exploit them as a means of self-realization or for the development of individuality.

In his later writings, it became increasingly clear that Simmel saw the solution of the cultural crisis, which also guided his political thinking (Moebius 2023), in the "law of the individual". In line with provenance to the educated bourgeoisie and as a result of an affirmative reception of life philosophy, Goethe and especially Nietzsche (Köhnke 1996: 489–514), with the search of the "law of the individual"

Simmel propagated the self-realisation of the individual over the masses (during World War I this was transformed in the struggle between the "self-realisation"/"individual law" of the German nation against the Western civilisations).

Max Weber too was concerned with the individual, especially with the question to which extent the individual conduct of life could be further guaranteed in the light of capitalist, rationalist, and bourgeois forces. Like Marx, Simmel, and Werner Sombart (1863–1941), Ernst Troeltsch (1865–1923), Weber was in his major work *Die Protestantische Ethik und der "Geist" des Kapitalismus* (1904–05/1920/2016) (Protestant Ethics and the "Spirit" of Capitalism), concerned with the rise of capitalism. Similar to Simmel's analysis of monetary culture, which sought to "construct a new storey beneath historical materialism" (Simmel 1900/1989: 13), Weber was concerned with an analysis of the central ideas that gave rise to capitalism. In his "Protestant ethics" he described the intellectual constitution that was decisive for the formation of capitalism. For Weber, the "Protestant ethic" was part of a general cultural process of rationalisation in Western societies. Rationalisation, Weber referred to it as the "destiny of our time", represented a general social process of systematisation, predictability, and effectiveness of all areas of society, social relationships, and actions. Weber described this rationalisation process as *Entzauberung der Welt* ("disenchantment of the world"), in the light of which each individual had to find their own way, following an "individual law", as he made clear to students in his lecture on *Wissenschaft als Beruf* (Weber 1917/1919/1994, "Science as Vocation").

Weber's understanding of science became particularly prominent in the course of the *Werturteilsstreit*, which can be literally translated as the "value judgement dispute", the first major controversy in German-speaking sociology (Albert 2010). The starting point of this controversy was a methodological dispute within the German Historical School of Economics, represented by Schmoller, and Carl Menger (1840–1921), representative of the Austrian *Grenznutzenschule* ("marginalist school"). In 1913 the dispute reached its peak in the *Verein für Sozialpolitik*. Opponents were Schmoller, Eduard Spranger (1882–1963) and Rudolf Goldscheid (1870–1931) on one side, and Menger, Lujo Brentano (1844–1931), Werner Sombart and Max Weber on the other. The latter group advocated that scholarship should abstain from value judgement. In 1909, so prior to the spectacular value judgement dispute of 1913, these tensions led to the founding of the German Sociological Association (GSA). The GSA explicitly avoided pursuing "practical" socio-political goals and instead pursued scholarship in the sense of abstaining from value judgement, free from power and religion, thus avoiding ideological conflicts as strictly as possible. In addition to Tönnies, Simmel and Weber, also the philosopher Paul Barth (1858–1922), the public lawyer Georg Jellinek (1851–1911), the historian Karl Lamprecht (1856–1915), the philosopher Theodor Lipps (1851–1914), the sociologist and economist Franz Oppenheimer (1864–1943), Werner Sombart, Ernst Troeltsch, the sociologists Alfred Vierkandt (167–1953) and Alfred Weber (1868–1858), and others, signed the invitation for the founding of the association.

The First German *Soziologentag* (literally "Sociologists' Day") took place in October 1910 in Frankfurt am Main. Tönnies, who was president of the GSA till 1934, opened the congress. He spoke about "ways and goals of sociology". Simmel gave a lecture on the "sociology of sociability". Finally, it ended in a scandal that again was caused by a dispute about value judgement. The second *Soziologentag* took place in Berlin in October 1912, on the theme of the nation. Alfred Weber delivered the keynote address, elaborating on the sociological concept of culture, as well as the distinction between culture and civilisation.

When World War I began in August 1914, like many intellectuals, the majority of German sociologists believed that the Germans now had the task and duty to defend the German "culture" against "civilsation". Nothing remained from the previously praised *Werturteilsfreiheit*. Max Scheler praised the "genius of war". For Simmel, the war marked a solution of the tragedy of culture, and socio-cultural disintegration. Max Weber immediately volunteered for military service and praised the "great and wonderful" "holy war of defense" (Kaube 2014: 350). Although Weber's initial enthusiasm for war declined over time, he did not develop a sociological analysis of the war and its effects. It was only the cultural sociologist Emil Lederer (1915/2014), who provided an objective sociological analysis of the war, the nationalist ideologies, and the interaction between military technology and the control of homogenized masses.

11.4 Sociology in the Weimar Republic

With Germany's defeat also the Reich ended.[2] The German people was shocked, and the mass euphoria of 1914 gave way to indignation over the course of the war as well as the economic, political, and social problems associated with its end. In November 1918 workers and soldiers sparked the German Revolution of 1918–1919 and began to found radical democratic councils. However, after numerous bloody battles and an enormous number of deaths, the conservative and reformist forces were able to prevail. The civil war-like conditions culminated on the 15th of January 1919 in the murder of Rosa Luxemburg (1871–1919) and Karl Liebknecht (1871–1919), maltreated and shot by secret Freikorps with the approval of the military of the governing Majority Social Democratic Party (MSPD). This was not exactly a radiant beginning of a new republic. The first elections to the National Assembly took place on the 19 of January, the women being allowed to vote for the first time. The new republic was faced with enormous problems like the conversion to a peace economy, reparation payments to the victorious powers, and further burdens from the Treaty of Versailles. Furthermore, the population lacked identification with the republic and parliamentarianism. This provided a breeding ground for radical parties. Social struggles were characteristic of the first years of the Weimar

[2] For the history of the Weimar Republic see Peukert (1987).

Republic. As it turned out, the military and judiciary were still riddled with conservative and right-wing staff.

Only in the second phase of the Republic (1923–1929) did the currency reform and the "Dawes" plan bring about an economic upswing. One social class gained a particular social importance: the employees. But this new middle class was deeply insecure, fearing to lose its social status, and open for nationalist attitudes (Geiger 1932). Despite—or perhaps because of—the massive economic, political, and social problems, the "Golden Twenties" were a time of an important cultural and artistic productivity. The content ranged from utopian ideas of a "new man" and an optimism about the progress of a "new world" to plain cultural criticism, and genres such as painting, music, dance, literature, theatre, photography, film, design, building and living. The Weimar period was a veritable "experimental field" and a "culmination point" of modernity (Becker 2018). Many of the topics that still concern us today, such as capitalism, democracy, community/society, mass culture, mass psychology, public opinion/press, new media, new art, or even relativism were already discussed then and also became central themes of the sociology of the interwar period (Moebius 2025a, 2025b; Acham and Moebius 2025, 2021, 2022, 2024).

It was especially the economic crisis of 1929 that changed the Republic. The unemployment rate rose rapidly from about 14,6% at the end of 1929 to roughly 22, 7% in 1930. In new elections in 1930, the KPD (Communist Party of Germany) and above all the NSDAP (National Socialist German Worker's Party, or short: Nazi Party) won additional votes. When in renewed elections at the end of 1932 it became apparent that the NSDAP was to receive fewer votes and the KPD more, General Paul von Hindenburg (1847–1934), the president of the German Reich, decided to make Adolf Hitler (1889–1945) chancellor, from the 30th of January 1933 on, in order to avert the danger from the left. The NSDAP became the state party und state terror began.

So much for the historical context. But let's go back to the development of sociology during the Weimar Republic. The initially war enthusiasm of sociologists 1914 weakened considerably till 1918 (Joas and Knöbl 2008: 142–143). Some sociologists like Tönnies, Max Weber, and Troeltsch became "republicans of reason", that means, that they supported the new democratic Weimar Republic less out of sympathy than for practical political reasons.

During the Weimar Republic, sociology as an academic discipline began to institutionalise and professionalise itself. Whereas in 1919 there were only three chairs of sociology in Germany, during the Weimar period "despite the fatal budgetary situation, professorships were established at 18 universities (…), whose titles included sociology (often in quotation marks and were held by 34 professors, 20 of whom were full professorships". (Dörk 2018: 843) This positive development of sociology at the beginning of the Weimar Republic was owed to Carl Heinrich Becker (1876–1933), a State Secretary and later Minister of Education in the Prussian Ministry of Culture. He promoted the establishment of new chairs of sociology.

The new role of sociology was the result of a much-discussed crisis in the humanities and their educational and cultural ideals. Since the old ideals of culture and education were at the political level regarded as part of the course that had led to total destruction during the war, some considered it necessary to "reorientate" and "change the vocabulary" (Bock 1994: 167). According to Becker, sociology offered the opportunity to contribute to a change of these orientations and synthesise the academic disciplines that were going through a process of increasing differentiation. He attributed an important cultural-political and moral educational task to sociology—much like Émile Durkheim did in France. In 1919, with the appointment of three professors at newly funded universities, the first steps towards an institutionalisation and professionalisation of sociology were taken: Franz Oppenheimer (1864–1943) was appointed to the first sociology chair in Frankfurt am Main, Leopold von Wiese (1876–1967) to the Chair of Economic Science of State, and Max Scheler (1874–1928) to the Chair of Philosophy and Sociology in Cologne.

Different centres of sociology developed during the Weimar Republic: In Cologne the first "Research Institute for Social Sciences" was founded in 1919 (Knebelspieß and Moebius 2019). The process of institutionalisation and professionalisation of sociology in Germany was advanced by the founding of the—still existing—sociological journal from Cologne (Moebius 2017): 1921 von Wiese founded the *Kölner Vierteljahrshefte* (*Cologne Quarterly*) that was renamed in 1923 and became the *Kölner Vierteljahreshefte für Soziologie* (*Cologne Quarterly of Sociology*). As *Cologne Journal for Sociology and Social Psychology,* it is still one of the most renowned sociological journals in Germany. The founding of this first purely sociological journal in Germany was an important milestone in von Wiese's efforts to institutionalise sociology within academia in the Weimar Republic, to establish it and to contribute to its reputation. The journal was also dedicated to the proclamation of von Wiese's own theory—the so-called *Beziehungslehre* (theory of relationships), which is sometimes also referred to as formal sociology. At the beginning of the twenties, it was still regarded as one of the most important sociological currents;in the following decades, however, it lost almost all of its relevance. Thus, von Wiese's significance for the professionalisation of sociology in the interwar period remained an organisational one. During the interwar period von Wiese increasingly advanced to become an important spokesman and "institution builder" (Käsler 1984: 462) of sociology. His "gate-keeper" position in the sociological field was strengthened by the fact that he was not only in charge of an institute with its own journal, but also he had an extraordinarily influential role in the *German Sociological Association* (GSA), where he served as first secretary and organiser of the *Soziologentage* (the meetings of the GSA) (Gorges 1986: 97).

Thanks to his "charismatic wealth of ideas" (Fischer 2008: 38), Max Scheler gathered circle of students and colleagues in Cologne, including Paul Honigsheim (1885–1963), Paul Ludwig Landsberg (1901–1944), Heinrich Lützeler (1902–1988) and Helmuth Plessner (1892–1985). As early as 1921 Scheler published an essay on the sociology of knowledge. Parallel to Karl Mannheim (1893–1947) he contributed to the development of the sociology of knowledge. He was, however, less rooted in sociology than in philosophy. Scheler came from the phenomenology of Edmund Husserl (1859–1938) and was considered a star philosopher. In 1928 Scheler founded a movement that he called/that was thereafter called Philosophical Anthropology with his monography *The Position of Man in the Cosmos*—that was published parallel to Helmuth Plessner's *Steps of the Organic and Man*. Both, Scheler and Plessner, had developed this line of thought independently. The Philosophical Anthropology, which assumes that man has always been a cultural being, is still a fundamental (and also unique) characteristic of sociology in the German-speaking area. The anthropologisation that could be observed in the inter-war period (Rehberg 1981) and to which Scheler's and Plessner's philosophical anthropology contributed significantly, can also be seen as one of many attempts to solve the "historism problem", meaning the relativism that accompanied historism (Acham 1998: 544; Moebius 2025b).

In 1924, before his magnum opus on philosophical anthropology, Plessner had written another important work: *Grenzen der Gemeinschaft. Eine Kritik des sozialen Radikalismus* (Plessner 2002 [1924], The Limits of Community. A critique of social radicalism). It was written as a reaction and criticism to the widespread longing for community in the Weimar Republic. The political right received Tönnies' concept of community with enthusiasm. While the political right received primarily the concept of the organic national community that was sacralised, the left received the concept of the production community.

Another centre of sociology was Heidelberg, where the Institute for Social and Political Sciences was founded in 1924 (cf. Blomert 1999). In contrast to von Wiese's theory of relations developed in Cologne, Heidelberg stood for a historically oriented sociology of culture, which was represented in particular by Alfred Weber (1868–1958). Characteristic for Alfred Weber's sociology of culture was an appreciation of the cultural as a spiritual-aesthetic movement (*Kulturbewegung*) opposed to technical-material phenomena of the civilisation process (capitalism, urbanisation, industrialisation, and rationalisation), and opposed to the social process (the economy, political power relations, and forms of organisation as well as the territorial division of the population).

Since 1924, Arnold Bergstraesser (1896–1964) was Alfred Weber's assistant. In Heidelberg he made an effort to promote international scientific exchange, especially with France. Talcott Parsons, a US-American sociologist, probably also came to Heidelberg via this newly founded exchange service. He wrote his doctoral thesis on capitalism with Sombart and Max Weber there in 1926/1927.

The second director of the Heidelberg Institute was Emil Lederer (1882–1939), who had been a staff member and editorial secretary of the *Archive for Social Science and Social Policy* since 1911. Lederer was professor for social policy. He was one of the very few sociologists who had written about the war from the perspective of his discipline. In Weimar Republic he wrote about the social classes, the social-psychic habitus, intellectuals, the sociology of revolutions, violence, art, public opinion, the new middle class of employees, work organization, and unemployment. Among Lederer's assistants were Hans Speier (1905–1990), Albert Salomon (1891–1966), Fritz Simon Croner (1896–1979), Svend Riemer (1905–1977), Hans H. Gerth (1908–1978), and Talcott Parsons (1902–1979). In 1933, Speier followed Lederer into exile in the USA, where Lederer played a central role as the first dean of the "University in Exile" of the *New School for Social Research.*

In 1924 Norbert Elias (1897–1990) came to Heidelberg and started to work on his habilitation, supervised by Alfred Weber. He soon belonged to the *inner circle of* the institute and became "unofficial assistant" (Elias 1990: 125) to Karl Mannheim (1893–1947). Born in Budapest, acquainted with Georg Lukàcs (1885–1971), participant in intellectual circles such as Simmel's circle in Berlin or Sonntag's circle in Budapest, Mannheim fled from right-wing extremist terror in Hungary in 1919 and came to Heidelberg in 1921, where he contributed significantly to the establishment of sociology of culture and sociology of knowledge. In 1928, Mannheim provoked the major sociological controversy of the Weimar Republic, the so-called "Sociology of Knowledge Dispute", which emerged at the 6th Congress of the German Sociological Association in Zurich (Meja/Stehr 1990). His book *Ideology and Utopia*, published in 1929, was widely discussed and became one of the most famous works of sociology in the Weimar Republic. His 1928 lecture on competition in the intellectual sphere, as well as his book, met with extensive criticism. Among those who, from different ideological standpoints, engaged critically with Mannheim's theses, were Alfred Weber, Werner Sombart (1863–1941), Otto Neurath (1882–1945), Max Horkheimer (1895–1973), Ernst Robert Curtius (1886–1956), Max Adler (1873–1937), Hannah Arendt (1906–1975), Herbert Marcuse (1898–1979), Karl August Wittvogel (1896–1988), and others. Their passionate and ideologically diverse responses, however, in a way confirmed Mannheim's own thesis about competition within the intellectual sphere; that all thinking, cognition, and knowledge are bound to their social and cultural existence. their

In Berlin, Becker's new university policy had a particularly positive effect on the social sciences. In addition to Werner Sombart, Alfred Vierkandt (1867–1953), Kurt Breysig (1866–1940), Richard Thurnwald (1869–1954), Karl Dunkmann (1868–1932) and Goetz Briefs (1889–1974) were appointed to chairs of sociology after 1918. Alfred Vierkandt was particularly important for the institutionalisation of sociology in Berlin because of his publication of the *Handwörterbuch der Soziologie* (1931), a "kind of codification" of the contemporary state of sociology in Germany. It shows that German-speaking sociology, despite its successive differentiation during this period, still considered itself primarily as cultural sociology. Most of the essays included in the dictionary were contributed by Sombart, followed by a number of contributions of Vierkandt and Theodor Geiger (1891–1952). Contemporary sociology of knowledge was also represented by a programmatic essay of Mannheim (1931).

Although there were some contacts with foreign sociologists, sociology in Germany remained largely isolated. One of the few exceptions was the ethnosociologist Richard Thurnwald, who taught in Berlin. By means of his reception of US-American sociology, social anthropology, and the foundation of the *Zeitschrift für Völkerpsychologie und Soziologie*in 1925 (together with Bronislaw Malinowski (1884–1942)), he fostered the internationalisation of sociology in the Weimar

Republic. From 1936 onwards he adapted to the NS regime and advocated German colonialism (Steinmetz 2015 [2009]).

Furthermore, Frida Wunderlich (1884–1965), a feminist researcher and activist, held a chair in social policy in Berlin. After she was forced to exile in 1933, she became the first female dean at the University in Exile in 1938/1939, succeeding Emil Lederer (Golesorkhi et al. 2023).

In Hamburg, sociology was represented by Andreas Walther (1879–1960), who had a clear empirical orientation, owed to a visit to the United States. There he became acquainted with empirical research techniques, which he subsequently used as a basis for his research. Conversely, he was also one of the first to make US American sociology known in Germany. From 1933 onwards, he placed his empirical social research in the service of National Socialist social-hygienic population policy (Waßner 1986: 407).

In Leipzig, Hans Freyer (1887–1969) was appointed to the Chair of Sociology in 1925 (Muller 1987: 122–222). Because of his important influence on younger sociologists, in particular national-revolutionary scientists, (Schäfer 1990; Muller 1987: 143 ff.), his circle is nowadays referred to as "Leipzig sociology" (Rehberg 2021). Particularly prominent representatives of Leipzig sociology, alongside Freyer were Arnold Gehlen (1904–1976) and Helmut Schelsky (1912–1984). After 1945, Schelsky became one of the most influential sociologists of West German sociology. All three "allied themselves with National Socialism—for motives going back to their engagement in the youth and national movement, for careerist reasons, for an anti-bourgeois (and therefore perhaps perceived as 'revolutionary') conservatism". (Rehberg 1999: 73) Other students and employees of Freyer included Gunter Ipsen (1899–1984), Karl-Heinz Pfeffer (1906–1971) and Karl Valentin Müller (1896–1963), all of whom became "convinced National Socialists", as can be seen from their work and political practice (Schäfer 1990: 158). Freyer described *Sociology as the Science of Reality,* which was the title of his main work published in 1930. Since every sociology is historically and culturally bound to a certain social reality, as he argued (Freyer 1930: 67), the different national sociologies also differ. Thus, sociology in his days had to be different to the one of the epoch of its emergence (liberalism, bourgeoisie and capitalism). The result is that every nation has its own sociology. Consequently, Freyer, adopted the political position that the time had come for a genuinely "German" sociology that would serve the German "Volkswerdung" ("Becoming a people"). Only one year later, in 1931, Freyer published a "Manifesto of the Conservative Revolution" (Schäfer 1990: 166, Fn. 61), the book entitled *Revolution von rechts* (Freyer 1931, revolution from the right). In this book he argued in favour of a unity of state, people and Führer. He reaffirmed his will to contribute to the Volkswerdungsprozess (process of "becoming a people") (Muller 1987: 186–266), and for the dissolution of society into a national "Volksgemeinschaft".

In Frankfurt, sociology developed absolute different. In 1919 Franz Oppenheimer received the first sociological professorship in Germany. He understood sociology as a historically universal science that should bring together economics, history, ethnology, law, art and religious studies. Oppenheimer retired in 1929. From

1934–35 onwards he taught in Palestine. In 1936 Oppenheimer was appointed honorary member of the *American Sociological Association.* He emigrated first to Japan, and finally, in 1940, to the USA, where he died impoverished in 1943 (Caspari and Lichtblau 2014). From 1921 to 1923 Fritz Sternberg (1895–1963) and Gottfried Salomon (1892–1964) were Oppenheimer's assistants. Since 1926, Salomon was the editor of the *Jahrbuch für Soziologie (Yearbook for Sociology)* that was aimed at achieving an internationalisation and alternative to a one-sided "German" sociology (Henning 2006: 77). He was also involved in the planning of the famous Davos university courses. Like many others, Salomon fled into exile in New York and became member of the *New School for Social Research.*

In 1930, Karl Mannheim succeeded Oppenheimer. When Mannheim left Heidelberg and went to Frankfurt, he was followed by some of his students such as Nina Rubinstein (1908–1996), Kurt Wolff (1912–2003), Norbert Elias und Hans Gerth, others, such as Gisèle Freund (1908–2000), Jakob Katz (1904–1998), Wilhelm O.H. Baldamus (1908–1991) and Wilhelm Carlé (1887–?), became his students in Frankfurt, so that a real "Mannheim circle" or Mannheim "research group" was formed in Frankfurt (Kettler et al. 2008). Elias became Mannheim's official assistant but had to share the position with the second assistant Hans Gerth. Since the mid-1930ies, Elias worked on his famous book on civilisation process. Among Mannheim's circle were several female students who regularly attended his courses (Honegger 1993): Besides Freund and Rubinstein these included Margarete Freudenthal (1894–1984), Frieda Haussig (1903–1982), Käthe Truhel (1906–1992), Toni Oelsner (1907–1981), Evelyn Anderson (1909–1977), Ilse Ziegellaub (1900–1984), and Natalie Halperin (1908–1974). This was also the case after Mannheim went into exile, when Charlotte Lütkens (1896–1967) and Viola Klein (1908–1973) were influenced by Mannheim's sociology of knowledge. Klein wrote one of the first books on sociology of gender, *The Feminine Character: History of an Ideology*, published 1946 in the series *International Library of Sociology and Social Reconstruction* founded by Mannheim.

Besides the Mannheim circle, Frankfurt became famous in sociology because of the Frankfurt Institute of Social Research (IfS), which was founded there in 1923/1924 (Wiggershaus 1988/2001; Demirović 1999). The first director was the Austrian Marxist Carl Grünberg (1861–1940). The initiative and idea to establish a Marxist research institute came from Felix Weil (1898–1975), the left-wing son of a millionaire. After Grünberg suffered a stroke in 1928, Max Horkheimer, a friend of Weil, became director of the IfS in the early 1930s. Unlike the economist Grünberg, Horkheimer aimed to connect Marxism with "bourgeois science" and philosophy, thereby laying the foundations for what would later be known as Critical Theory. This theoretical orientation was strongly influenced by the Marxist reception of Georg Lukács and Karl Korsch (1886–1961). With the recruitment of Erich Fromm, who had completed his doctorate under Alfred Weber, the Institute expanded its focus to include psychoanalysis and social psychology. The Institute's medium of publication was the *Zeitschrift für Sozialforschung (Journal for Social Research)* that first appeared in 1932 and continued to be published in exile until 1941. During the interwar period, the central figures of the IfS and the *Zeitschrift für Sozialforschung* included, among others, the economist Friedrich Pollock (1894–1970), the psychoanalyst Erich Fromm (1900–1980), the economist Otto Kirchheimer (1905–1965), the lawyer Franz Leopold Neumann (1900–1954), the economist Henryk Grossmann (1881–1950), the sociologist and philosopher Herbert Marcuse (1898–1979), the sociologist of literature Leo Löwenthal (1900–1993), the sociologists Siegfried Kracauer (1889–1966), the sociologists and philosophers Walter Benjamin (1892–1940) and Theodor W. Adorno (1903–1969).

The programmatic term "Critical Theory", which was later used to describe this group, goes back to Max Horkheimer's essay "Traditionelle und kritische Theorie" (*Traditional and Critical Theory*) published in 1937. In this essay Horkheimer attacked—long before the so-called "positivism dispute" of the 1960s—positivism and the orientation of traditional theory towards the model of mathematical-scientific procedures. Contrarily, he called for a philosophically trained and Marx-oriented critical theoretical perspective that consistently interprets the observable social facts in their historical development and in a social context.

From 1929 to 1931, an investigation led by Erich Fromm analysed the political attitudes of workers and employees. Despite some weaknesses in implementation and evaluation, certain results could be obtained, that were, however, depressing: the workers were less left-wing than expected and, even worse, they showed certain authoritarian tendencies.

At the beginning of the 1930s, as the National Socialists came to power, the resources of the IfS were taken abroad, and branches of the IfS were established in Geneva, London and Paris. This ensured the continuous existence of the IfS. This is when Horkheimer and Adorno began their work on the *Dialectic of Enlightenment* (1944).

Another important sociological centre was Braunschweig, with Theodor Geiger (1891–1952) as its key founding figure . René König (1906–1992), who had studied in Berlin in the 1920s and who became one of West Germany's leading sociologists after the Second World War, remarked in retrospect that towards the end of the Weimar period there had been a new, promising wave in sociology, represented in particular by Mannheim and Geiger. Geiger had been appointed to the chair of sociology at the Technical University of Braunschweig in 1928 following the recommendations of Tönnies and Vierkandt. During the interwar period, Geiger received international attention because of several publications: the monograph *Die Masse und ihre Aktion* (1926, *Mass and its action. A contribution to the sociology of revolutions*), articles on community, society, revolution and leadership in Vierkandts "Handwörterbuch" (1931, *Handbook of Sociology*) and the monograph *Die soziale Schichtung des deutschen Volkes* (1932, *The Social Stratification of the German People*). Geiger was one of the decisive critics of National Socialism, among the sociologist of his days. In *Die soziale Schichtung des deutschen Volkes* (1932) he presented his discovery of an increasingly popular approval of the politics of the NSDAP by the middle classes. Geiger was dismissed in September 1933. Despite certain opportunistic attempts to adapt to the ruling regime (Geißler and Meyer 2000: 280), he was threatened and subject to spying. He had, at the same time, impeded the planned sociological professorship for Adolf Hitler in Braunschweig.

He emigrated to Denmark, where he became the first Danish sociology professor at Aarhus University in 1938.

At the beginning of the 1930s, contemporaries such as Geiger had a favourable opinion of the epoch of Weimar sociology: "Sociology has at last also in Germany passed the epoch of the struggle for its recognition as an independent science." (Geiger 1931: 568) New research on the sociology of the interwar period clearly shows that at this time sociology in the German-speaking countries was highly differentiated, meaning that there were many theoretical approaches (from Marxist, positivistic, phenomenological, anthropological to more philosophical one's) and special sociologies (from sociology of culture, knowledge, economics, politics, religion, intellectuals, music, literature, media etc.) (Moebius 2025a, 2025b; Acham and Moebius 2025, 2021, 2022, 2024). But it was not only in the academic field that things seemed to develop positively for sociology. In addition to academic institutionalisation, the popular education movement and non-university institutions were of great importance for the recognition and dissemination of sociology, too.

In the Weimar Republic, most sociologists did not foresee the danger of fascism. National Socialism brought sociology, as an institutionalised discipline, to a provisional end. Already in 1933, when Hitler was appointed chancellor of the Reich, many sociologists were dismissed from their positions or forced into exile. The German Sociological Association was "closed down" in 1934 (Schnitzler 2018). The Nazis had no interest in sociology as an independent science. Nevertheless, even though sociology did not exists as an independent discipline between 1933 and 1945, there were individuals who continued to work in a sociological manner. It persisted, for example, in the form of so-called "völkische Soziologie". In particular, the empirical expertise of sociologists proved useful to the Nazis (Klingemann 1996). At the same time, numerous exiled sociologists—among them Karl Mannheim, Norbert Elias, Theodor W. Adorno, Max Horkheimer, René König, Franz Neumann, Alfred Schütz, and Siegfried Kracauer—continued their work abroad, especially in the United Kingdom and the United States. In exile, they preserved and further developed German sociological traditions, often under difficult, insecure, or marginal institutional circumstances. Some of their ideas, along with new methodological impulses from Anglo-American empirical research, later found their way back to Germany. Moreover, several of these scholars, such as Adorno and Horkheimer, eventually returned after 1945 and contributed significantly to the re-establishment of sociology in postwar Germany.

11.5 Sociology After 1945

After the collpase of the Nazi-Regime 1945, sociology had to be rebuilt. Journals were re-founded or newly founded, the German Sociological Association was restored, and sociology was re-established as a subject taught at universities. Different schools and regional centres of sociology emerged. The so-called Cologne School centred around René König. König, who was also president of the

International Sociological Association (ISA) in the early 1960s, was, first and foremost,shaped by the intellectual currents of Weimar sociology and by the Durkheimian School, but was also central for the reintroduction of empirical methods from the United States (Moebius 2015). "Cologne sociology" that has mostly been orientated towards empirical social research, is still mainstream in Germany today (Moebius/Griesbacher 2019). Other groups that importantly contributed to the re-establishment of sociology in Germany included the Frankfurt School, and the circle around Helmut Schelsky, as well as a Marxist oriented group around the former resistance fighter Wolfgang Abendroth (1906–1985), Werner Hofmann (1922–1969), and Heinz Maus (1911–1978) in Marburg (the Marburg School, see Peter 2019). In addition, Otto Stammer (1900–1978) in Berlin, Arnold Bergstraesser (1896–1964) in Freiburg, and Helmuth Plessner in Göttingen also played key roles in rebuilding the discipline. Despite their theoretical and political differences, up until the 1950s, they all shared a decisive commitment to political and social enlightenment concerning the post-war situation. By the late 1950s, however, field-specific dynamics began to gain momentum. The different schools and groups tried to secure and expand their positions within the sociological field, and their divergent research profiles became increasingly visible. Disciplinary struggles and camp-building processes during the first 20 years of post-war West German sociology revolved around the debate on role theory and the dispute over positivism (*Positivismusstreit,* cf. Strubenhoff 2017). By the end of the 1950s, an institutional and generational change can be observed. The so-called "post-war generation" that included Ralf Dahrendorf (1929–2009), Jürgen Habermas (*1929), Niklas Luhmann (1927–1998), Erwin K. Scheuch (1928–2003), Heinrich Popitz (1925–2002), Hans Paul Bahrdt (1918–1994), M. Rainer Lepsius (1928–2014), and Renate Mayntz (*1929), assumed central positions in organisations, editorial boards of journals, and universities. While the early schools and circles (König, Schelsky, Adorno and Horkheimer) initially focused on the sociology of the family and empirical research, the following generation concentrated foremost on industrial sociology, but also on topics of social structure and social stratification as well as on social mobility.

In 1968, massive student protests took place in Germany. The student movement brought sociology into the public limelight. The Frankfurt School and the more Marxist-oriented Marburg School became closely connected with the student movement. Sociology gained considerable importance as an academic discipline, as a subject of study, a development linked to the growing need for social reflection in all areas of life. Characteristic for sociology at this time was the increasing differentiation into specialised subfields. The number of academic positions in sociology, as well as the number of students, rose significantly, partly as a result of the founding of new universities and of reforms in higher education policy. The growing number of non-university research institutions complemented sociological research at the universities. This expansion, which coincided with a highly visible form of public sociology, also provoked counter-reactions: conservative sociologists such as Helmut Schelsky criticised the growing social influence of sociology and propagated an "anti-sociology."

As far as empirical social research is concerned, quantitative research had become more professional; interpretative social research had slowly developed, reinforced by the increasing reception of symbolic interactionism. The social "planning euphoria" of the 1960s and 1970s weakened. In the 1970ies many looked at 1968 with disappointment and some even turned away from sociology. There were debates, such as that between representatives of critical theory and systems theory (the so-called "Habermas-Luhmann debate"), the debate on "theory comparison," and controversies regarding "postmodernism" (Moebius 2021: 97–98; Moebius and Römer 2022).

Similar to other countries, in Germany too, research on gender and women's issues developed from the women's movement of the 1960s and 1970s (Paulitz 2019). During the first phase of research on gender and women's issues, it was mainly the social inequality affecting women that was criticised. The 1980s brought about a crucial change: Instead of the category of "women", the category of "gender" or "gender relations" became the primary focus of interest. Nevertheless, discussions often remained confined to a discourse that neither problematised the distinction between sex and gender nor the linking of the category of gender to certain ideas of reproduction, heterosexuality, and binary coding. It was not until the 1990s that this change occurred thanks to the publication and discussion of Judith Butler's *Gender Trouble* (1990), a book that also significantly changed feminism in Germany. The 1980s were a great era for sociological theory and history of sociology in Germany. Theories like Habermas' *Theory of Communicative Action* and Luhmann's System Theory were established. Even though these two independent (West) German "grand theories" were present, in the 1980s West German sociology continued to be influenced by another major theory: Parsons' theory synthesis. Other theoretical approaches were widely received and discussed, for example (post-)structuralism, works by Michel Foucault (1926–1984) and the Birmingham Centre for Contemporary Cultural Studies (CCCS), as well as symbolic interactionism and American pragmatism, a line of reception that was importantly advanced by Hans Joas (*1948) during the 1980s and 1990s. Works of Alfred Schütz (1899–1959) and Norbert Elias stated to be more widely received in the 1980s. A revival of the German tradition of historical sociology, which was also associated with the name Max Weber, became more pronounced among the new cultural sociologists. It was in the 1980s, too, that the reception of Pierre Bourdieu (1930–2002) intensified. Today, alongside Weber and Marx, he is one of the most renowned and "most received authors among students of sociology in Germany" (Schneickert et al. 2019). Beside this "theory-boom", a further increase in the differentiation and pluralisation of the sociological field could be observed.

In the GDR (German Democratic Republic), sociology did not emerge until the 1960s (Peter 2018). In 1963, the party programme of the *Sozialistische Einheitspartei Deutschlands* (SED, *Socialist Unity Party of Germany*, SUPG) explicitly called for the establishment of sociological research. In the GDR sociology developed under completely different conditions than in West Germany. In the former, its development was overshadowed by Marxist-Leninist philosophy and political economy and it was instrumentalised by economic policy. It focussed on the basic categories of work and production. The connection to economic policy and historical materialism promoted the tendency to economic reductionism in sociology. Methodologically, the main focus was on quantitative methods. Sociology had the function of confirming the social laws whose theoretical interpretation was at that time reserved for a specific, ideologically adapted form of historical materialism. It was not until the late 1980s that the situation stated to change, and—beyond economic determinism—a relative autonomy of the social came increasingly into focus. This also led to first approaches to study the social position of women and gender relations. Overall, sociology in the GDR remained committed to a canonising interpretation of Marxism-Leninism. In the 1980ies, when the social processes and dynamics could no longer be adequately described within the conventional ideological framework, certain changes happened within GDR sociology and it became open to the so called "bourgeois sociology".

While far reaching intellectual influences changed the profile of sociology in the 1980s, the development of sociology in the 1990s was first and foremost shaped by one concrete social and political transformation, the fall of the Berlin Wall. Soon after German reunification, East German sociology almost entirely disappeared, and West German sociology expanded to the East. The triumph of capitalist society fostered a short revival of modernisation theory. As the systemic transformation came along with severe social problems, theories and research projects focusing on social exclusion, precarious work, and xenophobia became core to sociological discourse. The first decade of the present century again brought about major changes for society and sociology. Market logic increasingly dominated social and education policy; economic thinking and its involvement in political affairs was on the rise and may have contributed to a marginalisation of the influence of sociology on policy making. Characteristic is a further specialisation and differentiation, that became visible through the multiplication of special sociologies. The landscape of sociological theory in Germany continued to change: Earlier, grand theories were dominant, whereas nowadays a trend towards sociological diagnoses of contemporary society can be observed. Overall, contemporary sociology in Germany can be characterised by the following features: (1) Historically and philosophically informed sociological theory has always been and still is important. (2) German sociology lacks self-confidence compared to US-American sociology. In recent decades sociology in Germany has come ever closer to the globally prevailing understanding of sociology as an empirically operating social science and has also contributed to this global mainstream of sociology. This means that a distinctive German sociology can be observed ever less. (3) German sociology has a critical attitude and a strong tradition of public sociology. (4) Self-critical debates and internal controversies have always existed and still persist in the field of German sociology.

References

Acham, Karl. 1996. Die 'kulturelle Krise' der Gesellschaft um 1900 und die Genese der Sozialwissenschaften. In *Vom Weltbildwandel zur Weltanschauungsanalyse. Krisenwahrnehmung und Krisenbewältigung um 1900*, ed. Volker Drehsen and Walter Sparn, 39–67. Berlin: Akademie Verlag.

Acham, Karl. 1998. Historische Umbrüche in dem halben Jahrhundert seit dem Ersten Weltkrieg. In *Erkenntnisgewinne, Erkenntnisverluste. Kontinuitäten und Diskontinuitäten in den Wirtschafts-, Rechts- und Sozialwissenschaften zwischen den 20er und 50er Jahren*, ed. Ders et al., 535–566. Stuttgart: Steiner.

Acham, Karl, and Stephan Moebius, eds. 2021. *Soziologie der Zwischenkriegszeit. Ihre Hauptströmungen und zentralen Themen im deutschen Sprachraum. Band 1*. Wiesbaden: VS.

Acham, Karl, and Stephan Moebius, eds. 2022. *Soziologie der Zwischenkriegszeit. Ihre Hauptströmungen und zentralen Themen im deutschen Sprachraum. Band 2*. Wiesbaden: VS.

Acham, Karl, and Stephan Moebius, eds. 2024. *Soziologie der Zwischenkriegszeit. Ihre Hauptströmungen und zentralen Themen im deutschen Sprachraum. Band 3*. Wiesbaden: VS.

Albert, Gert. 2010. Der Werturteilsstreit. In *Soziologische Kontroversen. Beiträge zu einer anderen Wissenschaft vom Sozialen*, ed. Georg Kneer and Stephan Moebius, 14–45. Berlin: Suhrkamp.

Blasius, Dirk. 1977. Lorenz von Stein als Geschichtsdenker'. In *Lorenz von Stein. Geschichts- und gesellschaftswissenschaftliche Perspektiven*, ed. Dirk Blasius and Eckart Pankoke, 3–76. Darmstadt: WBG.

Blomert, Reinhard. 1999. *Intellektuelle im Aufbruch. Karl Mannheim, Alfred Weber, Norbert Elias und die Heidelberger Sozialwissenschaften in der Zwischenkriegszeit*. München/Wien: Hanser.

Bock, Michael. 1994. Die Entwicklung der Soziologie und die Krise der Geisteswissenschaften in den 20er Jahren. In *Geisteswissenschaften zwischen Kaiserreich und Republik. Zur Entwicklung von Nationalökonomie, Rechtswissenschaft und Sozialwissenschaft im 20. Jahrhundert*, ed. Knut Wolfgang Nörr, Bertram Schefold, and Friedrich H. Tenbruck, 159–185. Stuttgart: Steiner.

Bollenbeck, Georg. 1994. *Bildung und Kultur. Glanz und Elend eines deutschen Deutungsmusters*. Frankfurt/M.: Insel.

Bruch, Rüdiger vom. 1989 [2005]. Kaiser und Bürger: Wilhelminismus als Ausdruck kulturellen Umbruchs um 1900. In *Bürgerlichkeit, Staat und Kultur im Kaiserreich*, ed. Rüdiger vom Bruch, 25–51. Stuttgart: Steiner.

Caspari, Volker, and Klaus Lichtblau. 2014. *Franz Oppenheimer. Ökonom und Soziologe der ersten Stunde*. Frankfurt/M.: Societäts-Verlag.

Demirović, Alex. 1999. *Der nonkonformistische Intellektuelle. Die Entwicklung der Kritischen Theorie zur Frankfurter Schule*. Frankfurt/M.: Suhrkamp.

Dörk, Uwe. 2018. Die Deutsche Gesellschaft für Soziologie (DGS) in der Zwischenkriegszeit (1918–1933). In *Handbuch Geschichte der deutschsprachigen Soziologie, Bd. 1*, ed. Stephan Moebius and Andrea Ploder, 829–848. Wiesbaden: VS.

Elias, Norbert. 1939 [1977]. *Über den Prozeß der Zivilisation. Soziogenetische und psychogenetische Untersuchungen, 1. Band: Wandlungen des Verhaltens in den weltlichen Oberschichten des Abendlandes*. Frankfurt/M.: Suhrkamp

Elias, Norbert. 1990. *Über sich selbst. Biographisches Interview. Notizen zum Lebenslauf*. Frankfurt/M.: Suhrkamp.

Fischer, Joachim. 2008. *Philosophische Anthropologie. Eine Denkrichtung des 20. Jahrhunderts*. Freiburg: Alber.

Freyer, Hans. 1930. *Soziologie als Wirklichkeitswissenschaft. Logische Grundlegung des Systems der Soziologie*. Leipzig/Berlin: Teubner.

Geiger, Theodor. 1931. Soziologie. In *Handwörterbuch der Soziologie*, ed. Alfred Vierkandt, 568–578. Stuttgart: Enke.

Geiger, Theodor. 1932. *Die soziale Schichtung des deutschen Volkes. Soziographischer Versuch auf statistischer Grundlage*. Stuttgart: Enke.

Geißler, Rainer, and Thomas Meyer. 2000. Theodor Geiger (1891–1952). In *Klassiker der Soziologie, Bd. 1*, ed. Dirk Kaesler, 278–296. München: Beck.
Golesorkhi, Lara-Zuzan, Ellen M. Freeberg, and Gina Luria Walker. 2023. Frieda Wunderlich: Feminist research and activism in Berlin. *Journal of Classical Sociology* 4:522–535.
Gorges, Irmela. 1986. *Sozialforschung in der Weimarer Republik 1918–1933*. Frankfurt/M.: Hain.
Henning, Christoph. 2006. Der übernationale Gedanke der geistigen Einheit. Gottfried Salomon(-Delatour), der vergessene Soziologe der Verständigung. In *Deutsch-jüdische Wissenschaftsschicksale*, ed. Amalia Barboza and Christoph Henning, 48–100. Bielefeld: transcript.
Joas, Hans, and Wolfgang Knöbl. 2008. *Kriegsverdrängung. Ein Problem in der Geschichte der Sozialtheorie*. Frankfurt/M.: Suhrkamp.
Käsler, Dirk. 1984. *Die frühe deutsche Soziologie 1909 bis 1934. Eine wissenschaftssoziologische Untersuchung*. Opladen: Westdeutscher Verlag.
Kettler, David, Colin Loader, and Volker Meja. 2008. *Karl Mannheim and the Legacy of Max Weber. Retrieving a Research Programme*. Aldershot/Burlington: Ashgate.
Knebelspieß, Stefanie, and Stephan Moebius. 2019. Programm, personelle und organisatorische Entwicklung des Forschungsinstituts für Sozialwissenschaften von 1918/19 bis zum heutigen Institut für Soziologie und Sozialpsychologie (ISS). *Kölner Zeitschrift für Soziologie und Sozialpsychologie* 71:515–552.
Köhnke, Klaus Christian. 1986. *Entstehung und Aufstieg des Neukantianismus. Die deutsche Universitätsphilosophie zwischen Idealismus und Positivismus*. Frankfurt/M.: Suhrkamp.
Köhnke, Klaus Christian. 1996. *Der junge Simmel in Theoriebeziehungen und sozialen Bewegungen*. Frankfurt/M.: Suhrkamp.
König, René. 1949. *Soziologie heute*. Zürich: Regio.
Koselleck, Reinhart. 1976 [1989]. 'Erfahrungsraum' und 'Erwartungshorizont'—zwei historische Kategorien. In *Vergangene Zukunft. Zur Semantik geschichtlicher Zeiten*, ed. Reinhart Koselleck, 349–375. Frankfurt/M.: Suhrkamp.
Lederer, Emil. 1915 [2014]. Zur Soziologie des Weltkriegs. In *Schriften zur Wissenschaftslehre und Kultursoziologie*, ed. Emil Lederer, 101–130. Wiesbaden: VS.
Lichtblau, Klaus. 1996. *Kulturkrise und Soziologie um die Jahrhundertwende. Zur Genealogie der Kultursoziologie in Deutschland*. Frankfurt/M.: Suhrkamp.
Lindenfeld, David F. 1997. *The Practical Imagination. The German Sciences of State in the Nineteenth Century*. Chicago/London: Chicago University Press.
Mannheim, Karl. 1931. Wissenssoziologie. In *Handwörterbuch der Soziologie*, ed. Alfred Vierkandt, 659–680. Stuttgart: Enke.
Marx, Karl. 1844 [2009] MEGA² I/2. Zur Kritik der Hegelschen Rechtsphilosophie. Einleitung. In *Gesamtausgabe (MEGA). Erste Abteilung. Werke, Artikel, Entwürfe. Band 2. März 1843 bis August 1844. Text, 2. Aufl.*, ed. Karl Marx, Friedrich Engels, 170–183. Berlin: Akademie.
Marx, Karl, and Friedrich Engels. 1844–1847/2017 MEGA² I/5. Deutsche Ideologie. Manuskripte und Drucke (1845–1847). In *Gesamtausgabe (MEGA). Erste Abteilung. Werke, Artikel, Entwürfe. Deutsche Ideologie. Manuskripte und Drucke. Band 5. Text*, ed. Karl Marx, Friedrich Engels, 1–643. Berlin: Dietz.
Mikl-Horke, Gertraude. 2001. *Soziologie. Historischer Kontext und soziologische Theorie-Entwürfe*. München/Wien: Oldenbourg.
Moebius, Stephan. 2017. Die Geschichte der Soziologie im Spiegel der Kölner Zeitschrift für Soziologie und Sozialpsychologie. In *Kölner Zeitschrift für Soziologie und Sozialpsychologie 56 (69)*, 3–44. Sonderheft: Soziologiegeschichte im Spiegel der Kölner Zeitschrift für Soziologie und Sozialpsychologie.
Moebius, Stephan. 2021. *Sociology in Germany. A History*. Cham: Palgrave.
Moebius, Stephan. 2023. Georg Simmel's political thought. Socialism and Nietzschean Aristocratism. *Journal of Classical Sociology* 23 (1): 20–62.
Mohl, Robert von. 1855. Die Geschichte und Literatur der Staatswissenschaften. In *Monographieen dargestellt. Erster Band*. Erlangen: Enke.

Mozetič, Gerald. 2018. Anfänge der Soziologie in Österreich. In *Handbuch Geschichte der deutschsprachigen Soziologie, Band 1*, ed. Stephan Moebius and Andrea Ploder, 37–64. Wiesbaden: VS.

Muller, Jerry Z. 1987. *The Other God That Failed. Hans Freyer and the Deradicalization of German Conservatism*. Princeton: Princeton University Press.

Pankoke, Eckart. 1970. *Sociale Bewegung—Sociale Frage—Sociale Politik. Grundfragen der deutschen "Sozialwissenschaft" im 19. Jahrhundert*. Stuttgart: Klett.

Pankoke, Eckart. 1984. Soziologie, Gesellschaftswissenschaften. In *Geschichtliche Grundbegriffe. Historisches Lexikon zur politisch-sozialen Sprache in Deutschland. Band 5*, ed. Otto Brunner, Werner Conze, and Reinhart Koselleck, 997–1032. Stuttgart: Klett.

Paulitz, Tanja. 2019. Feministische Soziologie, Gender Studies, Frauen-/Geschlechterforschung als Denkkollektiv: Soziologische Denkstile zu Geschlecht als sozialem Verhältnis und als soziale Kategorie. In *Soziologische Denkschulen in der Bundesrepublik Deutschland*, ed. Joachim Fischer and Stephan Moebius, 373–411. Wiesbaden: VS.

Peter, Lothar. 2018. Soziologie in der DDR. Legitimationsbeschaffung oder "machtkritische Subkulktur"? In *Handbuch Geschichte der deutschsprachigen Soziologie*, ed. Stephan Moebius and Andrea Ploder, vol. 1, 385–419. Wiesbaden: VS.

Peter, Lothar. 2019. *Marx on Campus. A Short History of the Marburg School*. Boston: Brill.

Peukert, Detlev J. K. 1987. *Die Weimarer Republik*. Frankfurt/M.: Suhrkamp.

Plessner, Helmuth. 2002 [1924]. *Grenzen der Gemeinschaft. Eine Kritik des sozialen Radikalismus*. Frankfurt/M.: Suhrkamp

Rehberg, Karl-Siegbert. 1981. Philosophische Anthropologie und die 'Soziologisierung' des Wissens vom Menschen. Einige Zusammenhänge zwischen einer philosophischen Denktradition und der Soziologie in Deutschland. In *Soziologie in Deutschland und Österreich 1918-1945. Sonderheft 23 der Kölner Zeitschrift für Soziologie und Sozialpsychologie*, ed. M. Rainer Lepsius, 160–198. Opladen: Westdeutscher Verlag.

Rehberg, Karl-Siegbert. 1999. Hans Freyer (1887–1960), Arnold Gehlen (1904–1976), Helmut Schelsky (1912–1984). In *Klassiker der Soziologie, Bd. 2*, ed. Dirk Kaesler, 72–104. München: Beck.

Rehberg, Karl-Siegbert. 2010. Das Unbehagen an der Soziologie. In *Soziologische Kontroversen. Eine andere Geschichte von der Wissenschaft des Sozialen*, ed. Georg Kneer and Stephan Moebius, 217–253. Frankfurt/M.: Suhrkamp.

Rehberg, Karl-Siegbert (2021) "Krisenwissenschaft", völkische Emphase und systemtheoretische Abkühlung: Die "Leipziger Schule" der Soziologie—Ein Mythos der Wissenschaftsgeschichtsschreibung? In: Karl Acham, Stephan Moebius (eds) Soziologie der Zwischenkriegszeit. Ihre Hauptströmungen und zentralen Themen im deutschen Sprachraum Wiesbaden: VS, pp. 383–411.

Riedel, Manfred. 1963. Der Staatsbegriff der deutschen Geschichtsschreibung des 19. Jahrhunderts in seinem Verhältnis zur klassisch-politischen Philosophie. In *Der Staat*, Band 2, Nr. 1, 41–63.

Schäfer, Gerhard. 1990. Wider die Inszenierung des Vergessens. Hans Freyer und die Soziologie in Leipzig 1925–1945. In *Jahrbuch für Soziologiegeschichte 1990*, ed. Heinz-Jürgen Dahme, Carsten Klingemann, Michael Neumann, Karl-Siegbert Rehberg, and Ilja Srubar, 121–175. Opladen: Leske & Budrich.

Schneikert, Christian, et al. 2019. The Sociological Canon, Relations between Theories and Methods, and a Latent Political Structure: Findings from a Survey of Sociology Students in Germany and Consequences for Teaching. *Teaching Sociology* 47:1–11. https://doi.org/10.1177/0092055X19865301. (accessed August 8, 2019).

Schnitzler, Sonja. 2018. Die Deutsche Gesellschaft für Soziologie zur Zeit des Nationalsozialismus. In *Handbuch Geschichte der deutschsprachigen Soziologie, Bd. 1*, ed. Stephan Moebius and Andrea Ploder, 849–865. Wiesbaden: Springer VS.

Stein, Lorenz von. 1842. *Der Socialismus und Communismus des heutigen Frankreichs. Ein Beitrag zur Zeitgeschichte*. Leipzig: Wigand.

Steinmetz, George. 2015 [2009]. Neo-Bourdieusche Theorie und die Frage wissenschaftlicher Autonomie: Deutsche Soziologen und der Imperialismus zwischen 1890 und 1945. In *Soziologiegeschichte. Wege und Ziele*, ed. Christian Dayé and Stephan Moebius, 336–399. Berlin: Suhrkamp.

Strubenhoff, Marius. 2017. The Positivism Dispute in German Sociology, 1954–1970. *History of European Ideas* 2:260–276.

Tönnies, Ferdinand. 1887. *Gemeinschaft und Gesellschaft. Abhandlung des Communismus und des Socialismus als empirischer Culturformen*. Leipzig: Fues.

Treitschke, Heinrich von. 1859 [1927]. *Die Gesellschaftswissenschaft. Ein kritischer Versuch*. Halle: Niemeyer.

Vierkandt, Alfred, ed. 1931. *Handwörterbuch der Soziologie*. Stuttgart: Enke.

Waßner, Rainer. 1986. Andreas Walther und das Seminar für Soziologie in Hamburg zwischen 1926 und 1945. Ein wissenschaftsbiographischer Umriss. In *Ordnung und Theorie. Beiträge zur Geschichte der Soziologie in Deutschland*, ed. Sven Papcke, 386–420. Darmstadt: WBG.

Weber, Max. 1904–5 [1920/2016]. *Die protestantische Ethik und der "Geist" des Kapitalismus*, hg. und mit einem Vorwort von Klaus Lichtblau/Johannes Weiß. Wiesbaden: VS.

Weber, Max. 1917/1919 [1994]. Wissenschaft als Beruf. In *Wissenschaft als Beruf 1917/1919. Politik als Beruf 1919. Studienausgabe*, ed. Max Weber, 1–23. Tübingen: Mohr.

Wehler, Hans-Ulrich. 1977. *Das Deutsche Kaiserreich 1871–1918*. Göttingen: Vandenhoeck & Ruprecht.

Wiggershaus, Rolf. 2001 [1988]. *Die Frankfurter Schule. Geschichte—Theoretische Entwicklung—Politische Bedeutung*, 6th ed. München: DTV.

Acham, Karl, and Stephan Moebius, eds. 2025. *Soziologie der Zwischenkriegszeit. Ihre Hauptströmungen und zentralen Themen im deutschen Sprachraum. Band 4*. Wiesbaden: VS.

Moebius, Stephan. 2025a. *Sociology in the Weimar Republic. Volume I*. Cham: Palgrave Macmillan.

Moebius, Stephan. 2025b. *Sociology in the Weimar Republic. Volume II*. Cham: Palgrave Macmillan.

Simmel, Georg. 1900 [1989]. *Die Philosophie des Geldes, Gesamtausgabe Vol. 6*. Frankfurt/M.: Suhrkamp.

Kaube, Jürgen. 2014. *Max Weber. Ein Leben zwischen den Epochen*. Berlin: Rowohlt.

Becker, Sabina. 2018. *Experiment Weimar. Eine Kulturgeschichte Deutschlands 1918-1933*. Darmstadt: WBG.

Freyer, Hans.1931. *Revolution von rechts*. Jena: Diederichs.

Honegger, Claudia.1993. Jüdinnen in der frühen deutschsprachigen Soziologie. In *Lektüren und Brüche. Jüdische Frauen in Kultur, Politik und Wissenschaft*, ed. Mechthild Jansen and Ingeborg Nordmann. 178-195. Wiesbaden: Hessische Landeszentrale für politische Bildung.

Klingemann, Carsten. 1996. *Soziologie im Dritten Reich*. Baden-Baden: Nomos.

Moebius, Stephan. 2015. *René König und die „Kölner Schule". Eine soziologiegeschichtliche Annäherung*. Wiesbaden: VS.

Moebius, Stephan, and Martin Griesbacher. 2019. Gab es eine „Kölner Schule" der bundesrepublikanischen Soziologie? Zu René Königs Professionalisierung der Soziologie Was There a "Cologne School" of West German Sociology? On René König's Professionalization of Sociology. In *KZfSS Kölner Zeitschrift für Soziologie und Sozialpsychologie 71(4)*. 553-592.

Moebius, Stephan, and Oliver Römer. 2022. "Die „wilden Siebziger" und ihre „gegnerischen Soziologien".: Historisch-soziologische Rekonstruktion einer Konstellation westdeutscher Soziologiegeschichte". In *Zeitschrift für Soziologie 51, no. 4*. 307-334.

Chapter 12
Three Methodological Ruptures in the History of Anthropology

Thomas Hylland Eriksen

Abstract Anthropology, described by Eric Wolf as the most scientific of the humanities and the most humanist of the sciences, is today a diverse and in many ways sprawling academic discipline. Yet, it is united through a shared genealogy and methodology. This chapter charts the development of anthropological theory and methodology by focusing on three historical breaks or ruptures, as well as the resulting bifurcations. The first break is conceptual and consists in the abandoning of unilinear evolutionist thinking about cultural change, and the main protagonist is Franz Boas, whose historical particularism and methodological relativism contributed to a less biased and more rigorous approach to the diverse empirical materials. The second break is methodological and is associated with Bronislaw Malinowski. He emphasised the need to move beyond superficial survey work and distorting interviews, and laid out the principles of participant observation as the most accurate methodology for gaining knowledge about any society. The ethnographic method has subsequently become the most significant defining trait of social and cultural anthropology. The third, more recent break was expressed through the epistemological critique retrospectively known as 'the Writing Culture turn', where fundamental questions were raised about the validity of anthropological concepts, methods of comparison and theoretical constructions.

Each of these historical ruptures are accompanied by frictions, disagreements in the professional community and theoretical bifurcation. After Boas, neo-evolutionists tried to recover big questions formerly raised in cultural history; the fieldwork revolution, though less controversial, led to its own counterreactions in the shape of theoretical and comparative work; and the 'postmodernism' from the 1980s has similarly led to counterreactions among scholars who insist on the veraciousness of their objects of study.

Thomas Hylland Eriksen was died at the time of publication.

T. H. Eriksen (Deceased) (✉)
University of Oslo, Oslo, Norway
e-mail: t.h.eriksen@sai.uio.no

Y. Shan (ed.), *History and Philosophy of the Social Sciences*,
https://doi.org/10.1007/978-3-032-08167-4_12

Constructing the chapter around a few dramatic moments enables the author to demonstrate that anthropology is a community of disagreement, advancing through internal friction, and at the same time a unitary discipline based on fundamental, shared principles of a theoretical, epistemological and methodological nature. Since anthropology is a comparative, empirical discipline rooted in local realities, theoretical and methodological developments vary geographically. Some of this variation, though not all, are discussed in the chapter.

Keywords Boas · Malinowski · Evolutionism · Cultural relativism · Writing culture

12.1 Introduction

Early attempts at accounting for cultural variation and human universals, the subject matter of anthropology, can be traced back to ancient Greek geography and history, Renaissance and early modern travelogues, comparative sociology from the fourteenth century Tunisian scholar Ibn Khaldun, and philosophical treatises from the European Enlightenment. They generally fell short of being scientific because of inherent bias, lack of theoretical sophistication, and, frequently, poor empirical material leading to irresponsible speculation. In the nineteenth century, theorists such as Lewis Henry Morgan, Henry Maine, Auguste Comte, and Edward Burnett Tylor developed ambitious evolutionist theories of culture, implying a hierarchical view of cultures. This was a time when ideas of progress and development were powerful and almost ubiquituous, not just in the Anglophone world, but across Europe and the colonies. Ideologies emphasising progress were bolstered by the success of Darwin's theory of evolution, published in *The Origin of Species* in 1859. In an era when the British and French empires were expanding, turning millions into allegedly inferior subject peoples, at the same time as the USA relied for its prosperity on the marginalisation of indigenous Americans and the maintenance of a strict racial hierarchy, both during and after slavery, social evolution was taken for granted by most early anthropologists, although many did not subscribe to racism. Rather, most of the Victorian-era anthropologists sought explanations for cultural diversity in history and ecological circumstances. Yet, unilinear social evolution was assumed by the majority of nineteenth century anthropologists, and some, like Morgan, developed typologies describing the stages through which evolution progressed.

Modern anthropology emerged in the early decades of the twentieth century, when a new generation of intellectual leaders, foremost among them Franz Boas in the USA, A.R. Radcliffe-Brown and Bronislaw Malinowski in Great Britain, and Marcel Mauss in France, rejected evolutionism, replacing it with cultural relativism as a conceptual device, also refining field method and methods of comparison. In the decades following World War II, new approaches such as structuralism, cultural ecology, and neo-Marxism would develop as anthropology grew and diversified, branching into various specializations including psychological and medical anthropology. Different European countries and regions developed their own approaches, although in dialogue with French and Anglophone scholars. The German *Volkskunde* tradition, which studied domestic folk culture, was an influential and often a-theoretical contribution to modern nation-building (Gellner 1983), and its influence stretched from Scandinavia to Russia and Bulgaria. A similar divide could be observed between the English-speaking and Afrikaans-speaking South African anthropologists until the end of apartheid (Hammond-Tooke 1997). The most powerful scholarly communities, increasingly Anglophone, would move, in the post-war decades, towards the study of modern phenomena such as nationalism, migration, consumption, and digitalization, yet retaining its commitment to the intensive, fieldwork-based study of social life in local settings. Others would soon follow suit.

Social or cultural anthropology can be defined, loosely and broadly, as the comparative science of culture and society, and it is the only major discipline in the social sciences that has concentrated most of its attention on non-Western societies. Although many of the classic problems investigated by anthropologists are familiar to the European history of ideas, the subject as it is known today emerged only in the twentieth century, became institutionalized at universities in the Western world in mid-century, and underwent a phenomenal growth and diversification in the latter half of the century.

The man who is often hailed as the founder of modern British social anthropology, and whose influence went well beyond the Anglophone world, was a Polish immigrant, Bronislaw Malinowski (1884–1942), whose more than 2 years of fieldwork in the Trobriand Islands set a standard for ethnographic data collection that is still largely unchallenged. Malinowski argued the need to learn the local language properly and to engage in everyday life, in order to see the world from the actor's point of view and to understand the interconnections between social institutions and cultural notions. He placed an unusual emphasis on the acting individual, seeing social structure not as a determinant of but as a framework for action, and he wrote about a wide range of topics, from garden magic, economics, and sex to the puzzling *kula* trade.

The transition from global theories of social evolution and grand syntheses to specific, detailed, and empirically systematic work, in reality amounted to an intellectual revolution. The works of respected Victorian anthropologists such as Tylor and Morgan had been relegated to the mists of history, and in the leading countries—France, the UK and the USA—the discipline had been taken over by small groups of researchers, typically with an education in a different subject, who saw intensive fieldwork, cultural relativism, the study of single societies, and rigorous comparison as its essence. The academic institutions, the conferences, and the learned journals of the discipline continue to build on the early-twentieth century anthropology of Boas, Malinowski, Radcliffe-Brown, and Mauss. This is to a great extent also true of the anthropological traditions of other countries (Vermeulen and Roldán 1995, Ribeiro and Escobar 2006, Boskovic 2008), including India, Australia,

Mexico, Argentina, Brazil, Japan, South Africa, The Netherlands, Spain, and Scandinavia. As noted, Soviet/Russian and Central/East European anthropologies have followed different itineraries, retaining a connection with the *Volkskunde* tradition.

This chapter describes three major ruptures in the history of the subject. The first two are foundational, while the third, beginning in the 1980s, led to internal controversies and disagreements over methodology and theory which are still simmering. Other examples could have been invoked, notably regarding postcolonialism and current bids to decolonize anthropology, or the sometimes bitter controversies over 'scientific' approaches, often based on Darwinist evolutionism, versus hermeneutic and symbolic ones. However, the following three cases show both how anthropology came into its own as an academic discipline, and some of the perennial epistemological issues concerning comparison, representation and cultural translation.

12.2 From Evolutionism to Cultural Relativism

In 1886, Franz Boas found himself in New York. He was 28 and meant to be on his way home to Germany. Boas (1858–1942) held a doctorate from Kiel and good prospects for an academic career in Germany. He had spent a year alone with the Baffin Land Inuit and participated in several ethnographic expeditions to northern and western Canada. He chose to stay in New York, partly because it brought him closer to the Native American peoples that fascinated him. He worked first as the editor of a scientific journal, subsequently as an academic at Clark University, and in 1899 he was appointed professor of anthropology at the prestigious Columbia University in New York, where he remained until his death in 1942. During the intervening 43 years, he was teacher and mentor to several generations of American anthropologists.

Boas' message to his students was simple, but fundamental. He had been trained by German scholars who were critical of evolutionism, and remained until his death sympathetic to diffusionist ideas, although Fritz Graebner's influential *Kulturkreis* theory was too general and therefore unacceptable to him (Boas 1911). Graebner and other German diffusionists held that cultures normally changed owing to the adoption of influences from outside (diffusion). In Boas's view, this perspective was empirically simplistic and underestimated human, and by extension cultural, creativity. Against global theories about social evolution or diffusion, Boas held that culture was extremely complex and that social theories were dependent on a firm and reliable empirical base, rather than neat armchair theories that reduced reality to schemata. The first task of anthropology was therefore to provide the data. Only then could one discuss theory. Boas' insistence on detailed empirical facts was related to his *cultural relativism*. Each people, each nation, each tribe had its destiny, its unique character, and it was the task of the anthropologist, as Boas saw it, to document and defend it. In this he was a loyal follower of the early Romantic philosopher Johann Gottfried Herder and the anthropologist Adolf Bastian, who had briefly taught Boas. Bastian insisted on the psychic unity of mankind, in other words that people everywhere were born with the same mental capacities. This insight would be decisive in Boas's subsequent work, and it is indeed a requirement for cross-cultural comparison to be possible without recourse to racist views.

In the USA, the German influence from the humanities (*Geisteswissenschaften*) was retained through Boas and many of his students, the first generation of whom were immigrants with German origins like himself. The discipline became known as *cultural* rather than social anthropology. Here, E. B. Tylor's famous definition of culture from 1871 remained canonical. In the German-American (and Tylorian) sense, culture is a far wider concept than society. If society is made up social norms, institutions and relationships, then culture consists of everything that humans have created, *including* society: material phenomena (a field, a plough, a painting), social conditions (marriage, households, the state), body techniques (dance, manual skills, sense perception), and symbolic meaning (language, ritual, belief). Anthropology was *the* humanist science, the science of humanity, and was, quite literally, concerned with everything human. Boas recognised that no individual could contribute equally to all parts of this vast subject. Accordingly, advocated a 'four-field approach', dividing anthropology into anthropological linguistics, physical anthropology, archaeology and cultural anthropology (which included social anthropology). Students were trained in all four fields, later to specialise in one of them. This broad approach to human life was thus an integral part of American anthropology from its inception, while in Britain and France a more sociological approach remained the ideal. As early as the 1930s, there existed well-established research teams in the United States, specialising, for example, in Native North American languages.

Boas' own writings covered all four fields broadly, but his main contributions were in cultural anthropology. He had carried out long-term field research among the Inuit in 1883–84, and later worked for extended periods with research assistants who collected material on the north-west coast Kwakiutl and other Native American peoples. During fieldwork, he made use of linguistically proficient members of the tribe under study, who would record, discuss and interpret interlocutors' statements. Some of these collaborators, notably the prodigious George Hunt (1854–1933), who co-wrote several of Boas' books on the Kwakiutl and supplied him with thousands of pages of ethnographic documentation, were later recognised as anthropological scholars in their own right.

Fieldwork, as Boas saw it, was often collaborative, and did not presuppose a lone researcher subjected to the long-term stresses of continuous 'immersion' in the field. Individual field stays were often short, and even during longer stays, he travelled extensively between local groups (some 5000 km during his year among the Inuit). Projects was frequently 'long-term' in another sense as well, since field trips in American anthropology were repeated many times over the years, often by a number of people who collaborated on the project (Foster et al. 1979). This methodological strategy had precursors in the German tradition, and was also a reasonable approach given that 'the field' was close at hand in the USA, not in distant areas, as in the case of Britain and France, where most fieldwork took place in the colonies.

Unlike his British contemporaries, Boas was not hostile to historical reconstruction. This perspective was shared by researchers influenced by the Volkskunde tradition, but also in Chinese and Japanese anthropology. Native American cultural history was an integral part of Boas's research agenda, and he included physical anthropology and archaeology as parts of the holistic anthropological enterprise. His basic theoretical orientation remained that of a cultural historian, and the perspective he developed is often loosely referred to as *historical particularism*. Like his erstwhile teacher Bastian, he held that each culture had its own values and unique history, which anthropologists could attempt to reconstruct. He saw intrinsic value in the plurality of cultural practices and was deeply sceptical of any attempt, political or academic, to undermine this diversity. Writing about Kwakiutl dance, for example, he says it illustrates the culture's approach to rhythm, and cannot be reduced to a mere 'function' of society. One must ask what the rhythm *is* for the people who dance to it, and the answer can only be found by exploring the emotional states that generate and are generated by the rhythm (Boas 1927). In other words, Boas saw a hermeneutic methodology as being superior to causal explanation in the study of culture.

A tireless critic of racism and the science inspired by it, Boas was a highly visible presence in American public debate. Against scholars who claimed that 'races' had different innate potentials for cultural development, Boas argued that culture was *sui generis*—its own source—and that inborn differences could not account for the impressive range of cultural variation that anthropologists had already documented by the early twentieth century. The term *cultural relativism*, to which I have referred above, was coined by Boas. In our own day, it is often asked whether relativism should be understood as a methodological or a moral imperative, and the short answer is that in anthropological research, it is a methodological device. In order to understand a different society on its own terms, your own prejudices, moral and epistemological assumptions have to be bracketed.

Boas, fondly nicknamed 'Papa Franz' by his younger colleagues, dominated American anthropology through four decades, but left no grand theory or monumental work that is read by succeeding generations of anthropologists. This reflects his distrust of lofty generalizations. During his time with Bastian, he was warned against the perils of lofty theorising, and he took these warnings to heart. In his writings, he tries to identify the unique circumstances that generate particular cultures or cultural patterns, hardly ever jumping to general conclusions about humanity or human nature. He was cautious in his use of comparison, which all too easily established spurious similarities between circumstances that were fundamentally different. There is thus an important sense in which Boas remained loyal to the principles of natural science in which he had been trained, in that he sought the particular instance rather than the general scheme. His life-long scepticism of Émile Durkheim's sociology confirms this.

Several of Boas's students and associates nevertheless did develop general theories of culture, notably Ruth Benedict, Alfred Kroeber and Robert Lowie. His most famous student was Margaret Mead (1901–78). Although her best-selling books on Pacific societies have been criticized for being superficial, she used material from non-Western societies to raise questions about gender relations, socialization, and politics in the West, and Mead's work indicates the potential of cultural criticism inherent in the discipline. This was also the case in his Brazilian student Gilberto Freyre's (1900–87) work. Freyre was among the first to argue that racial mixing, or 'miscegenation', was a positive force for the cultural creativity in a diverse country such as Brazil (Freyre 1933). While this view did not follow logically from Boas's cultural particularism, it was—in Freyre's view—a possible implication of it. Freyre's work has later been severely criticised for not taking racism and inequality properly into account.

One of Boas' most remarkable associates, the linguist Edward Sapir (1884–1939), formulated, with his student Benjamin Lee Whorf, the Sapir–Whorf hypothesis, which posits that language determines cognition. Consistent with a radical cultural relativism, the hypothesis implies that, for example, the Hopi perceive the world in a fundamentally different way from Westerners, due to differences in the structure of their respective languages. Although heavily criticised, the Sapir–Whorf hypothesis refuses to die, and in comparative studies of cognitive systems, it keeps returning in various guises, sometimes under the label of linguistic determinism, but also as an implication of the 'ontological turn' where a transnational group of anthropologists under the leadership of the Brazilian Eduardo Viveiros de Castro (2012) emphasise the uniqueness of small-scale life-worlds, often with an ethnographic focus in the Amazon.

Although it would be an exaggeration to claim that Boas single-handedly created modern American cultural anthropology, it is difficult to overestimate his influence in the early decades of the twentieth century. He took a clear position against the dominant grand theories of his time—evolutionist and diffusionist—and insisted on detailed studies of entire cultures and their internal workings instead. He also replaced causal explanation with interpretation as the superior method for understanding cultures. Moreover, he demolished racist pseudoscience on empirical grounds, arguing forcefully that at people's intelligence, creativity and other traits are shaped by their environment, not by inborn characteristics. In contemporary American anthropology, the legacy of Boas remains visible, as I will indicate in the third case discussed in this chapter.

12.3 Malinowski and the Ethnographic Method

> I have spent over 8 months in one village in the Trobriand and this proved to me, how even a poor observer like myself can get a certain amount of reliable information, if he puts himself into the proper conditions for observation.
>
> —*Bronislaw Malinowski (letter to A.C. Haddon, May 1916)*

Twenty-four years after Boas made his decision to stay in the USA, a young Polish intellectual from an upper middle-class academic family in the venerable city of Kraków, Bronislaw Kasper Malinowski (1884–1942), moved from Leipzig to London. A few years earlier, he had defended a highly praised doctorate in physics and philosophy in Krakow, then part of the Austro-Hungarian Habsburg Empire (now in Poland). In Leipzig, he had studied under the psychologist Wilhelm Wundt, who convinced him that society had to be understood *holistically*, as a unity of integrated parts, and that this was best achieved by synchronous (rather than historical) analysis. Moreover, his reading of James Frazer's massive comparative work on myth, *The Golden Bough* (Frazer 1974 [1890]) had inspired him to study so-called primitive peoples, and he emigrated to study under Charles Seligman at the London School of Economics, which had a reputation for offering good conditions for exotic fieldwork. Seligman had been a member of the interdisciplinary group undertaking the important Torres Straits Expedition in 1898–99.

In 1914, Malinowski travelled to Australia, intending to spend about 6 months there. However, during his stay, war broke out in Europe, and as a citizen of the Habsburg Empire, Malinowski was classified as an alien, and could not return to Britain. Owing largely to interventions by academic supporters, he was nevertheless allowed to remain in Australia. His first Melanesian fieldwork, on Mailu and Woodlark Island, lasted from August 1914 to March 1915, and Malinowski considered it a failure since he had not engaged sufficiently with the local community—like many anthropologists at the time, he lived with missionaries—and failed to learn the language properly. This shortcoming would be overcome during his next fieldwork, this time in Kiriwina, one of the Trobriand Islands off the coast of New Guinea, which were under Australian rule, where he spent altogether 2 years doing fieldwork between 1915 and 1918. After the war, Malinowski returned to London, producing a string of ethnographically detailed monographs from the Trobriand Islands, the first of which was *Argonauts of the Western Pacific* (Malinowski 1922), possibly the single most revolutionary work in the history of social anthropology. The book was prefaced by Sir James Frazer, who lavished praise on the young Pole, probably unaware that he, in an academic sense, was signing his own death warrant. *Argonauts* is a long book with an unprecedented degree of ethnographic detail. Following a much quoted introductory chapter which outlines the aims and methods of social anthropology (or ethnology, as it was still often called), the author leads us through a tightly focused and detailed examination of a single institution among the Trobriand islanders, namely the *kula* trade, where symbolic valuables circulate over a large area between the islands of Melanesia. Malinowski describes the planning of expeditions, the routes followed, the rites and practices associated with them, and traces the connections between the *kula* trade and other Trobriand institutions, such as political leadership, domestic economics, kinship and rank. Although the goods exchanged are merely of symbolic value, Trobrianders invest a great deal of time and energy into their *kula* expeditions.

A younger contemporary and countryman of the novelist Joseph Conrad, Malinowski brought home news from 'the heart of darkness', in the form of nuanced and naturalistic images of the Trobriand islanders, who ultimately emerge as neither spectacular, exotic nor 'radically different' from Westerners, but simply as distinctive.

At the London School of Economics, Malinowski attracted a small group of accomplished and devoted students of varying backgrounds. They would contribute massively to making the Malinowskian fieldwork ideals the very trademark of social anthropology, its impact being felt across the discipline. Malinowski was in the middle of a new ethnographic study of social change among peasants in Mexico when he suddenly died of a stroke in New York aged 58.

In the words of his student Raymond Firth, later an important anthropologist in his own right, Malinowski 'was a thorough and systematic ethnographer, with an unusual capacity for acquiring languages and outstanding faculties of observation' (Firth 1957). To him, it was essential to stay long enough in the field to be able to use the local vernacular as a working language. Only thus could one gain direct access to the culture. Interpreters cut you off from the informal aspects of communication, in the worst case giving access only to information the interpreter consciously decides to give. Most anthropologists before Malinowski had been in that position, as was he himself on his first, less successful fieldwork.

Malinowski's concept of *participant observation* set a new standard for ethnographic research, expanding the methodology to include nonverbal phenomena. No fact was too trivial to be recorded. Formal interviews and social aloofness would no longer suffice. As far as practically possible, the ethnographer ought to take active part in the ongoing flow of everyday life, avoiding asking questions that might divert the natural stream of events, and without restricting attention to particular parts of the scene.

Fieldwork following the standards defined by Malinowski in the first chapter of *Argonauts* can be extremely demanding. The ethnographer has to adjust to local ways of living, frequently deprived of creature comforts such as electricity and tap water. They must also make themselves available socially at most times, leaving little room for privacy. The ethnographer him- or herself is the most important instrument of data collecting, implying that they are obliged to invest themselves personally and existentially into the research, which is anything but a nine-to-five job.

Malinowski wrote that he lived alone in a tent in the middle of a Trobriand village for months on end—though photos from the field indicate that he kept his tropical suit and hat immaculately white, and his posthumously published diaries (Malinowski 1967) reveal that he often felt homesick, despondent and sick and tired of 'the natives'. At the same, Malinowski was too good a fieldworker to restrict himself to unstructured methods. He collected precise quantitative data on yam yields, land rights, gift exchange, trade patterns and political conflicts, and many other aspects of Trobriand society, and carried out structured interviews whenever deemed necessary.

What Malinowski did not do to any significant extent was to contextualize the Trobriand islanders within a wider historical and regional context. By contrast, his contemporary, the great French anthropologist and sociologist Marcel Mauss, positioned the *kula* ring in a wider regional system of exchange in his seminal essay on gift exchange (Mauss 1954 [1923]). Mauss was a specialist on the Pacific, and had a broader and deeper knowledge of the region's cultural history than Malinowski, but he had never actually been there, and had to draw on Malinowski's ethnography and that of others.

Virtually everything Malinowski published drew on his Trobriand data. He wrote about economics and trade, marriage and sex, magic and worldviews, politics and power, human needs and social structure, gardening and aesthetics. (He promised to write one on kinship, but never did.) His published Trobriand material runs over several thousand pages, and demonstrates convincingly the potential of long-term intensive fieldwork. The sheer number of Trobriand institutions, beliefs and practices showed beyond doubt that a 'primitive', 'simple' society, near the bottom of any evolutionist ladder, could be a highly complex and multifaceted universe in itself. Malinowski's work revealed, more convincingly than any theoretical argument, the absurdity of comparing individual traits. From now on, context and interconnection would be essential qualities of any anthropological account.

Considering his starting point in a single society with which he became intimate, Malinowski demonstrated a method of generalisation which was very different from that of the previous generation, with their more piecemeal local knowledge and grand comparative ambitions. Malinowski regarded all institutions of a society as being intrinsically linked to each other, and stressed that every social or cultural phenomenon ought to be studied in its full context. He also believed that inborn human needs were the driving force in the development of social institutions, and his brand of functionalism is often described as 'biopsychological functionalism'.

Malinowski's theoretical views were less enthusiastically received than his methods and ethnography. He referred to his theoretical stance as *functionalism*, though in practice he is rather eclectic. Social practices and institutions were functional in the sense that they fit together in a functioning whole, which they contributed to maintaining. But unlike the structural-functionalists who followed Durkheim and Radcliffe-Brown, Malinowski saw individuals, not society, as the system's ultimate motivating engine. Institutions existed for people, and it was their needs, ultimately their biological and psychological needs, that were the prime movers of social stability and change. This was another expression of methodological individualism, and in a collectivist academic climate dominated by Durkheimians, it was not favourably received in British anthropology. For a few decades after his death, Malinowski's star faded, until disillusionment with 'Grand Theory' set in during the 1980s, leading to his rehabilitation on both sides of the Atlantic. Malinowski's attention to detail and the native's point of view may seem to exclude high-flying theories, and there is a clear similarity to Boas here, an indication of their common Germanic training. Malinowski differed from Boas, however, in his reluctance to engage in historical reconstruction. With his contemporary and rival Radcliffe-Brown, he waged an anti-evolutionary—and anti-historical—campaign that was so successful that historical perspectives was more or less absent from British anthropology for half a century (Kuper 1996; Eriksen and Nielsen 2012).

For the Durkheimian structural-functionalists, the individual was an epiphenomenon of society and of little intrinsic interest—what mattered was to elicit the elements of social structure. These two lineages of British social anthropology—biopsychological functionalism and sociological structural-functionalism—highlight a basic tension in the discipline and indeed in the social sciences more generally, between *agency* and *structure*. The individual has agency in the sense that he or she is a creator of society. At the same time, society imposes structure on the individual and limits his or her options. The two viewpoints are, as Giddens (1979) points out, complementary. But in interwar British anthropology this was not seen. Malinowski and Radcliffe-Brown were considered diametrical opposites.

Before the Malinowskian revolution, the quality of the data used by anthropologists was variable. Many of the most famous ones, such as Tylor and Frazer, never did fieldwork themselves, but relied on written sources, ranging from missionaries' accounts to travelogues of varying accuracy. The need for more reliable data began to make itself felt in the Anglophone world, where the ignorance of ethnographic studies carried out by Russian, Dutch and other anthropologists was almost complete. Expeditions and systematic surveys increasingly provided researchers around the turn of the twentieth century with improved knowledge of cultural variation, which eventually led to the downfall of the ambitious theories of unilineal evolution characteristic of nineteenth-century anthropology.

Along with the increasing lionising of Malinowski as a founding father of modern social anthropology came critical interrogations of his actual significance. How original was he, and to what extent can his position be seen as a result of his own successful marketing campaigns?

This question has been raised by many anthropologists, perhaps most convincingly and thoroughly in the edited volume *Ethnographers Before Malinowski* (Rosa and Vermeulen 2022). In this book, consisting of essays introducing early ethnographers, the editors include a bibliography consisting of 365 entries produced by 220 ethnographers between 1870 and 1922. None of them were 'armchair theorists' although, arguably, the most famous and influential of the Victorian anthropologists were. Here is a piece of methodological advice for the novice ethnographer:

> A typical piece of intensive work is one in which the worker lives for a year or more among a community of perhaps four or five hundred people and studies every detail of their life and culture; in which he comes to know every member of the community personally; in which he is not content with generalized information, but studies every feature of life and custom in concrete detail and by means of the vernacular language.

This does sound like something Malinowski or one of his students might have said in the 1920s. In fact, the quotation precedes the 'Malinowskian revolution' by about a decade and was written by a major immediate predecessor of Malinowski, namely W. H. R. Rivers (1864–1922), who made major contributions to both psychiatry and anthropology, and was Chair of the Royal Anthropological Institute at the time of his untimely death, at the age of 58, in 1922. The quotation is from Rivers's book on kinship studies (Rivers 1913: 7). It is known that Malinowski availed himself of Rivers's methodological advice, which distinguishes sharply between superficial survey work and intensive fieldwork. Rivers had himself carried out intensive fieldwork among the Toda people in south India, though he did not live in their village nor acquire fluency in their language.

Although the principles of fieldwork had been laid out, and many ethnographers had produced thorough studies of single people before Malinowski, his great influence is not undeserved. Grand theories and speculative historical reconstructions were still hegemonic when he began to teach at the LSE. The clarity with which he formulates the principles of fieldwork also gave him many readers, and unlike Rivers's detailed essay on methodology, *Argonauts* does not stop at a discussion of fieldwork methodology, but shows, in the bulk of the book, why these methods—participant observation, living in the community, using the vernacular, a holist approach to institutions and details—are superior. In conclusion, Malinowski and his students did revolutionize the discipline, slimming and sharpening it at the same time as society at large was carrying out a similar project by attempting to leave the Victorian era resolutely behind.

12.4 *Writing Culture* and the Literary Turn

Boas and Malinowski were contemporaries, and in complementary ways, they made major contributions towards making anthropology more precise, more scientific and more useful for the cumulative growth of knowledge and theory. Their epistemological and methodological principles were by and large taken up, though often in modified forms, by anthropologists across the world.

By the 1980s, theoretical currents had flourished and withered—structuralism, Marxism, structural Marxism, transactionalism, cultural materialism and so on. The impact of feminist anthropology, especially noticeable in the Anglosphere from the 1970s onwards, was significant not only by focusing on gender and women in ways that were unheard of a generation earlier, but it also influenced methodology and epistemology. If 'Man' was a hunter (Lee and DeVore 1968), then 'Woman' was a gatherer (Dahlberg 1983) whose economic contributions to society were crucial. In a restudy of the Trobriand Islands, Anette Weiner (1983) showed that Malinowski's economic anthropology was incomplete owing to his neglect of valuables produced and distributed by and among women. By including women as equal partners in society, anthropologists could depict societies in a more holistic, naturalistic and accurate way than earlier, when women had rarely featured as independent actors. The postcolonial turn, inspired by Edward Said's *Orientalism* (1978) but also older critiques, accused anthropology of reproducing a Western gaze on the rest of the world. A further destabilizing tendency was postmodernism, which began to make inroads in the discipline a few years after the feminist and postcolonial critiques. Eschewing 'grand narratives', postmodernist philosophy seemed to be compatible with the legacies from Boas, Malinowski and others. However, postmodernism also seemed to relativise knowledge as such, leading anthropologists to question the validity of their ethnographic findings.

What could retrospectively be called the postmodernist movement in anthropology, especially in the USA, was associated with the work of a fairly small group of scholars. The core included non-anthropologist James Clifford, a historian of anthropology with leanings towards literary studies; and anthropologists Stephen Tyler (a convert to postmodernism from ethnoscience), George Marcus, Michael

Fischer, Renato Rosaldo and Paul Rabinow. Others associated with the movement included Johannes Fabian, Richard Handler (who studied nationalism as a form of discourse), Lila Abu-Lughod (an Arabist), as well as Akhil Gupta and James Ferguson, who co-wrote articles and edited an important work on the discursive construction of space and narrative (Gupta and Ferguson 1997). In spite of mutual differences (regarding, for example, the possibilities and limitations of ethnography), these and other scholars shared a number of concerns and were seen, from outside, as constituting a movement and proposing a radical shift. They were uncomfortable with the reified 'othering' typical of classical modernist anthropology, and sought to redress this, often by advocating 'experimental ethnographies', where the informants participated as equal partners in the production of knowledge (Marcus and Fischer 1986). They were critical of the Boasian idea of cultures as integrated wholes with deep historical roots. In this, they targeted Clifford Geertz, who was the most influential torch-carrier of Boas's legacy. Inspired by the philosopher Paul Ricœur, Geertz (1973) advocated a hermeneutical model for reading cultures, assuming them to be intrinsically meaningful and integrated. Against Geertz, the *Writing Culture* group rejected the idea that cultures were somehow integrated and stable. Inspired by Foucault and cultural Marxists like Antonio Gramsci (1891–1937), they were also concerned with modes of representation and the power implied by particular styles of writing.

A blanket term covering everything that was meaningful to the people under study, the concept of culture was being criticized for being too broad to be useful, too bounded, static and rigid, and for being the construction of the anthropologist, not of the communities under scrutiny. Alternative ways of conceptualising bundles of collective meaning were proposed (e.g. unbounded 'cultural flows' or 'fields of discourse', or 'traditions of knowledge'), and some even wanted to abolish the concept altogether (Clifford and Marcus 1986; James et al. 1997; Ortner 1999). Interestingly, the concept of society has been subjected to similar critiques (see Urry 2000), but problematic as they may be, both concepts still form part of the conceptual backbone of anthropology. In his book-length review of the culture concept in American cultural anthropology, Kuper (1999: 226) notes that '[t]hese days, anthropologists get remarkably nervous when they discuss culture—which is surprising, on the face of it, since the anthropology of culture is something of a success story'. The reason for this 'nervousness' is not just the contested meaning of the term 'culture', but also the fact that culture concepts that are close kin to the classic anthropological one are increasingly being exploited politically in exclusionary and often xenophobic identity politics. For example, several of the intellectual architects behind South African apartheid were trained as anthropologists.

The year 1986, in many ways an *annus mirabilis* for the postmodern movement, saw the publication of two important books and the launching of a new journal edited by Marcus, entitled simply *Cultural Anthropology*. The first book was Marcus and Fischer's aforementioned *Anthropology as Cultural Critique*, subtitled *An Experimental Moment in the Human Sciences* (Marcus and Fischer 1986) Arguing that the discipline suffered from 'a crisis of representation', they went on to present several of the problems outlined above, and to affirm the importance of reflexivity (positioning the anthropologist's knowledge) and wider systemic concerns (incorporating an understanding of world history and economics in ethnographic analyses). They argued that a main objective of the discipline should be to engage in cultural criticism 'at home', and that an appropriate way of achieving this was by *defamiliarisation*—creating a sense of 'strangeness' by pointing out the similarity of the readers' own world to distant and 'exotic' cultures. In their view, the 1980s were a period of unique potential for the fulfilment of anthropology's promise as an instrument of cultural critique. The discipline was in disarray; the broad postwar consensus had broken down on both sides of the Atlantic; the postcolonial movement had generated uncertainty; grand theories had lost their appeal. In this situation, 'experimental ethnographies' could contribute substantially to critical self-reflection on Western society.

Anthropology as Cultural Critique, which emphasised continuity with the concerns of anthropologists like Mead, Sahlins and Douglas, was less radical than the edited volume *Writing Culture* (Clifford and Marcus 1986). Although the latter book contained about a dozen chapters written by different scholars, who represented various positions in the postmodernist debate, it was perceived as a single, collective assault on the dominant concept of culture. The contributors unanimously distanced themselves from the idea of culture as an 'integrated whole', questioned the validity of 'scientific' anthropology, and argued the merits of both dialogic methods (the main inspiration here being the Russian literary critic Mikhail Bakhtin), and historical contextualisation in the increasingly problematic art of anthropological representation.

The hugely influential *Writing Culture* was followed 2 years later by Clifford's *The Predicament of Culture* (Clifford 1988), which could be summarised as a long, historically based argument against essentialism. In the same year, Geertz published a small, elegant book called *Works and Lives: The Anthropologist as Author* (Geertz 1988). This was a collection of essays about famous anthropologists that focused on the rhetorical and literary aspects of their writings, and was based on a lecture series which antedated *Writing Culture*. The anthropological quest now seemed to have been brought to a close, since it no longer dealt with living people, but with texts; not with cultures, but with their representations.

This view had its limitations. Never had so much fieldwork been done, in so many different places, as in the 1980s. In most parts of the world, anthropologists continued with business as usual. In deeply unequal societies like Brazil, India and South Africa, anthropologial research continued to be engaged and oriented towards social problems. In China and Japan, the accuracy of the ethnographic description was crucial, and many Australian anthropologists were—and are—engaged in land rights issues on behalf of the Aboriginal communities they studied.

These anthropologies had modest immediate benefits from the postmodern turn, and so it may seem to have been a limited impact on the discipline as a whole. Yet it remains a fact that some of the most hotly debated writings of the last decades of the twentieth century were reflexive texts, which questioned ethnographic authority, questioned the ethical legitimacy of turning 'natives' into empirical data, and ultimately challenged the validity of ethnographic representation as such. These problems would also seep into and contribute to shaping other anthropologies as well.

The connection between American postmodernist anthropology and literary studies was strong. Both looked to recent French philosophy for inspiration, and the young anthropologists, steeped in Geertzian hermeneutics, were predisposed to see cultures as texts. Their break with Geertz consisted largely in problematising the subject–object relationship between anthropologist and informant, and in not seeing cultures ('cultural texts') as integrated wholes. However, these problems were not foreign to Geertz himself. He once compared an integrated culture to an octopus, a loosely coordinated animal with a weak brain that does not always know what each of its arms is doing; and in *Works and Lives*, he effectively deconstructs classical anthropological texts, as imaginative, historically situated works. Indeed, seen from a distance, American postmodern anthropology might resemble an extreme form of Boasianism, nothing entirely new.

Geertz is the key figure here. Though the younger anthropologists, who congregated around *Cultural Anthropology*, had an ambivalent relationship to Geertz' hermeneutics, he was their favourite discussion partner. There was continuity between his view of anthropology and the radical deconstruction advocated in *Writing Culture*, but Geertz himself felt that the younger generation had gone too far, and coined the term 'epistemological hypochondria' to describe the excessive self-criticism that prevented people from doing good ethnographic work.

Many anthropologists would agree with Geertz or even go further in their criticism than he did. Some of the more explicit critics were Steven Sangren (1988), who saw the 'reflexive turn' as a retreat from anthropology's proper mission, and Jonathan Spencer (1989), who argued that anthropology might be better viewed as a style of working than as a form of writing. The cultural materialist Marvin Harris, in one of his many position papers, thoroughly thrashed (some would say parodied) the likes of George Marcus, who seemed to favour an increase in 'the number of experimental, personalistic, and idiosyncratic field studies carried out by untrained would-be novelists and ego-tripping narcissists afflicted with congenital logodiarrhea' (Harris 1994: 64).

Ernest Gellner, in a little book devoted to the defence of rational science in the face of the two-pronged threat of closed-world fundamentalism and boundless postmodern relativity, berated the American postmodernists, singling out Clifford and Rabinow as his main targets, for their sloppy thinking and poorly defined concepts, and ultimately for being more interested in their own interpretations than in understanding the world (Gellner 1992). While Gellner saw Geertz as an important precursor of the postmodernist movement, he noted that Geertz, at least, was still trying to 'say something about something', as Geertz himself once put it. Against the critique of Great Divide (or 'Big Ditch') theories (that contrast 'us' with 'them', 'modern' with 'primitive', and so on), Gellner confirmed his position as a modernist. Elsewhere, Gellner (1993) also voiced misgivings about postcolonialism, for undermining scientific truth claims, confusing ideology and analysis, and not understanding that the 'problem of power and culture ... is too important to be left to lit[erary] crit[icism]' (Gellner 1993: 4). Finally, Gellner hinted that there was an element of careerism in the postmodern movement, remarking that '*Sturm und Drang and Tenure* might well be their slogan' (Gellner 1992: 27). While there may be something to this, the same might be said of all innovative movements in anthropological history, from the Malinowskian–Radcliffe-Brownian 'functionalist revolution', via neo-evolutionist and transactionalist 'revolutions', to the 'revolution' in nationalism studies that Gellner himself had made major contributions to.

It was also often noted that many of the ideas inspiring American postmodern anthropology had originated in France, and that the Americans tailored the French masters to fit their anthropology, often trivialising their views in the process. Foucault, in the American reading, became a theorist of discourse; Derrida a philosopher of relativism. While not entirely off the mark, such characterisations simplify complex positions, but there were reasons for these (mis-) interpretations, and it is important to situate American anthropology itself in its broader social context. Boas, a Jewish German, had in his day understood very well the potential for ethnic and racial conflict in the sprawling, multinational United States, and events since his death have not proven him wrong. Clifford and Marcus may seem crude from the vantage-point of Parisian intellectuals, but relativism, even extreme relativism, was an understandable stance in a country split apart by a history of Afro-American slavery, Native American genocide and immigration from every part of the world.

Some critics doubted that postmodernism was as radical a departure from anthropological tradition as it claimed to be. However, this argument was double-edged, and was also used by some of the postmodernists to legitimise their project. Thus, Kirsten Hastrup, whose work has always been relentlessly anti-positivist, and has in some ways represented a European counterpart to American postmodernism, argued that anthropology *had always been* a postmodern science, ever since it started to confront the West with images of other lifeworlds (Hastrup 1995). Though Hastrup may have overstated her case, there are definite affinities between the deconstructive efforts of the postmodernists and earlier trends in anthropological history. Questioning the objectivity of ethnographic method had been a recurring anthropological theme on both sides of the Atlantic at least since the Second World War, and the problem of translation between local world-views and anthropological discourse is a constitutive problem in the discipline. Above all, of course, there is continuity with the historical particularism of Boas and the German Romantic tradition. In general, American anthropologists, who were steeped in this tradition, were more favourably disposed to postmodernism than their European colleagues, who were the heirs of confirmed positivists such as Radcliffe-Brown and had a stronger grounding in the social sciences (Kuper 1996: 189). There were other important precursors to the postmodern turn in cultural antropology. David Schneider (1984) argued against the reification of kinship as a comparatively useful concept. His work, questioning the cross-cultural validity of the kinship concept, had little to do with postmodernism, but was the work of a devoted Boasian. Schneider's work would be cited approvingly in Great Britain, by the Czech-British anthropologist Ladislav Holy in his textbook on kinship (Holy 1996), but Holy's point of reference is not Boas, but the methodological individualism of the British anti-structural-functionalist movement of the 1950s and 1960s. This movement can also be seen as a precursor of postmodernism. When Fredrik Barth, in the mid-1960s, critically interrogated the concept of social structure and posited that stable social forms were just unintentional aggregate outcomes of individual decisions, this was (in a sense) a deconstructive argument which closely paralleled the postmodernists' deconstruction of Boasian and Geertzian concepts of integrated cultural wholes.

In spite of all this, the postmodern critique of anthropology, with occasional support from feminist and postcolonial scholars, did represent something new, though its originality was often exaggerated at the time. The newness, as far as anthropology was concerned, lay mainly in the reflexive emphasis on styles of writing, in the rejection of a neutral and non-positioned authorial voice, and (most fundamentally) in the application of reflexivity to anthropology itself. After postmodernism, anthropology could no longer see itself as a privileged discourse with access to objective truth about the peoples it studied.

12.5 Conclusion

In this chapter, I have chosen three controversies, or ruptures, in the history of anthropology in the last century. Others could have been discussed instead. The most famous controversy in anthropology may be that involving Margaret Mead and Derek Freeman (Mead 1978 [1928]: Freeman 1983). Both did fieldwork in Samoa at different times, and Freeman, an adept of evolutionary anthropology, criticised the hermeneutician Mead not only for her theoretical approach, but also for having carried out sloppy fieldwork and misrepresented her interlocutors. ther tensions and controversies

The impact of feminism in the 1970s and postcolonial thought in the 1980s, briefly discussed above, has left a perceptible legacy in anthropology, both as regards methodology, research ethics and theoretical assumptions. More recently, the increased interest in ecology, human-nature relations and the Anthropocene more generally, has divided anthropologists over methodological questions (such as the possibility of doing multispecies fieldwork) and epistemological assumptions (should humanity and nature be separated or joined analytically?). It should also be kept in mind that the reception and impact of these three shifts has varied significantly as the wider context in which anthropologists work is shaped by their socio-political environment.

The three cases I have chosen nonetheless highlight some of the fundamental questions raised within the discipline: On what grounds is it defensible to make empirical generalisations? Are theoretical perspectives and analytical concepts cross-culturally valid or fictions devised by Western academics? Is society mainly a product of individual agency, or is it the other way around?

In spite of fairly deep differences along these and other lines, social and cultural anthropologists share important commonalities. There is by and large agreement about the importance of ethnography and requirements for fieldwork. Cultural relativism, as a methodology, is taken for granted by anthropologists, who suspend moral judgement for the sake of neutrality or even objectivity. Finally, doubt about the validity of comparative concepts such as culture, is constitutive of the discipline. The fact that these and other fundamental quesions are being discussed should be seen as a healthy sign since it inoculates the discipline against smug complacency. The broad agreement about the relevance of a cluster of methodological and theoretical debates—although there is not agreement concerning the answers—indicates that notwithstanding its sprawling character, social and cultural anthropologists form a community, however loosely integrated, namely a community of disagreement.

References

Boas, F. 1911. Review of Graebner, *Methode der Ethnologie*. Reprinted in *Race, Language, and Culture*. New York: Macmillan, 1940.

Boas, F. 1927. *Primitive Art*. New York: Dover.

Boskovic, A., ed. 2008. *Other People's Anthropologies*. Oxford: Berghahn.

Clifford, J. 1988. *The Predicament of Culture*. Berkeley: University of California Press.

Clifford, J., and G. Marcus, eds. 1986. *Writing Culture: The Poetics and Politics of Ethnography*. Berkeley: University of California Press.

Dahlberg, F., ed. 1983. *Woman the Gatherer*. New Haven: Yale University Press.

Eriksen, T. H., and F. S. Nielsen. 2012. *A History of Anthropology*. 2nd ed. London: Pluto.

Firth, R., ed. 1957. *Man and Culture: An Evaluation of the Work of Bronislaw Malinowski*. London: Routledge & Kegan Paul.

Foster, G. M., T. Scudder, E. Colson, and R. V. Kemper. 1979. *Long-Term Field Research in Social Anthropology*. London: Academic Press.

Frazer, J. G. 1974 [1890]. *The Golden Bough*. New York: Macmillan.

Freeman, D. 1983. *Margaret Mead and Samoa: The Making and Unmaking of an Anthropological Myth*. Cambridge, MA: Harvard University Press.

Freyre, G. 1933. *Casa-Grande e Senzala*. Rio de Janeiro: Livraria José Olympio Editora.

Geertz, C. 1973. *The Interpretation of Cultures*. New York: Basic Books.

Geertz, C. 1988. *Works and Lives: The Anthropologist as Author*. Cambridge: Polity.

Gellner, E. 1983. *Nations and Nationalism*. Oxford: Blackwell.

Gellner, E. 1992. *Postmodernism, Reason and Religion*. London: Routledge.

Gellner, E. 1993. The mightier pen? Edward Said and the double standards of inside-out colonialism. *TLS*, 19 February 1993.

Giddens, A. 1979. *Central Problems in Social Theory*. London: Macmillan.

Gupta, A. and J. Ferguson, eds. 1997. Anthropological Locations. Berkeley, CA: University of California Press.

Hammond-Tooke, W. D. 1997. *Imperfect Interpreters: South Africa's Anthropologists, 1920–1990*. Johannesburg: Wits University Press.

Harris, M. 1994. Cultural materialism is alive and well and won't go away until something better comes along. In *Assessing cultural anthropology*. Edited by Robert Borofsky, 62–75. New York: McGraw-Hill.

Hastrup, K. 1995. *A Passage to Anthropology: Between Experience and Theory*. London: Routledge.

Holy, L. 1996. *Anthropological Perspectives on Kinship*. London: Pluto.

James, A., J. Hockey, and A. Dawson, eds. 1997. *After Writing Culture: Epistemology and Praxis in Contemporary Anthropology*. London: Routledge.

Kuper, A. 1996. *Anthropology and Anthropologists: The Modern British School*. 3rd ed. London: Routledge.

Kuper, A. 1999. *Culture: The Anthropologists' Account*. Cambridge, MA: Harvard University Press.

Lee, R., and I. DeVore, eds. 1968. *Man the Hunter*. Chicago: Aldine.

Malinowski, B. 1967. *A Diary in the Strict Sense of the Term*. London: Routledge & Kegan Paul.

Malinowski, B. 1984 [1922]. *Argonauts of the Western Pacific*. Prospect Heights, IL: Waveland.

Marcus, G., and M. Fischer. 1986. *Anthropology as a Cultural Critique: An Experimental Moment in the Human Sciences*. Chicago: University of Chicago Press.

Mauss, M. 1954 [1924]. *The Gift*, tr. Ian Cunnison. Glencoe: Free Press.

Mead, M. 1978 [1928]. *Coming of Age in Samoa*. Harmondsworth: Penguin.

Ortner, S., ed. 1999. *The Fate of 'Culture': Geertz and Beyond*. Berkeley: University of California Press.

Ribeiro, G. Lins, and A. Escobar, eds. 2006. *World Anthropologies: Disciplinary Transformations Within Systems of Power*. Oxford: Berg.

Rivers, W. H. R. 1913. *Kinship and Social Organisation*. London: Taylor & Francis.

Rosa, F. D., and H. F. Vermeulen, eds. 2022. *Ethnographers Before Malinowski: Pioneers of Anthropological Fieldwork, 1870-1922*. Oxford: Berghahn.

Said, E. 1978. *Orientalism*. New York: Pantheon.

Sangren, S. 1988. Rhetoric and the authority of ethnography: 'postmodernism' and the social reproduction of texts. *Current Anthropology* 29 (3): 405–435.

Schneider, D. 1984. *A Critique of the Study of Kinship*. Ann Arbor: University of Michigan Press.

Spencer, J. 1989. Anthropology as a kind of writing. *Man* 24 (2): 145–164.

Urry, J. 2000. *Sociology Beyond Societies*. Cambridge: Polity.

Vermeulen, H. F., and A. A. Roldán, eds. 1995. *Fieldwork and Footnotes: Studies in the History of European Anthropology*. London: Routledge.

Weiner, A. 1983. *The Trobrianders of Papua New Guinea*. New York: Holt, Rinehart & Winston.

Chapter 13
History and Philosophy of Market Design: Mathematical Politics of Resource Allocation

Ivan Boldyrev

Abstract The standard economic problem of optimally allocating scarce resources has, over recent decades, evolved into a set of practical and consequential policy interventions on the part of economists. Key fields of (mostly game-theoretic) microeconomics—social choice and mechanism design—have joined forces with experimental economics and, more recently, computer science to help 'build economic machines' (Guala 2001). In fact, a substantial share of today's theoretical microeconomic research is directly or indirectly related to market design. It would be fair to say that the field became a key driver in the development of 'theory' in economics. The chapter explores the emergence of market design and its major conceptual and policy achievements. It argues that this field prompts us to challenge the traditional notion of economic models as epistemic artifacts devised by scholars to solely understand the world. Market design models also serve as instruments for intervention and accompany the actual design process in a way that could be 'performative'. The chapter reviews existing critical work on market design, which attempts to make sense of the specific cases, contexts, and conceptual/epistemic significance of market design interventions. It also reflects on research challenges and offers some general insights drawn from the survey.

Keywords Market design · Performativity · History of microeconomics · Philosophy of economics · Economic models · Auctions · Matching

I am extremely grateful to Philippe van Basshuysen, Béatrice Cherrier, Sascha Füllbrunn, Edward Nik-Khah, Yafeng Shan, and an anonymous referee for the comments on an earlier draft of this paper. Of course, all remaining errors or omissions are solely my responsibility.

I. Boldyrev (✉)
Radboud University, Nijmegen, The Netherlands
e-mail: ivan.boldyrev@ru.nl

Y. Shan (ed.), *History and Philosophy of the Social Sciences*,
https://doi.org/10.1007/978-3-032-08167-4_13

13.1 Introduction

The standard economic problem of optimally allocating scarce resources has, over recent decades, evolved into a set of practical and consequential policy interventions on the part of economists. Key fields of (mostly game-theoretic) microeconomics—social choice and mechanism design—have joined forces with experimental economics and, more recently, computer science to help 'build economic machines' (Guala 2001). The roots of market design in the socialist calculation controversy, its entanglement with an algorithmic vision of rationality, its promise of directly applying pure microeconomic theory to real-world policymaking, its multiple and complex interdisciplinary interactions, and its growing significance within contemporary mainstream economics[1]—all these aspects make it a particularly intriguing field for historians and philosophers of economics. In fact, a substantial share of today's theoretical microeconomic research is directly or indirectly related to market design. It would be fair to say that the field became a key driver in the development of 'theory' in economics.[2]

The chapter explores the emergence of market design and its major conceptual and policy achievements. It argues that this field prompts us to challenge the traditional notion of economic models as epistemic artifacts devised by scholars to solely understand the world. Market design models also serve as instruments for intervention and accompany the actual design process in a way that should 'perform' what Morgan (2012) refers to as the 'model world.' The chapter reviews existing critical work on market design, which attempts to make sense of the specific cases, contexts, and conceptual/epistemic significance of these interventions. It also reflects on research challenges and offers some general insights drawn from the survey.

13.2 Market Design: Blueprints for a History

The emergence of market design[3] followed several interconnected tendencies in economics. General equilibrium theory began to lose popularity, increasingly complemented in both graduate curricula and theoretical work by game theory (Rizvi 1994). However, from a conceptual standpoint, many insights from general equilibrium theory remain relevant to market design. Crucially, both fields share a focus on exploring and *finding* equilibria—whether equilibrium prices or allocations. The

[1] As of today, the work of 9 Nobel Memorial prize winners (Leonid Hurwicz, Eric Maskin, Paul Milgrom, Roger Myerson, Alvin Roth, Lloyd Shapley, Vernon Smith, William Vickrey, and Robert Wilson) and 3 Clark medalists (Susan Athey, Parag Pathak, Philipp Strack) is directly related to mechanism/market design.

[2] Otherwise theory in economics is receding in importance and giving way to more atheoretical empirical approaches (Backhouse and Cherrier 2017).

[3] Sometimes the field is called *marketplace design*, to emphasize its more local focus and the fact that the designed marketplaces are often embedded in broader markets (Vulkan et al. 2013).

broader theoretical discipline that addresses institutions-building is known as *mechanism design*. This field examines how to create or modify institutions with desirable properties by structuring the strategic interactions of economic agents (see Lee 2006 for a more detailed history). Additional connections between general equilibrium and mechanism design range from the central metaphor of *tâtonnement* to the metatheoretical notion that mechanism designers must consider the relationship between the model worlds they construct and the larger economic and social systems in which these mechanisms operate.

Regarding this latter point: One way to distinguish between earlier mechanism design (which featured more non-cooperative elements) and later market design (which incorporated insights from cooperative game theory) is that earlier mechanism design models tended to be self-contained, merging the simplified world of the model with the real-world contexts. However, market designers later recognized that this implicit identification might not hold: agents might have options outside of a given mechanism, and these alternatives must be accounted for and not assumed away (Milgrom 2011; Roth and Wilson 2019). Here is a nice general characterization of market design (as opposed to broader theoretical literature in mechanism design):

> Market designers typically do not try to design a market *all* of whose equilibria accomplish something, but rather try to design a marketplace with a good equilibrium, and then try to achieve that equilibrium. If unanticipated behavior develops, the market can be modified (Vulkan et al. 2013: 3).

Game-theoretic and mechanism design models (in which institutions themselves are conceived as games) became the dominant language of economic theory during the shift toward viewing economics as the science of optimal resource allocation. This transformation is largely attributed to Hayek's (1945) influential conception of markets as information processors (see also Myerson 2009; Mirowski and Nik-Khah 2017). Economic theorists began considering situations where resource allocation is complicated by missing markets (markets that therefore needed to be *created*) and by various incentive constraints, whether informational or strategic. As a result, *economics evolved into the science of incentives*.[4] General mechanism design theory, as developed by Hurwicz (1960), formalized the economic problem as one of information exchange.[5]

From the outset, market design developed in two major directions—auctions and matching. While the two fields are naturally connected (see e.g., Hatfield and Milgrom 2005), most researchers tend to specialize in one or the other. Unlike auctions, matching as an allocation mechanism typically does not involve prices (with the notable exception of labor markets). Matching also places a somewhat stronger emphasis on efficiency considerations and preference elicitation, and often

[4] This became particularly clear in the 1970s when incentive analyses became widespread—ranging from optimal taxation (Mirrlees 1971) to the theory of the firm (Jensen and Meckling 1976).

[5] The issues of information and communication loom large in market design till today and account for its increasing collaboration with computer science.

operates, at least initially, on a more localized scale than auctions (Roth and Wilson 2019).

What follows is a bird-eye view of the major conceptual developments and innovations in these fields. It is necessarily brief and cannot do full justice to the vast and ever-expanding body of literature. Its purpose is to provide a broad overview of how the field developed and to highlight the *synchronicities* among various theoretical contributions (hence the—otherwise unhelpful—chronological structure of my narrative). I hesitate to call it a history, as that would require a much more extensive discussion of contexts and applications, a more detailed examination of conceptual subtleties and challenges, and a more specific analysis of institutional sites and affiliations of market design.[6] Such a comprehensive account cannot be provided here, and the proper history of market design remains to be written.

Some of the key insights that shaped market design were already formulated in the 1960s. A number of them were associated with the study of existing institutions, primarily auctions. Notably, Vickrey (1961), in his work on the independent private value auctions,[7] came up with the idea of 'revenue equivalence.' He could establish the conditions under which any auction design yields the same revenue for the seller, assuming the auction is optimal.[8] Building on his work from the 1950s, Smith (1962) pioneered the experimental study of markets with a focus on auctions. About the same time, Gale and Shapley (1962) characterized a stable matching[9] in a two-sided 'marriage market' and showed that the so-called 'deferred acceptance algorithm' finds a stable matching. Another algorithm—the 'top trading cycle' one—was suggested by Gale to Shapley and Scarf (1974): along with an existence theorem, their work showed how, in the case of indivisible commodities (anticipating, for example, kidneys in the kidney exchange mechanisms developed later), an efficient outcome can be achieved.

Wilson (1967, 1969) made significant advances in the study of auctions by modeling auctions with 'common values,' where the ultimate value of an object is the same for all participants, though each has different information about it. This was one of the earliest game-theoretic formalizations of auctions. A breakthrough in terms of analytical tools came almost immediately from Harsanyi's (1967, 1968a, b) analysis of games with incomplete information, introducing the concept of Bayesian Nash equilibrium. Harsanyi's framework encapsulated agents' lack of

[6] One recent study in this latter vein is Cherrier and Saïdi (2020) on Stanford.

[7] Its participants have independent and privately known valuations of the auctioned object.

[8] That is, the lot goes to the agent who values it the highest. One of the most general revenue equivalence theorems was proven by Myerson (1981). Vickrey's (1945) early analysis of marginal utility measurement also included the discussion of optimal taxation trying to find a 'compromise between equality and incentives' (329) in view of asymmetric information.

[9] In the stable matching, 'no man or woman is matched to an unacceptable mate, and no man and woman who are not matched to each other would both prefer to be' (Roth 2008: 538). Interestingly, Gale and Shapley (1962) precedes Gary Becker's theory of marriage. Becker (1993: 128) explained that his vision of marriage markets as full-fledged markets involving output and prices makes the underlying preferences endogenous and dependent 'on how the outputs produced with different mates were divided,' whereas in Gale and Shapley (1962) the rankings are given.

knowledge about others' beliefs by modeling these beliefs as privately known types, represented as random variables. This allowed theorists to work with the probability distribution of types to address behaviorally significant information asymmetries.

Hurwicz (1972) then introduced the concept of incentive compatibility, a condition necessary for the optimal functioning of economic mechanisms in the presence of asymmetric information. In an incentive-compatible mechanism, it is every agent's dominant strategy to truthfully report their private information (preferences). This contribution responded both to Hayek's (1945) view of markets as information processors and to Samuelson's (1954) assertion regarding the inherent inefficiency in the production of public goods. Hurwicz's contribution paved the way for a coherent framework to analyze informational and strategic incentives—specifically, what agents know (and may wish to conceal) and what they intend to do (which may diverge from what they are expected to do) within various social arrangements.

Almost immediately, the so-called 'revelation principle' was first formulated (Gibbard 1973), with more general versions developed later by others, including Myerson (1979). This principle implied that the set of all possible mechanisms could be reduced to a mathematically more manageable subset of incentive-compatible 'direct mechanisms,' in which agents' actions consist solely of reporting their 'types.' Around the same time, a key theoretical advancement involved the application of Vickrey's idea in formulating the Vickrey-Clarke-Groves (VCG) mechanism for the allocation of public goods. In this mechanism, agents are rewarded or taxed based on the opportunity costs of their contributions (or bids), which are interpreted as externalities (Clarke 1971; Groves 1973). This was an influential formalization of a social technology designed to incentivize agents to truthfully report their preferences.

The importance of incentive compatibility was further underscored by a famous (and negative) result in the mechanism design literature: the Gibbard-Satterthwaite theorem (Gibbard 1973; Satterthwaite 1975), a development of Arrow's impossibility theorem for voting systems. Under certain reasonable assumptions, it demonstrated that any collective choice procedure—or incentive-compatible mechanism—can only be consistent if it operates as a 'dictatorship.' The key issue behind this impossibility was the potential for manipulability, or the strategic misrepresentation of preferences to influence the outcome in one's favor.

An important next step was Wilson's (1977) model of auctions with 'common value' emphasizing the 'winner's curse' problem: In the presence of asymmetric information, auction winners find themselves discouraged if they have reasons to believe others did not bid as much because of some negative information they possessed about the auctioned unit.[10] Wilson (1977), however, examined this issue in the context of 'perfect competition' demonstrating that the winning bid converges

[10] As McAffee and McMillan (1987: 721) explain in an influential survey, 'sophisticated bidders, when deciding their bidding strategies, take into account the fact that winning reveals to the winner that his estimate of the item's value was the highest estimate; as a result, they bid more cautiously than if they adopted naive strategies.'

to the 'true value' of the object as the number of bidders increases—an issue underlying much work in microeconomic theory at that time.

Maskin (1999, circulated in the end of the 1970s)—inaugurated the *theory of implementation* seeking to determine which mechanisms/social choice rules, and under which conditions, align equilibrium with optimal outcomes. Maskin (1999) introduced a key condition—called Maskin monotonicity—guaranteeing that all Nash equilibria generated by a mechanism are optimal with respect to a given social choice rule. Myerson's (1981) general theory of auctions 'as noncooperative games with imperfect information' (58), applied the broader insights of mechanism design to a rigorous analysis of auction optimality.[11] This research, like the Arrow-Debreu model, provided a benchmark that later scholars would use to relax its assumptions and address more realistic scenarios.

These years witnessed a remarkable upsurge of groundbreaking work in the field. Around the same time Roth (1982) demonstrated the strategy-proofness of the Gale-Shapley deferred acceptance algorithm; Miglrom and Weber (1982) showed how, under certain assumptions, one can compare revenues from different auction formats (pace the revenue equivalence theorem) and actually mitigate the 'winner's curse' through more efficient information transmission; Kelso and Crawford (1982) emphasized the importance of the substitutability assumption in labor markets (and thus for indivisible goods), which in some sense connected general equilibrium, auctions, and matching literatures; Rassenti et al. (1982) conducted the first experimental analysis of the combinatorial auctions[12]; while Smith (1982), in an influential paper, linked the agendas of mechanism design and experimental economics research.[13] Around the same time, an important impossibility theorem emerged in auctions theory ((Myerson and Satterthwaite 1983): in a model of bilateral trade with asymmetric information, no efficient mechanism exists[14]), while Roth (1984) proposed the new design for allocating doctors among hospitals in the US and thus inaugurated the whole new field of research.

Following these contributions, market design theory grew rapidly in popularity and began to develop much more intensively. The remarkable surge in theoretical work during the early 1980s appears to have been driven by an increasing interest in designing new markets, partly associated with the wave of decentralization that began in the 1970s (Mirowski and Nik-Khah 2017).

Several notable applications of market design ideas stand out. In 1994, auction theory was famously put to work when the Federal Communication Commission

[11] Parallel to Myerson (1981), optimal auctions have been studied by Riley and Samuleson (1981) who also generalized Vickrey's (1961) results and proved a revenue equivalence theorem.

[12] That is, the auction that allows bidding for packages, and not only for single items. For the history, see Svorenčík (2017).

[13] On the links between mechanism design and experimental economics communities see Lee (2016).

[14] This result emphasized the need to ensure efficient allocation from the very beginning, so that no further trade is needed, and in this sense made economists rethink the ideas behind the Coase theorem, see Medema (2020).

(FCC) introduced new auction procedures for allocating electromagnetic spectrum rights. In designing and implementing this simultaneous *multi-unit* auction design, both theorists (Paul Milgrom, Robert Wilson and Preston McAfee) and experimentalists (Charles Plott) played important roles.[15] Almost simultaneously, in 1995, Alvin Roth began contributing to the redesign of the *National Resident Matching Program*, the largest clearinghouse for matching doctors to hospitals in the US (Roth and Peranson 1999). This effort led to a broader application of market design ideas, with varying degrees of success (see, for example, Klemperer 2013 on auctions and Hitzig 2020 on matching).

The literature on market design has since grown so vast that it would require much more space than is available here to give an accurate account of the most recent developments. These include the redesign of school choice (Abdulkadiroğlu and Sönmez 2003); kidney exchange programs (Roth et al. 2004); new auction formats (Ausubel 2004); online ad auctions (Varian 2009) and the 'incentive auction' designed to reallocate spectrum licenses from TV broadcasters to internet companies providing wireless services. The latter auction involved solving large-scale computational problems and developing exceptionally complex designs (Milgrom 2017; Hitzig 2019).

What lessons can we draw from this history? Market design has clearly transformed economics as a body of knowledge and as an intellectual community/thought collective. It has redefined the discipline's priorities (game-theoretic work received an additional boost; experimental economics was increasingly mobilized for policy interventions; economics has engaged in closer interaction with computer science). The problems market designers were tackling have grown increasingly complex—partly due to evolving societal concerns. For example, foundational assumptions have been renegotiated: one needed to accommodate selling more complex objects or being able to match not individual doctors, but the families of doctors to hospitals (which challenged the assumption of substitutability). These practical involvements, in turn, drove further theoretical developments. To become part of the ideational infrastructure of markets and organizations, economic models and concepts had to be sufficiently adapted and modified to tackle certain issues (Milgrom 2017).

The status of economists has changed as well. There has been a long journey from Coase (1959)'s initial armchair suggestion to use auctions for distributing licenses to the proposed auction designs based on sophisticated game-theoretic models and the constant hands-on involvement of 'economic engineers' (see on that Hitzig 2019; van Basshuysen 2021, who distinguishes designers, engineers, and plumbers; and Sönmez, who adds the identity of 'minimally invasive surgeons').

At the same time, the work in market design was closely connected to broader developments in formal economic theory. For instance, Wilson's early research on auctions was done in parallel with Harsanyi's general theory of Bayesian Nash

[15] The history is critically discussed e.g. in Guala (2001), Nik-Khah (2008), Alexandrova and Northcott (2009), and Mirowski and Nik-Khah (2017).

equilibrium, while Milgrom's analysis of complementarities was informed by the theory of supermodular games and monotone comparative statics (Milgrom and Roberts 1990; Milgrom and Shannon 1994).

Experimental economics accompanied these developments all along. Auctions and other allocation mechanisms were particularly well-suited for experimental investigation (Kagel 1995; Roth 2016), allowing experimentalists to manipulate the components of these 'economic machines' and contribute significantly to practical applications (Guala 2001).

13.3 Epistemologies of Market Design

It has become clear that market design calls for a rethinking of our methodological views on economics and economic modeling. Roth predicted already in 1991 that

> [i]n the long term, the real test of our success will be not merely how well we understand the general principles which govern economic interactions, but how well we can bring this knowledge to bear on practical questions of microeconomic engineering [...] [,] the extent to which [microeconomics] becomes the source of practical advice, solidly grounded in well tested theory, on designing the institutions through which we interact with one another (Roth 1991: 113).[16]

Interestingly, although market design as a discipline was not a major focus in the philosophy of economics, one of the earliest critical reflections on the *application* of auction theory emerged from this very field (Guala 2001). Guala's paper was the first comprehensive and philosophically informed reflection on the roles of theory and experiment in auction design. Focusing on the FCC auction for electromagnetic spectrum rights, it highlighted the constructedness of markets and the continuous need for experimental calibration to make the 'economic machines' function effectively. Drawing on the accounts of economists Guala (2001) demonstrated that auction design—particularly the design of the FCC auctions—was not a straightforward application of (rational choice/game) theory in a strict sense, but rather a systematic use of game-theoretic concepts and partial models to develop practical intuition.

Following up on Guala's work Alexandrova (2008) and Alexandrova and Northcott (2009) argued that the FCC case compels us to reconsider the understanding of how models in economic theory are used to explain phenomena and inform policy. Specifically, this case revealed the limitation of several influential philosophical accounts of modeling. These are the accounts claiming that theoretical models in economics

- provide *ceteris paribus* claims and explanations on the basis of some assumptions[17] being satisfied in the real world (Hausman 1992);

[16] See Janssen (2021: 183–184) for a similar evaluation and Boldyrev (2012) on the general implications for economic methodology as a field.

[17] More exactly: the assumptions that cannot be de-idealized (i.e. replaced by more realistic ones).

- reveal the 'stable capacities' of auction designs and identify disturbing factors (following Cartwright 1999);
- create and explore parallel 'credible worlds' in which certain causal effects are observed (Sugden 2000).

The first—'de-idealization'—account is criticized for being unhelpful in assessing the application of auction design models to real-world auctions, with all their complexity. These models relied on numerous assumptions that were not met in practice *and* could not be relaxed. In fact, Alexandrova and Northcott (2009) contend that no such analysis was conducted when applying auction theory to the FCC auctions design.

The stable capacities account is also problematic, because 'the effects of different features of the auction on its outcomes are not stable' (Alexandrova 2008: 394). Various auction formats were experimentally tested to see how different elements of auction settings produce different outcomes. Alexandrova and Northcott (2009) interpret Plott's (1997) understanding of auction rules and experimental testbeds as a holistic search for the best design, in which results stem from the *interaction* of different elements. While theory can suggest what effects certain elements (e.g., closed vs. open auction formats) should have on desired outcomes (e.g., efficiency), Alexandrova and Northcott (2009) argue that these effects were not predetermined (as it might seem when looking at the theorists' accounts of the auctions). Instead, they were treated skeptically in the actual design process.

Similarly, the credible worlds account offers only limited insight, as the credibility of parallel worlds in auction theory models—something that could validate these models as useful for explaining auction dynamics and guiding design choices—is not to be found in any prior theoretical knowledge and has to be demonstrated by actually *running* the auctions.

All these philosophical accounts of models are thus portrayed as too abstract/narrow to fully capture the real-world application of theoretical models in economics. What are these models, then? According to Alexandrova (2008), they are 'open formulae' with a structure like 'an *F* causes a *B* in a situation of type *S*.' The framework is used for formulating hypotheses that do the explanatory work, meaning that explanation might occur outside the model itself. This formula is open, because the situation *S* may not exist and the list of its relevant characteristics may be not unique. In other words, there is a loose connection between the specific assumptions of each model and the actual causal claims made on the basis of the models. This openness, Alexandrova (2008) argues, is an attractive feature for institutional designers who often do not know whether the assumptions of their models are actually satisfied. They may also recognize that certain assumptions are unrealistic *and* lack the control over real-world conditions to better align them with the model's premises. Interestingly, the practice of auction design is described as finding the constructive ways out of this impasse by searching for alternative conditions under which the causal claims made in the models hold true. On the open formula account, the process of explanation is linked to the material realization of an actual situation in which the causal effect described in the model holds. This material realization 'closes' the open formula.

The open formula perspective tries to capture situations where, as with the FCC auctions, certain outcomes emerged unexpectedly, were not accounted for in the abstract models (which therefore cannot, strictly speaking, be considered as explanations of the auction outcomes). These outcomes were only discovered through an *experimental material realization* of the auction itself.[18] The approach thus follows the holistic perspective of experimentalists who, unlike game theorists, allegedly 'have tended to explore the properties of the machine as a whole' (Mirowski and Nik-Khah 2017: 188).

Mirowski and Nik-Khah (2017) make some more sweeping claims about market design epistemologies. They argue that market design exemplifies economists' evolving role as guardians of truth—the only source of which is increasingly believed to be—in the wake of Hayek—the working of the market.

> If markets indeed validate truth, then the cadre that gets to construct the markets gets the final say on the nature of truth. The visible hand that fashions the auction believes it can govern the world. (Mirowski and Nik-Khah 2017: 7f.).

In their narrative, market design became influential due to a growing preoccupation with information, the attitudes to which were shifting, and because it 'constitutes the precepts of neoliberalism taken to their logical conclusion' (128). While the connection is at times less convincing, I believe they are correct in stating that the shift toward market design represents 'an epoch-making departure in the history of the praxis of economics' (154).

Mirowski and Nik-Khah (2017) classify different 'schools' of market design based on their approach to the role of information and knowledge. Their classification is also derived from the stylized evolution of Hayek's thought. The 'Walrasian school' treats information as a thing (echoing Shannon's ideas) or commodity and insists on the importance of the knowledge possessed by individual agents (Hurwicz's early models of mechanism fall under this category). The 'Bayes-Nash school,' following what they term the 'Blackwell doctrine,' treats information gathering as an inductive partitioning of the possible states of the world in a suitable space. Finally, information as a result of the work of markets as computational machines is associated with more algorithmic experimental approaches. There,

---

[18] I should add here that I would complement Alexandrova and Northcott's (2009) conclusions with regard to the ticklish issue of the progress in economics, something they infer from the auction theory case. They limit progress primarily to applications—the 'engineering' aspect of economic theory. However, I would argue that market design theorists have developed their own internal notions of progress, drawing not only on purely theoretical advancements (weakening the assumptions etc.), but also on practical applications (such as the practice of auction designs or school choice experiments). The complex and multifaceted understanding of progress, with all its vagueness, enables the discipline to evolve (and the field is currently an immensely active one, perhaps more so than it was in the 2000s). However, this complexity also precludes any straightforward assessment of progress: many theoretical innovations may prove practically irrelevant in specific designs, while an obscure theorem might unexpectedly provide the key to solving a major conceptual or practical challenge. In this sense, I believe market design theorists still value what Alexandrova and Northcott (2009: 334) describe as 'mathematics-type, purely abstract theoretical development.'

Mirowski and Nik-Khah argue, the knowledge of an individual agent is not actually revealed, as in the Bayes-Nash approach, but rather becomes irrelevant. While they acknowledge that individual economists may have differing implicit understandings of information and epistemic agency, Mirowski and Nik-Khah maintain that their classification sheds light on significant changes within the profession. Ultimately, they emphasize the supposed epistemic capacities of markets, a concept that resonated with both economists and policymakers.[19]

Overall, this classification might be useful for shedding light on the ongoing interplay between 'agency' and 'structure,' which underpins some important methodological debates in economics (Boldyrev 2019). However, in certain respects, it is less helpful. Specifically, the algorithmic and experimental dimensions that Mirowski and Nik-Khah make central in characterizing the experimentalist 'school' were prominent, indeed sometimes decisive, in other 'schools.'[20] The paternalistic tendencies Mirowski and Nik-Khah associate with neoliberalism is a separate issue, to be discussed below.

13.4 Performing Theory?

The performativity of economics is a radical approach to economic modeling, rooted in Marxist, pragmatist, and sometimes 'constructivist' ideas, which challenges the traditional separation between 'theory' and 'practice.' According to this view, not only does practice inform and guide theory, but theory also shapes and transforms its object of study—practice itself. In other words, performativity calls into question the distinction, on which most models—and their philosophical interpretations—are based.[21] In its most controversial form, the performativity thesis suggests that the major success of market design lies not in explaining institutions, but in (re)constructing them according to the precepts of the theory, thereby making the theory true as a practical endeavor. The relevant question concerning modeling assumptions, then, is not whether they *are* realistic, but whether they *can be made* realistic.

To explore this idea, it is essential to note that, from the outset, the models of mechanism/market design were the models of *institution building*. Economic

[19] On this important, but underappreciated topic see also the recent paper by Herzog (2024).

[20] Thus, Walrasian *tâtonnement* and other adjustment processes are inherently algorithmic; the communities of mechanism designers and experimentalists were closely interacting; Bayesian design theories are systematically tested through experiments; practical designs within the Bayes-Nash tradition inevitably face all sorts of computational constraints (such as the solvability and computational efficiency of solving combinatorial optimization problems).

[21] The early and provocative definition of performativity was given by an actor-network theorist and economic sociologist Michel Callon (1998: 30) who argued that the economy "is embedded not in society but in economics." On this topic in general see MacKenzie et al. (2007), Muniesa (2014), Boldyrev and Svetlova (2016), and Brisset (2019).

engineering is, thus, an element of their modeling architecture, much like it was in earlier planning models. Importantly, most problems of market design are more clearly defined compared to other economic contexts. In auctions or other designed marketplaces—such as systems for school allocations—agents operate under clear, well-defined rules, much like in game theory.

The epistemologies of market design discussed above are, in various ways, connected to the performativity account. Guala (2001) argues that in defining a 'mechanism,' there is persistent confusion between the actual institution and its (mathematical) representation. However, the confusion is not accidental, and this actually follows from Guala's paper. From the performativity perspective, we are dealing with a representation that *becomes* the institution itself. (Of course, the reverse is also true: as demonstrated by the history outlined earlier, intellectual progress was decisively shaped by practical work in designing auctions or matching mechanisms.) When looking at an open formulae view of models (Alexandrova 2008), one cannot help but notice the formal similarity between the open formula—'an *F* causes a *B* in a situation of type *S*'—and Searle's (2005) famous representation of institutions: '*X* counts as *Y* in *C*.' In market design, *models are institutionalized*. The explanatory power of market design models is explicitly linked to the constructive—performative!—effort of making them function by adjusting decision-making environments to fit the theoretical frameworks.

Mirowski and Nik-Khah (2017) dedicate some attention to the critique of the performativity literature, though they ultimately agree with it in singling out market design as an important constructivist endeavor (cf. Callon 1998, 2007 and Mirowski and Nik-Khah 2017: 180). At times, their language overlaps with the one of actor-network theory, e.g., when they describe markets as hybrid computational devices (185, 188).[22] A helpful point they raise is the potential problem of conflating different approaches to 'economics' when discussing performativity. This critique should be taken seriously, as different research programs may vary significantly in how they aim to enact their theoretical visions. The critique is also voiced by Santos and Rodrigues (2009) who have reasons to be skeptical about the 'strong'—or, as they argue, the only potentially meaningful and interesting—version of performativity in the case of the FCC auction design. Indeed, no single theory was 'performed' in that context.[23]

What can we actually learn about market design by drawing on the performativity program? It is broadly agreed that theory—and theorists—began to play an

[22] And of course some further claims of theirs would be quite compatible with performativity accounts (cf. this interpretation of market designers' position: "It is not merely markets that are tasked by market designers to replace politics, but some combination of markets and *economics*" (Mirowski and Nik-Khah 2017: 218)).

[23] When discussing the case of the FCC auctions, they write: 'Possibly both economics and the interests of the telecoms contributed to the choice of the simultaneous–multiple-round–independent auction. What evidence does not show, however, is that the auctions constitute an instance of performativity in the strong sense. They did not render auction theory, or any other theory for that matter, true by construction in any meaningful way' (Santos and Rodrigues 2009: 996f.)

active role in shaping institutions by influencing the conversation on the rules and aims associated with market design. While the final outcomes might have deviated from the ideal visions economists initially conceived, it is reasonable to believe—despite certain doubts (Nik-Khah 2008)—that economists aimed to implement these visions and that their framing efforts were driven by their theoretical, analytic concerns. At the very least, theory provided a conceptual *language* and a set of exemplary—paradigmatic—cases that had to be considered in real-world designs. And of course numerous specific ideas from market design literature—like *strategy-proofness*, *exposure problem* or *unraveling*—now structure the expert discourse.

The influence of economic ideas is visible not only in the language of the debate, but also, for example, in the very specific rules developed and defended by auction theorists *and* experimentalists (for the performativity of market design, it is not essential what type of economic knowledge was mobilized). For example, beginning with Hayek and Hurwicz, it was economic theory that emphasized the importance of information—its transmission and asymmetries—in determining economic outcomes. Therefore, it is no surprise that many adjustments proposed by auction designers focused on revealing new information. The phenomena and the complexities identified by theory directly guided the design efforts.

The intricate system of rules created by the market designers demonstrates that, while real-world designs were never direct implementations of a single auction theory model, the *efforts* of designers were directed at transforming the incentive structures of market participants. In doing so they effectively constructed rational agents[24] capable of producing the desired outcomes.

A useful way to examine this construction in more detail would be to look at the issue of *assumptions*—a classic theme in economic methodology. The work of market designers highlights a key aspect of assumptions: the intentional work towards actually bringing the assumptions into being.[25]

> [D]ue to the partly self-referential character of social concepts, for the construction of a stable and reliable socio-economic mechanism to be possible at all, adequate self-perpetuating mechanisms must be set in place (Guala 2001: 475; with reference to Barnes 1983).

These constructive mechanisms, or 'self-implementing technologies' (Boldyrev and Ushakov 2016, see also Tee 2019), are crucial components of market design models. In these models, the simple idea that *assumptions represent social facts* is taken more seriously than in other areas of economic research—to the extent that, again, the distinction between representation and the represented collapses. The

---

[24] Examples of this construction include the stopping rules and penalties for withdrawing bids in the 1994 FCC auctions (the rules helped ensure the auction remains simultaneous without overwhelming participants); the constraints on bidding in the 2017 forward auctions (Milgrom 2017); or framing households as agents with explicit preferences for schools (Hitzig 2020). In this sense, Wilson's observation that, in the future, market participants themselves will be designed (Roth and Wilson 2019: 138), can be equally applied to past designs.

[25] In defending the power of abstraction, Milgrom (2021) argues that the variability of contexts in market design actually justifies the use of *less*, rather than more, realistic assumptions.

rules used by market design experimenters to elicit the 'right' behavior constitute, Guala (2001: 464) argues, a specific kind of 'wisdom.' These guidelines 'are not *part of* economic theory, but rather tell us (roughly) how to construct real economic systems faithful to our abstract models,' which is true for more recent designs as well. For instance, the repacking rules in the incentive auction helped enact the important economic assumption of *substitutability* (Milgrom 2017; Kominers and Teytelboym 2020).

Another promising aspect of the performativity program is the recognition that the 'performation struggles' between different ideas and other parties willing to shape new institutions are always *materially mediated*. The materiality of economic engineering, particularly in market design, has not been sufficiently studied (with notable exceptions such as research on electricity markets by Breslau 2013 and Rilinger 2023). The new entities: new rules, algorithms, clearinghouses, systems of evaluation—*embody* theoretical ideas and do much of the 'performative work' (Beunza and Ferraro 2019). As the platform economy continues to expand, the role of technical infrastructures in market design will only become more prominent.

To sum up, beyond the provocative claim that theory is made 'true' by construction, the performativity approach highlights the pragmatic aspects of modeling, the social nature of assumptions, the uncertainties in neatly distinguishing 'theory' from 'practice', and the materiality of technical infrastructures mediating between design ideas and their practical implementation. As Hitzig (2020: 424) helpfully put it,

> the power of performativity as a lens through which to understand market designers' activities lies in precisely identifying how and at what cost economists' definitions of value-laden terms get transposed onto the realities they describe.

Some interpretations of performativity might overstate the role of economic engineering. However, the analysis of 'performation struggles'—where economic *framings* shape realities and contend with *overflows* (the unintended consequences and complexities that arise from these framings)—remains as relevant as ever.

13.5 Challenges and Blindspots

If one were to identify the single most significant theme in the critical discussion of market design, it would certainly be its role in society and policy. Broadly, this involves examining how market design efforts are embedded in societal and political contexts and exploring the social roles of both theory and theorists. Both the designers and their critics are preoccupied with incorporating these broader contexts, reflecting on market design's successes and failures (Klemperer 2013).

The major issue is that market design exemplifies what Eyal (2019) calls 'regulatory science' or what Halsmayer and Hounshell (2025), drawing partly on Joseph Vogl, describe as 'interventionist knowledge'—a science closely tied to, and shaped by, the 'collective concerns' (Frankel et al. 2019) of various social groups and stakeholders. Much of the critique centers on the role these groups play and the potential for designed markets to favor some participants at the expense of others. The wide applications of market design, therefore, demand additional ethical reflection (Li 2017). As van Basshuysen (2023) argues, this reflection must extend beyond familiar criticisms of 'marketization.'

Critiques of market design are diverse. Mirowski and Nik-Khah (2017) note that the current tendency in market design is to construct the agents as cognitively challenged and to rely instead too much on the 'structure' of a market (thus abandoning the Enlightenment vision of individual rationality[26] and risking to skew some markets in favor of large and powerful players). Sönmez (2023) expresses doubts about the radicality of social engineering in rebuilding institutions. Spiegler (2024) laments the overemphasis on the designer's perspective in market design models. Many critiques thus echo Milgrom's (2011: 317) assertion: 'The game is bigger than you think'.

In fact, the challenges most frequently discussed in the critical literature stem from the debates surrounding performativity. One important lesson of these debates is that, during implementation, certain assumptions within the design may contradict each other and prove impossible to reconcile. This infeasibility may only come to light *once the designs are put to practice*, thus remaining a 'blindspot' (for more on the concept and a case study, see Rilinger 2023).

The difficulties market designers encounter when analyzing relevant markets and mechanisms can be grouped under the broad concept of 'complexity.' In fact, the history of auction design is one of increasing complexity, which has posed significant challenges to theoretical models. The two most famous auction designs, that of the FCC auctions and the Incentive auction, have both been described as something uniquely complex. Kominers and Teytelboym (2020) classify the complexities and distinguish the difficulties of defining and balancing design objectives, optimally exchanging information between designers and market participants, and solving computational problems.

The *goals* of design, and its broader normative dimension, remain in the center of debate. Li (2017: 717) suggests that 'market designers should study the connection between designs and consequences, and should not attempt to resolve fundamental ethical questions.'[27] In contrast, Hitzig (2020) argues that often there is a normative vision market designers enact, which does not necessarily coincide with that of other stakeholders. Market design involves not only proposing optimal solutions for predefined normative goals but also reconfiguring those goals, thereby

[26] It would be instructive in this respect to recall Nietzschean critique of Enlightenment for hailing 'malleability of men' as progress (see the discussion in Adorno and Horkheimer's *Dialectic of Enlightenment* (1989 [1947]: 44)).

[27] Curiously, this perspective reminds of an ages-old 'Robbins-Parsons consensus', with sociologists discussing normative goals and values and economists focusing on technical issues of their optimal implementation (Parsons 1935).

embedding a specific, idealized normative theory within the process.[28] Drawing on Sen's critique of transcendental theory of justice, Hitzig further argues that enacting the 'ideal theory' might distract from the actual demands of justice emphasized by non-ideal, comparative theory. This latter approach would prioritize addressing the existing inequities, rather than devising an algorithm that satisfies the whole set of abstract 'transcendental' conditions. Such a non-ideal perspective is closely tied to democratic deliberation. It is the lack of broader discussion, or in other words, *depoliticization* associated with market design efforts, that is at the root of many ethical concerns. Moreover, the ambiguities surrounding the very concept of a 'market' further complicate the ethics of market design (Frankel 2015; Nik-Khah and Mirowski 2019).

An important aspect of market design, as discussed earlier, is the creation of new entities—sometimes referred to as machines or devices (Hirschman and Popp Berman 2014)—that embody economic principles. Examples include algorithm-based school allocation systems (Hitzig 2020), electricity demand curve (Breslau 2013), emission trading schemes (MacKenzie 2009), labor market clearinghouses (Roth 2008), auction infrastructures (Milgrom 2017), and infrastructures for kidney exchange (Sönmez 2024). These entities, prefigured already in Lange's market socialist projects[29] or in Hurwicz's (1960) notion of a mechanism aggregating information, mediate between individual agency and public policy. Once established, such entities become difficult to contest without specialized expertise, raising concerns about democratic deliberation and impartiality. This issue of *opacity* (van Basshuysen 2023) suggests that the mechanisms created by economic engineers may be so intricate that the participants cannot always assess whether these new institutions truly serve their needs. In some subtle ways, the mechanisms might even hinder democratic participation, replacing it with what Hitzig (2020), drawing on Porter's (1995) concept of 'mechanical objectivity,' describes as technical and impartial evaluation. The critical questions of how to enhance democratic participation and what forms it should take, however, remain unresolved.

Different models of engagement with design stakeholders have been proposed. Callon (in Callon and Roth 2021) advocates for 'enlightened engineering,' while Sönmez (2024) supports what he terms 'minimalist market design.' The minimalist approach focuses on improving existing institutions by addressing stakeholders' concerns, rather than engineering entirely new ones. However, participation might be a persistent challenge. Roth argues that even designs intended to be highly

[28] In the case of the Boston public school system Hitzig (2020) considers, the theory in question defined the policy goal as equal opportunity of welfare.

[29] In discussing the notion of social optimum, Lange argued that the optimal solution "can be carried out by a mathematical machine. This opens up to a socialist society the possibility of automation of administrative processes. In agreement with the postulates of socialist democracy, all members of the society would register their preferences (this too could be mechanized), and the decision would be computed by 'electronic brains' on the minimax principle. Such decisions would be the objective logical outcome of valuations by individuals, free from arbitrariness and subjectivism, and from possible distortions by bureaucratic decision making agencies" (cit. in: Zauberman 1967: 126).

collaborative—such as when economists were designing a labor market for themselves—faced issues with inadequate participation (Callon and Roth 2021).

Several related challenges may arise. Individuals may lack a full understanding of the mechanisms (Li 2024), or new markets may be seen as 'repugnant,' leading agents to refuse participation (Roth 2007; Brisset 2021). In any case, the logic of inclusion necessitates addressing these concerns. New approaches to modifying designs and fostering communicative spaces are emerging. In these spaces, dispersed economic knowledge, including theoretical models, may be 'redistributed' and, it is hoped, harnessed for emancipation rather than domination.

13.6 Conclusion

Marx's 11th thesis on Feuerbach:

> The philosophers have only interpreted the world, in various ways; the point is to change it
> – is, curiously, and perhaps inadvertently, echoed by Robert Wilson:
> I foresee more economists improving the allocation of scarce resources rather than (just) studying it (Roth and Wilson 2019: 139).

Market designers approach markets as specific infrastructures for resource allocation to be either created from scratch or reconfigured with certain normative goals in mind. In order to understand this field of economics we need to rethink the traditional Millian view of economic analysis as armchair deduction from assumptions to conclusions. The new generation of 'worldly philosophers' are not here to only study optimal patterns of resource allocation. They are actively involved in shaping relevant institutions.

Beyond familiar areas of market design interventions–such as auctions, organ transplantation, clearinghouses for medical doctors, and school choice–new domains are coming into focus (Kominers et al. 2017; Cramton et al. 2020; Hadad and Teytelboym 2022). Historians and philosophers of economics should explore these areas in greater depth, paying closer attention to the specific motivations of market designers and to the complexities of their contexts. This paper has touched on some of the critical work that, in fact, closely follows the concerns within the discipline. It also demonstrated that the societal embeddedness of design ties together many epistemic and ethical considerations with issues raised by performativity theory. To be clear, these issues are not unique to market design (one could consider other applied policy-driven fields within economics). However, market design stands out. It has become a focal point for much work and innovation in seemingly abstract *microeconomic theory*. Its emphasis on constructing institutions in the context of normative objectives and on using the strategic analysis of incentives lends itself to practical application.

Potential design failures, the interplay between intellectual and policy priorities, the creation of relatively autonomous entities, unintended consequences of implementation and, finally, possibilities for inclusive deliberation—all these aspects figure prominently in market design as a distinct *practice* of interventionist knowledge. This practice will undoubtedly give rise to new epistemic challenges and ethical debates. By reflecting on the designers' efforts, by conversing on the terms of these debates and by bringing together the insights from various fields, historians and methodologists of economics can help illuminate important issues that often remain invisible to many stakeholders and society at large.

References

Abdulkadiroğlu, A., and T. Sönmez. 2003. School Choice: A Mechanism Design Approach. *American Economic Review* 93 (3): 729–747.

Adorno, Th. and M. Horkheimer. 1989 [1944]. *Dialectics of Enlightenment*. N.Y.: Continuum.

Alexandrova, A. 2008. Making Models Count. *Philosophy of Science* 75 (3): 383–404.

Alexandrova, A., and R. Northcott. 2009. Progress in Economics: Lessons from the Spectrum Auctions. In *The Oxford Handbook for Philosophy of Economics*, ed. Harold Kincaid and Don Ross, 306–336. New York: Oxford University Press.

Ausubel, L. M. 2004. An Efficient Ascending-Bid Auction for Multiple Objects. *American Economic Review* 94 (5): 1452–1475.

Backhouse, R., and B. Cherrier, eds. 2017. *The Age of the Applied Economist: The Transformation of Economics since the 1970s. Annual Supplement to History of Political Economy*. Durham, NC and L.: Duke University Press.

Barnes, B. 1983. Social Life as Bootstrapped Induction. *Sociology* 17:524–545.

Becker, G. 1993. *A Treatise on the Family*. Cambridge, MA: Harvard University Press.

Beunza, D., and F. Ferraro. 2019. Performative Work: Bridging Performativity and Institutional Theory in the Responsible Investment Field. *Organization Studies* 40 (4): 515–543.

Boldyrev, I. 2012. Philosophy of Science or Science and Technology Studies? Economic Methodology and Auction Theory. *International Studies in the Philosophy of Science* 26 (3): 289–307.

Boldyrev, I. 2019. New Wine into Old Wineskins? *Methodenstreit*, Agency, and Structure in the Philosophy of Experimental Economics. In *Diversity of Experimental Methods in Economics*, ed. T. Kawagoe and H. Takizawa, 177–183. Singapore: Springer.

Boldyrev, I., and E. Svetlova, eds. 2016. *Enacting Dismal Science: New Perspectives on the Performativity of Economics*. New York: Palgrave Macmillan.

Boldyrev, I., and A. Ushakov. 2016. Adjusting the Model to Adjust the World: Constructive Mechanisms in Postwar General Equilibrium Theory. *Journal of Economic Methodology* 23 (1): 38–56.

Breslau, D. 2013. Designing a Market-Like Entity: Economics in the Politics of Market Formation. *Social Studies of Science* 43 (6): 829–851.

Brisset, N. 2019. *Economics and Performativity: Exploring Limits, Theories and Cases*. L. and N.Y.: Routledge.

Brisset, N. 2021. What Do We Learn from Market Design? On the Moral Foundations of Repugnance. *Revue de philosophie économique* 22:29–53.

Callon, M., ed. 1998. *The Laws of the Markets*. Oxford: Blackwell.

Callon, M. 2007. What Does It Mean to Say That Economics Is Performative? In *Do Economists Make Markets? On the Performativity of Economics*, ed. D. MacKenzie, F. Muniesa, and L. Siu, 311–357. Princeton: Princeton University Press.

Callon, M., and A. E. Roth. 2021. The Design and Performation of Markets: A Discussion. *AMS Review* 11: 219–239.

Cartwright, N. 1999. *The Dappled World: A Study of the Boundaries of Science*. Cambridge: Cambridge University Press.

Cherrier, B., and A. Saïdi. 2020. A Century of Economics and Engineering at Stanford. *History of Political Economy* 52 (S1): 85–111.

Clarke, E. H. 1971. Multipart Pricing of Public Goods. *Public Choice* 11: 17–33.

Coase, R. H. 1959. The Federal Communications Commission. *Journal of Law and Economics* 2 (1): 1–40.

Cramton, P., A. Ockenfels, A. E. Roth, and R. B. Wilson. 2020. Borrow Crisis Tactics to Get COVID-19 Supplies to Where They Are Needed. *Nature* 582: 334–336.

Eyal, G. 2019. *Crisis of Expertise*. N.Y.: Wiley.

Frankel, C. 2015. The Multiple-Markets Problem. *Journal of Cultural Economy* 8 (4): 538–546.

Frankel, C., J. Ossandón, and T. Pallesen. 2019. The Organization of Markets for Collective Concerns and Their Failures. *Economy and Society* 48 (2): 153–174.

Gale, D., and L. S. Shapley. 1962. College Admissions and the Stability of Marriage. *The American Mathematical Monthly* 69 (1): 9–15.

Gibbard, A. 1973. Manipulation of Voting Schemes: A General Result. *Econometrica* 41: 587–602.

Groves, T. 1973. Incentives in Teams. *Econometrica* 41 (4): 617–631.

Guala, F. 2001. Building Economic Machines: The FCC Auctions. *Studies in History and Philosophy of Science Part A* 32 (3): 453–477.

Hadad, J., and A. Teytelboym. 2022. Improving Refugee Resettlement: Insights from Market Design. *Oxford Review of Economic Policy* 38 (3): 434–448.

Halsmayer, V. and E. Hounshell. 2025. Introduction: Historicizing Interventionist Social Knowledge, 1950s–1990s. *Science in Context*. https://doi.org/10.1017/S0269889725100872

Harsanyi, J. C. 1967. Games with Incomplete Information Played by "Bayesian" Players, Part I-III. Part I. The Basic Model. *Management Science* 14: 159–182.

Harsanyi, J. C. 1968a. Games with Incomplete Information Played by "Bayesian" Players. Part II. Bayesian Equilibrium Points. *Management Science* 14: 320–334.

Harsanyi, J. C. 1968b. Games with Incomplete Information Played by "Bayesian" Players. Part III. The Basic Probability Distribution of the Game. *Management Science* 14: 486–502.

Hatfield, J. W., and P. R. Milgrom. 2005. Matching with Contracts. *American Economic Review* 95 (4): 913–935.

Hausman, D. 1992. *Inexact and Separate Science of Economics*. Cambridge: Cambridge University Press.

Hayek, F. A. von. 1945. The Use of Knowledge in Society. *American Economic Review* 35(4): 519–530.

Herzog, L. 2024. Are Financial Markets Epistemically Efficient? In *The Philosophy of Money and Finance*, ed. Joakim Sandberg and Lisa Warenski, 91–110. Oxford: Oxford University Press.

Hirschman, D., and E. Popp Berman. 2014. Do Economists Make Policies? On the Political Effects of Economics. *Socio-Economic Review* 12 (4): 779–811.

Hitzig, Z. 2019. *Market Design and Recentralization in the FCC Incentive Auction*. Unpublished draft.

Hitzig, Z. 2020. The Normative Gap: Mechanism Design and Ideal Theories of Justice. *Economics and Philosophy* 36(3): 407–434.

Hurwicz, L. 1960. Optimality and Informational Efficiency in Resource Allocation Processes. In *Mathematical Methods in the Social Sciences*, ed. Kenneth Arrow et al., 27–46. Stanford: Stanford University.

Hurwicz, L. 1972. On Informationally Decentralized Systems. In *Decision and Organization: A Volume in Honor of Jacob Marshak*, ed. C. B. McGuire and R. Radner, 297–336. Amsterdam: North-Holland.

Janssen, M. C. W. 2021. Reflections on the 2020 Nobel Memorial Prize Awarded to Paul Milgrom and Robert Wilson. *Erasmus Journal for Philosophy and Economics* 13 (2): 177–184.

Jensen, M. C., and W. H. Meckling. 1976. Theory of the Firm: Managerial Behavior, Agency Costs and Ownership Structure. *Journal of Financial Economics* 3 (4): 305–360.

Kagel, J. H. 1995. Auctions: A Survey of Experimental Research. In *The Handbook of Experimental Economics*, ed. Alvin E. Roth and John H. Kagel. Princeton: Princeton University Press.
Kelso, A. S., Jr., and V. P. Crawford. 1982. Job Matching, Coalition Formation, and Gross Substitutes. *Econometrica* 50:1483–1504.
Klemperer, P. 2013. Using and Abusing Auction Theory. In *Handbook of Market Design,* ed. by Nir Vulkan, Alvin E. Roth, and Zvika Neeman, 62-89. Oxford etc.: Oxford University Press
Kominers, S. D., A. Teytelboym, and V. P. Crawford. 2017. An Invitation to Market Design. *Oxford Review of Economic Policy* 33 (4): 541–571.
Kominers, S. D., and A. Teytelboym. 2020. The Parable of the Auctioneer: Complexity in Paul R. Milgrom's Discovering Prices. *Journal of Economic Literature* 58 (4): 1180–1196.
Lee, K. S. 2006. Mechanism Design Theory Embodying an Algorithm-Centered Vision of Markets/ Organizations/Institutions. In *Agreement on Demand: Consumer Theory in the Twentieth Century*, History of Political Economy 38 (supplement), ed. Philip Mirowski and D. Wade Hands, 283–304.
Lee, K. S. 2016. Mechanism Designers in Alliance: A Portrayal of a Scholarly Network in Support of Experimental Economics. *History of Political Economy* 48 (2): 191–223.
Li, S. 2017. Ethics and Market Design. *Oxford Review of Economic Policy* 33 (4): 705–720.
Li, S. 2024. *Designing Simple Mechanisms*. https://arxiv.org/pdf/2403.18694.pdf.
MacKenzie, D. 2009. Making Things the Same: Gases, Emission Rights and the Politics of Carbon Markets. *Accounting, Organizations and Society* 34 (3–4): 440–455.
MacKenzie, D., F. Muniesa, and L. Siu, eds. 2007. *Do Economists Make Markets? On the Performativity of Economics*. Princeton, NJ: Princeton University Press.
Maskin, E. 1999. Nash Equilibrium and Welfare Optimality. *The Review of Economic Studies* 66 (1): 23–38.
McAfee, R. P., and J. McMillan. 1987. Auctions and Bidding. *Journal of Economic Literature* 25 (2): 699–738.
Medema, S. G. 2020. The Coase Theorem at Sixty. *Journal of Economic Literature* 58 (4): 1045–1128.
Milgrom, P. 2011. Critical Issues in the Practice of Market Design. *Economic Inquiry* 49: 311–320.
Milgrom, P. 2017. *Discovering Prices*. N.Y.: Columbia University Press.
Milgrom, P. 2021. Auction Research Evolving: Theorems and Market Designs. *American Economic Review* 111(5): 1383–1405.
Milgrom, P., and J. Roberts. 1990. Rationalizability, Learning, and Equilibrium in Games with Strategic Complementarities. *Econometrica* 58(6): 1255–1277.
Milgrom, P., and C. Shannon. 1994. Monotone Comparative Statics. *Econometrica* 62(1): 157–180.
Milgrom, P. R., and R. J. Weber. 1982. A Theory of Auctions and Competitive Bidding. *Econometrica* 50(5): 1089–1122.
Mirrlees, J. A. 1971. An Exploration in the Theory of Optimum Income Taxation. *Review of Economic Studies* 38(2): 175–208.
Mirowski, P. and E. Nik-Khah. 2017. *The Knowledge we Have Lost in Information: The History of Information in Modern Economics*. Oxford etc.: Oxford University Press.
Morgan, M. 2012. *The World in the Model: How Economists Work and Think.* Cambridge: Cambridge University Press.
Muniesa, F. 2014. *The Provoked Economy: The Provoked Economy Economic Reality and the Performative Turn*. L. and N.Y: Routledge.
Myerson, R. 1979. Incentive Compatibility and the Bargaining Problem. *Econometrica* 47(1): 61–74.
Myerson, R. B. 1981. Optimal Auction Design. *Mathematics of Operations Research* 6(1): 58–73.
Myerson, R. B., and M. A. Satterthwaite. 1983. Efficient Mechanisms for Bilateral Trading. *Journal of Economic Theory* 29(2): 265–281.
Myerson, R. B. 2009. Fundamental Theory of Institutions: A Lecture in Honor of Leo Hurwicz. *Review of Economic Design* 13: 59–75.
Nik-Khah, E. 2008. A Tale of Two Auctions. *Journal of Institutional Economics* 4(1): 73-97.
Nik-Khah, E., and P. Mirowski. 2019. On going the Market One Better: Economic Market Design and the Contradictions of Building Markets for Public Purposes. *Economy and Society* 48(2): 268–294.

Parsons, T. 1935. Sociological Elements in Economic Thought II. The Analytical Factor View. *Quarterly Journal of Economics* 49(4): 646–667.
Plott, C. R. 1997. Laboratory Experimental Testbeds: Application to the PCS Auction. *Journal of Economics & Management Strategy* 6: 605–638.
Porter, T. M. 1995. *Trust in Numbers: The Pursuit of Objectivity in Science and Public Life*. Princeton, NJ: Princeton University Press.
Rassenti, S. J., V. L. Smith, and R. L. Bulfin. 1982. A Combinatorial Auction Mechanism for Airport Time Slot Allocation. *Bell Journal of Economics* 13(2): 402–417.
Riley, J. G., and W. F. Samuelson. 1981. Optimal Auctions. *American Economic Review* 71(3) 381–392.
Rilinger, G. 2023. Conceptual Limits of Performativity: Assessing the Feasibility of Market Design Blueprints. *Socio-Economic Review* 21 (2): 885–908.
Rizvi, S. A. T. 1994. Game Theory to the Rescue? *Contributions to Political Economy* 13 (1): 1–28.
Roth, A. E. 1982. The Economics of Matching: Stability and Incentives. *Mathematics of Operations Research* 7 (4): 617–628.
Roth, A. E. 1984. The Evolution of the Labor Market for Medical Interns and Residents: A Case Study in Game Theory. *Journal of Political Economy* 92(6): 991–1016.
Roth, A. E. 1991. Game Theory as a Part of Empirical Economics. *The Economic Journal* 101 (404): 107–114.
Roth, Alvin E. 2007. Repugnance as a Constraint on Markets. *Journal of Economic Perspectives* 21 (3): 37–58.
Roth, A. E. 2008. Deferred Acceptance Algorithms: History, Theory, Practice, and Open Questions. *International Journal of Game Theory* 36: 537–569.
Roth, A. E. 2016. Experiments in Market Design. In *Handbook of Experimental Economics*, ed. J. H. Kagel and A. E. Roth, vol. 2, 290–346. Princeton: Princeton University Press.
Roth, A. E., and E. Peranson. 1999. The Redesign of the Matching Market for American Physicians: Some Engineering Aspects of Economic Design. *American Economic Review* 89 (4): 748–780.
Roth, A. E., T. Sönmez, and M. Utku Ünver. 2004. Kidney Exchange. *The Quarterly Journal of Economics* 119(2): 457–488.
Roth, A. E., and R. B. Wilson. 2019. How Market Design Emerged from Game Theory: A Mutual Interview. *Journal of Economic Perspectives* 33 (3): 118–143.
Samuelson, P. A. 1954. The Pure Theory of Public Expenditure. *The Review of Economics and Statistics* 36(4): 387–389.
Santos, A. C., and J. Rodrigues. 2009. Economics as Social Engineering? Questioning the Performativity Thesis. *Cambridge Journal of Economics* 33 (5): 985–1000.
Satterthwaite, M. A. 1975. Strategy-proofness and Arrow's conditions: Existence and Correspondence Theorems for Voting Procedures and Social Welfare Functions. *Journal of Economic Theory* 10(2): 187–217.
Searle, J. R. 2005. What Is an Institution? *Journal of Institutional Economics* 1 (1): 1–22.
Shapley, L., and H. Scarf. 1974. On Cores and Indivisibility. *Journal of Mathematical Economics* 1 (1): 23–37.
Smith, V. L. 1962. An Experimental Study of Competitive Market Behavior. *Journal of Political Economy* 70 (2): 111–137.
Smith, V. L. 1982. Microeconomic Systems as an Experimental Science. *American Economic Review* 72 (5): 923–955.
Sönmez, T. 2024. *Minimalist Market Design: A Framework for Economists with Policy Aspirations*. https://arxiv.org/abs/2401.00307.
Spiegler, R. 2024. *The Curious Culture of Economic Theory*. Cambridge, MA: MIT Press.
Sugden, R. 2000. Credible Worlds: The Status of Theoretical Models in Economics. *Journal of Economic Methodology* 7 (1): 1–31.
Svorenčík, A. 2017. Allocating Airport Slots: The History of Early Applied Experimental Research. *History of Political Economy* 49 (Supplement): 240–263.
Tee, S.-H. 2019. Constructing Reality with Models. *Synthese* 196(11): 4605–4622.

van Basshuysen, P. 2021. How to Build an Institution. *Philosophy of the Social Sciences* 51 (2): 215–238.
van Basshuysen, P. 2023. Markets, Market Algorithms, and Algorithmic Bias. *Journal of Economic Methodology* 30 (4): 310–321.
Varian, H. R. 2009. Online Ad Auctions. *American Economic Review* 99 (2): 430–434.
Vickrey, W. 1945. Measuring Marginal Utility by Reactions to Risk. *Econometrica* 13(4): 319–333.
Vickrey, W. 1961. Counterspeculation, Auctions and Competitive Sealed Tenders. *Journal of Finance* 16(1): 8–37.
Vulkan, N., A. E. Roth, and Z. Neeman. 2013. Introduction. In *The Handbook of Market Design*, ed. N. Vulkan, A. E. Roth, and Z. Neeman, 1–4. Oxford: Oxford University Press.
Wilson, R. B. 1967. Competitive Bidding with Asymmetric Information. *Management Science* 13(11): 816–820.
Wilson, R. B. 1969. Competitive Bidding with Disparate Information. *Management Science* 15(7): 446–448.
Wilson, R. B. 1977. A Bidding Model of Perfect Competition. *Review of Economic Studies* 44(3): 511–518.
Zauberman, A. 1967. *Aspects of Planometrics*. L.: Athlone Press.

Chapter 14
A History of Methodological Debates in Political Science

Sharon Crasnow

Abstract This chapter addresses two areas in which the interests of political scientists overlap with those of philosophers of science: how political science should be understood as a science and causal inference. The examination of the history of political science reveals that these are closely connected. During the twentieth century political scientists often strove to make the discipline "more scientific", evoking what Karl Popper called "the demarcation problem". One prominent answer to that question was that science required establishing causal relations among the phenomena. This, in turn, led to methodological debates about how to produce the best evidence for causal inference, since different methods would seem to produce different sorts of evidence. The chapter explores how current philosophical discussions about evidence for causal claims have intersected with those methodological debates, looking particularly at multi-method research and evidential pluralism.

Keywords Political science · Causal inference · Demarcation · Evidential pluralism · Multi-method research · Detachability ideal

14.1 Introduction

Political science has intersected with philosophy throughout its history. Most directly it shares its origins in philosophy along with the other sciences, a legacy that is most evident in political theory. In addition, debates about the nature of the discipline—its aims and methodologies—have been a key part of the evolution of the discipline. This chapter limits the discussion to two areas where the interests of political scientists have explicitly intersected with philosophy of science: how

S. Crasnow (✉)
Norco College, Norco, CA, USA

Y. Shan (ed.), *History and Philosophy of the Social Sciences*,
https://doi.org/10.1007/978-3-032-08167-4_14

political science should be understood as a science and causal inference. The history of modern political science shows these two concerns to be closely connected.

To clarify the first of these, it is the question of what it means to study politics *scientifically*. While answers to this question have had their own trajectory in the discipline during the twentieth century and since, they have at times been influenced by the philosophy of science literature when it sought to clarify the nature of science and distinguish it from pseudoscience (what Karl Popper referred to as "the demarcation problem") (Popper 1962). Political scientists rarely address demarcation directly, but the language that they use often reflects those philosophical debates. Part 1 explores how the conversation about the scientific status of political science has shifted over time and the extent to which philosophy is relevant to that question. While it is sometimes claimed that changes in focus and methods in the discipline amount to scientific revolutions in the Kuhnian sense, I argue that the persistence of diverse methodological approaches in the discipline counts against that interpretation. However, the identification of causal relationships has emerged as an important goal in the discipline. The question of whether and how the multiple methodological approaches embraced by the discipline can contribute to that goal is an ongoing debate.

Consequently, Sect. 14.2 looks more closely at such methodological debates in political science, focusing on the question of what sort of evidence best supports causal inference.. Political methodologists have sometimes appealed to philosophical work on causality to clarify their thinking on these issues. Recently, philosophers have begun to look at causal reasoning in the social sciences (including political science) to better understand how social scientists treat causality and what kinds of evidence are thought to be needed for causal inference. Since different methods would seem to produce different sorts of evidence, Ddebates about causal inference are directly tied to questions of methodology. Section 14.3 explores some issues in the current discussion about causality that are relevant to both disciplines.

Until recently, philosophers of science have rarely engaged with political science directly. Bruno Verbeek and Lee McIntyre note this in the final chapter of *The Routledge Companion to the Philosophy of Social Science*, where they ask "Why is There No Philosophy of Political Science?". They point out that if you look at the "big six" social sciences –economics, psychology, sociology, anthropology, history, and political science—it is only philosophy of political science that does not have journals, conferences, or audiences clearly associated with it (Verbeek and McIntyre 2017, 433). Harold Kincaid and Jeroen van Bouwel pick up this point in their introduction to *The Oxford Handbook of Philosophy of Political Science* where they state as one of their goals the desire to increase interaction between political scientists and philosophers of science because there are indeed philosophical issues to be addressed that are of interest to both (Kincaid and van Bouwel 2023). Their volume, along with work in the last decade, seeks to begin to fill this lacuna.[1]

[1] There is also some work in political science which has engaged with philosophical questions. Dowding (2016) is an example.

The present chapter does not aim to provide a comprehensive history of philosophical thought in or about political science. This is in part because of the complexity of the discipline. The American Political Science Association has 55 organized sections, reflecting the diversity of the field. Some of these sections are focused on particular regions (e.g., South Asian Politics; American Political Economy; Canadian Politics; European Politics and Society). Others reflect interests spanning regions (e.g., Political Economy; Human Rights; International Relations Theory; Democracy and Autocracy; Sexuality and Politics; Internationality Security). There are also sections that are more theoretically oriented (Foundations of Political Theory; Politics and History; Politics, Literature, and the Arts; Ideas, Knowledge, and Politics). And still others are organized around methodological approaches (Comparative Politics, Political Methodology; Qualitative and Multi-method Research; Experimental Research; Formal Theory). Some of these subdivisions have histories that differ from others. As a result, generalizations about the intellectual history of the discipline are difficult to support.

A further caveat is that this account is based primarily on political science in the United States. This is in part because it is the political science with which I am most familiar, but also because prior to World War II, the United States had the only institutionalized professional political science organizations (Gunnell and Easton 1990, 1) and consequently offered a model for professionalizing political science. After World War II, that model was transported to other parts of the world and the political and economic dominance of the United States was at least partially responsible for American political science shaping the discipline globally. While political science had its roots in European social science, the periods that I focus on are ones in which political science in the United States separated itself from that tradition.

14.2 A Science of Politics

14.2.1 The Founding of the Discipline

Political science began to distinguish itself as an empirical science in the modern sense in the late nineteenth and early twentieth century. At the time of the founding of the American Political Science Association (APSA) in 1903 one of the stated motivations for the establishment of the organization was the desire to distinguish political science from history, philosophy, economics, and law. Although *The Annals of the Academy of Social and Political Science* (first published in 1890) and the *Political Science Quarterly* (first published in 1886) were already in existence, the creation of a new journal associated with the new organization, the *American Political Science Review* (APSR), signaled support for a different approach to studying the politics (Gunnell 2006). Discussing APSA's founding and the motivations behind it, W. W. Willoughby, the first editor of the APSR, notes approvingly that APSA's constitution declares that the aim of the organization is to promote the "*scientific* study of politics, public law, administration and diplomacy" (Willoughby 1904, 109, emphasis added).

John Gunnell, writing about the history of the discipline, explores why there was a felt need for this separation and identifies as one stated motivation the desire to alter the relationship of the discipline to political life and policy (Gunnell 2006, 479). He cites Goodnow's presidential address at the first APSA meeting in 1904, in which Goodnow suggests that the annual meetings of the APSA will provide an opportunity for the "closet philosopher" to interact with the "man of action" and that "the ideal and the practical may meet . . . ought to be of benefit to both" (Goodnow 1904, 45–46). The idea that political science should be more than an academic study—that it should be relevant and serve as a guide to political life—reemerges repeatedly in the history of the discipline, usually around calls for reform in the discipline, as we shall see.

Such sentiments expressed by the founders of APSA are consistent with the desire to distinguish the scientific study of politics from a historical, theoretical, and philosophical tradition of the previous century. Mark Bevir describes this new approach as a modernist empiricism that grows to prominence after World War I when the progressive narrative that was common in the nineteenth century becomes more difficult to sustain (Bevir 2022, 12). The changes in the discipline not only involved new approaches—a shift away from long-arc narratives of the previous century—but also alterations to what was thought to be the appropriate subject matter of the discipline. The nineteenth century historicist tradition took as its subject matter "the nation, the state, and freedom" (Adcock et al. 2007). Political science during the interwar period began to turn its attention to the study of political behavior and was characterized by research on parties, public opinion, and pressure groups (Bevir 2022, 19).

14.2.2 The Behavioral Revolution

Gunnell describes this interwar period as foreshadowing what came to be known as "the behavioral revolution" of the 1950s and 1960s.[2] The idea that this was a "revolution" evokes Thomas Kuhn's account of scientific theory change (Kuhn 1962). That political scientists would think of the history of the discipline in these terms is not surprising given that Kuhn's ideas had made their way into discussions of many of the sciences as well as into popular culture. In describing the behavioral revolution, Gunnell argues that political scientists understood it "as effecting a paradigmatic shift in political studies and as transforming the discipline into a truly scientific enterprise." Here he again uses Kuhnian terminology (Gunnell 1990, 14). This framing of behavioralism 40 years after its heyday reflects a shared understanding

[2] The behavioral movement is not at all related to behaviorism in psychology, in spite of the common root for the terms (see Easton 1990, 278).

by members of the discipline of this history. As Adcock notes "The behavioral movement of the 1950s and 1960s is central to the way most contemporary American political scientists envision their discipline's past" (Adcock 2007, 180).

What was the behavioral revolution and was it indeed a Kuhnian revolution? There is general agreement that the seeds for the behavioral revolution are present in changes in the discipline that took place in 1920s and 30s, among American political scientists. During the midcentury, the movement gained momentum, and by the early 1960s behavioralists held prominent positions in APSA. As Robert Adcock describes, behavioralism highlighted political behavior and the absence of research on the legal structures that had dominated research in the past (Adcock 2007, 182–183).[3]

In addition to new subject matter, behavioralism also urged new methodologies. Earlier political science research had been mostly qualitative—field interviews and case studies (Adcock 2007, 184). Adcock goes on to say that until the 1940s "most political scientists were uninterested in, or even actively hostile to, neopositivist crusades seeking to tighten the standards of social science. They thought of themselves as scientists, and they did so with little anxiety or, for that matter, self-reflection. A low-key notion of science as fact gathering and objective reporting prevailed" (Adcock 2007, 187).[4] The methodological shift associated with behavioralism involved a call for "a new kind of theoretical work and for the use of more sophisticated empirical techniques" (Adcock 2007, 189).

The "more sophisticated empirical techniques" were primarily quantitative, large-*N* analyses, and an increasing preference for quantitative methods dates from this period. Quantitative techniques for research were already well-established in the other social sciences, particularly psychology and sociology, and behavioralists transported them into political science. In doing so they transformed the discipline from one that was loosely empirical to one that sought to connect theory and empirical evidence. In doing so, they saw themselves as adopting approaches that were described by philosophers as scientific and those in the profession sometimes identified them as positivist. This was positivism to the extent that it was an empiricism

[3] APSA had organized groups to assess the state of the 10 areas of the discipline and the volume *Research in Political Science*, Ernest S. Griffith (ed.) published in 1948 resulted from the work of those groups

[4] Adcock refers to "neopositivism" here and so prompts a comment on how the term "positivism" is used by the various histories of political science I have relied upon for this chapter. In most of the history of political science literature "positivism" is not used to refer to the philosophical position Logical Positivism.. But it is not always clear what is intended when the term is used. Bevir claims that we shouldn't think of rational choice theorists as positivists in spite of rational choice theorists using of formal models to deductively predict empirical outcomes (Bevir 2008, 53). In doing so, he seems to be making that connection. Easton, on the other hand, does explicitly identify rational choice theorists as positivist (Easton 1990, 288). In post-behavioral political methodology, "positivism" is often used as an epithet to indicate old-fashioned behavioralism that is associated with some form of modeling the social sciences on the natural sciences either as a search for laws or for mechanisms. In any case, the term does seem to refer to some, not clearly specified, empiricist approaches that seek to link theory to phenomena through something like the hypothetico-deductive method.

that sought to relate theory and empirical phenomena, but not strictly speaking the philosophical positivism of early twentieth century philosophy.[5] Often, the term "positivism" (or "neopositivism") are used as criticisms aimed at those in the discipline who were perceived as trying to make political science more like the natural sciences, or rather more like what they believed to the natural sciences to be.

Behavioralism adopts something like a hypothetico-deductive model of science in that it calls for empirical observations to be deduced from theory, thus providing a means of theory testing. However, the failure to agree on a theoretical framework for behavioralism was a weakness in the approach, that ultimately led to its demise (Adcock 2007). Adcock describes a variety of theoretical frameworks that were proposed, including systems theory, favored by David Easton, a prominent behavioralist. Easton advocated the use of systems theory as a means of organizing data and sought universal theories through which to give accounts of political behavior (see Easton's *The Political System*). Disagreements about what theories were to be adopted and what level of theory was appropriate (universal and abstract, or middle-range) was not agreed upon by behavioralists, although the importance of theory (rather than what they saw as the "hyperfactualism" of previous work) was agreed upon. In addition, the sorts of theories that were advocated differed from the political theory frameworks that had dominated the historicist approaches that behavorialism sought to replace. "Behavioural theory … is empirically oriented. It seeks to help us explain, understand, and, if possible, predict the way in which people behave politically and the way political institutions operate" (Easton 1990, 279).

Although quantitative research represented the majority of publications in the major disciplinary journals from the 1950s onward, qualitative research never disappeared from the discipline and many political scientists pursued a plurality of approaches throughout this period.[6] Even during behavioralism's heyday there were those in the profession who continued to see the importance of history and criticized behavioral approaches for decontextualizing knowledge in their search for theories through which to understand political behavior. Bevir describes much of this work as a continuation of the earlier institutionalism which favored inductive, small-*N* comparative studies and sought middle-range theories as opposed to the more general theories most often associated with behavioralism and supported through large-*N* techniques (Bevir 2008, 52–53). Dowding (1994) notes that a goal of behavioralists was to study human behavior *within* different institutional settings and so these approaches should not be seen as at odds with each other. Rather than

[5] Lawler and Waldner have recently argued that the legacy of this positivism is the causal inferentialism that dominants current conceptions of the goals of the discipline as discussed in Part 2 (Lawler and Waldner 2023).

[6] There are two senses in which quantitative research might be thought to dominate. One is in terms of prestige. This is hard to document but appears to be a shared impression in the discipline. There is more concrete evidence both that quantitative and formal work is more frequently published in top journals and that graduate schools offer fewer courses in qualitative methods than quantitative methods. See Bennett et al. (2003).

thinking of them as emblematic of different paradigms they should be understood as seeking to answer different sorts of questions.

Interestingly, as political science entered the late 1960s, a criticism that the behavioral movement had used against traditional historicism—that the discipline was not relevant—reemerged as one challenge to behvioralism during the Viet Nam era.

14.2.3 Post-behavioralism

In his 1969 APSA Presidential Address, Easton describes a new, "post-behavioral", revolution spurred by "a deep dissatisfaction with political research and teaching, especially of the kind that is striving to convert the study of politics into a more rigorously scientific discipline modelled on the methodology of the natural sciences" (Easton 1969, 1051). Easton sees post-behavioralism as a coalition of conservation (historicist) and liberal (politically progressive) forces against the political science of the time.

There were other reasons for the decline of behavorialism. As mentioned, the failure to agree on a theoretical underpinning of political behavior made it difficult to construct accounts that fulfilled an understanding of science as composed of theoretically based empirical accounts—an ideal closely associated with the version of positivism absorbed by behavioralist in the discipline.

In the early 1960s, William Riker began using game theory to explain the formation of coalitions (Riker 1962). While his interest was primarily in political coalitions, he urged the use of rational choice models in political science more generally. He argued that the approach was more scientific, in part, because the use of formal models allowed for the deduction of results which could then be interpreted empirically. Such empirical results could then serve as tests of the success of such models (Quack and Herfeld 2023, 556). This characterization of the approach seemed to fit well this the behavioralist understanding of science as following a hypothetico-deductive pattern of reasoning.

Although rational choice approaches were not widely adopted until the 1970s, when they were, it was in part because they offered the sort of theoretical framework that behavioralist had been seeking. However, there were disputes about the assumptions of rationality that were embedded in such models. In addition, there was resistance to the approach from political scientists who continued to be committed to historical comparative approaches as well as those less concerned about the scientific status of political science (constructivists and interpretivists), at least if that status was understood to be based on natural science as the exemplar. Nonetheless, by the 1990s, game-theoretic approaches had been broadly embraced and continue to be used, although they no longer are as dominant as they once were. They have remained appealing, in part, because models can be understood as models of mechanisms and so may play of role in causal reasoning. The persistence of multiple approaches in contemporary political science should not be surprising given the diversity of the discipline. Even acknowledging a primary goal of seeking evidence that supports causal claims a variety of research tasks are part of that process. Particular political events are understood as events of some type and the regularites that those types exhibit are sought. Political scientists also seek theoretical frameworks through which to explain such regularities. Different approaches are focused on different apsects of research that are all seen as relevant to the overarching aims of the discipline, and so that a variety of approaches remain active is not particularly surprising.[7]

And yet, the sociology of knowledge point that at particular periods particular approaches are likely to garner more prestige and be considered more important is also true. Describing these shifts as revolutions reflects the awareness of Kuhn's work and the debates about how science progresses (or not) that surrounded it. When coupled with the desire to understand the discipline as a science, it tempting to misdescribe such sociological shifts as revolutions. The title of Easton's 1969 Presidential Address, "The New Revolution in Political Science," shows that he embraced this framing, however, the history of the discipline does not fully support such an understanding. While it is true that both behavioralism and post-behavioralism elevated new approaches to prominence, qualitative, historical, and interpretivist methods were never abandoned. Adcock's account disputes the characterization of the behavioral revolution as a rejection of historical research, and indeed, the persistence of multiple methodological traditions within the discipline would seem to support his claim (Adcock 2007). There is no clear replacement of one paradigm by another in political science. There are however periods where particular approaches are strongly preferred and become dominant. The shift to research supported through quantitative (statistical) research has remained prominent. Game theoretic, formal, approaches are still widely used. Most recently, experimental methods have come to prominence.

14.2.4 The Perestroika Movement

In the fall of 2000, another change in the discipline was instigated through an anonymous email that circulated through the APSA membership. The email subject line reads "On the irrelevance of APSA and APSR to the study of Political Science!" and it is signed by Mr. Perestroika.[8] The body of the email consists of 11 rhetorical questions that, in various ways, call out elitism in the choice of APSA officers, methodological favoritism in APSR publication, and the recurring concern about the relevance of the discipline. For the first, both the dominance of east coast institutions and the lack of diversity—racial, gender, and sexual orientation—are mentioned specifically. But it is the lack of methodological diversity that stands out. An

[7] Dowding (1994) makes a point similar to this.

[8] The authorship of this email is still unknown.

aspect of the critique included charges that the skills of those who use quantificational and game-theoretic approaches fall far short of the economists who they are accused of trying to emulate and so both the dominance of these methods and the lack of expertise in their use were targeted. The email names a number of qualitative researchers and their exemplary work and charges APSA leadership with ignoring the achievements of these researchers and excluding them from leadership roles.

Mr. Perestroika's email came at a time when there were other discontents in the disicipline. Six years earlier, debates about the merits of qualitative research had been reinvigorated with the publication of *Designing Social Inquiry: Scientific Inference in Qualitative Research* (King et al. 1994). According to King, Keohane, and Verba's (KKV) account, all methodological approaches should follow the same underlying pattern of reasoning. "Our main goal is to connect the traditions of what are conventionally denoted "quantitative" and "qualitative" research by applying a unified logic of inference to both. …The same underlying logic provides the framework for each research approach" (King et al. 1994, 3). Among those who did primarily qualitative work, KKV was hotly debated and an alignment with what came to be known as the Perestroika movement seemed natural. Among the changes sparked by this movement were the election of the officers of APSA (in contrast, to their appointment which had previously been the norm). In addition, efforts were made to publish more qualitative work in the APSR as well as the founding of a new APSA journal *Perspectives in Politics* in 2003. The Qualitative and Multi-method Research Section—a new section of APSA—was formed and has been active in training researchers in qualitative research techniques at its Summer Institute of Qualitative and Multi-Method Research (IQMR). Changes in the APSR have been slow but recent appointments of editorial teams signal an ongoing commitment to reform and relevance.[9]

While behavioralism marks the beginning of a widely (although not universally) shared understanding of the primary goal of political science research: the identification of causes, KKV staked out a position about how to accomplish that goal qualitatively. For many qualitative researchers what KKV described did not seem to be what they were doing in their research, nor what they thought they should be doing. Most particularly, KKV did not capture how case studies might be used to produce evidence supporting causal claims. Part 2 explores how the methods debates in political science have come to focus on causality and consequently how they have intersected with work in philosophy of science.

[9]The 2020–2024 team stated as among their goals "we intend to publish problem-driven scholarship that is well-conceptualized, ethically-designed, and well-executed; research on topics and by scholars the discipline has been slow to engage; and work that uses a range of methods and approaches to address both timely and timeless questions about power and governance that are central to the study of politics everywhere" ("APSA Announces the New Editorial Team for the American Political Science Review" *politicalsciencenow.com*. 26 July 2019).

14.3 Methods and Causality

14.3.1 *Causal Inferentialism*

The Perestroika movement resulted in disciplinary self-reflection, but it did not stem the emerging dominant understanding of the goal of science (and so political science) as the identification of causes. Janet Lawler and David Waldner describe this view as "causal inferentialism" (Lawler and Waldner 2023). In political science, this causal inferentialism includes the idea that the best methods for establishing causal claims are experimental. However, since experimentation is difficult (or even impossible) in the social sciences, methods that approximate experimental reasoning must be relied upon.[10] Techniques for providing evidence of causality that either are or are proxies for experiments, such as econometrics and other statistical techniques, are thus favored. These techniques all involve what Derek Beach and Rasmus Brun Pedersen call "variance-based" reasoning. Variance-based designs are distinguished from case-based research designs (Beach and Pedersen 2016). The latter are typically single case studies involving qualitative methods.

The distinction between these two types of reasoning is a rejection of the KKV strong methodological unity thesis consisting of three key elements: (1) there is one logic of inference; (2) there is one goal of science; and (3) science is unified by method (King et al. 1994, 9). The one goal of science that KKV refer to is establishing causal claims. The logic of inference is thus a logic of causal inference and the unified method through which to establish such claims is a form of variance-based reasoning.

KKV's claim that method should be unified is cashed out for qualitative research in a variety of ways. They call for qualitative research to be "more rigorous". By this they mean specifically that it should be more self-conscious about concept formation, research design, and data collection. They argue that the way to achieve this is to follow the example of quantitative research. This might seem puzzling at first given that qualitative research is usually case-based and so focused on the close examination of a single or a few cases—small-*N*—whereas quantitative research is large-*N*. However, they argue that the many observations that qualitative research produces can serve as the basis for observational studies that make use of large-*N* statistical techniques.[11] In this way, both quantitative and qualitative approaches could be understood as sharing a commitment to an inferential logic of variance-based approaches.

---

[10] Experiments first appear in more psychologically oriented spheres of political science—for example, in the study of voting behavior. Experimental research design is increasingly popular in political science, having traveled to the discipline via development economics and has rapidly become the new gold-standard methodology for the discipline. However, experiments are not always possible and qualitative (observational) data still plays an important role in the discipline.

[11] To do so involves coding cases or observations within cases. This in itself is an interesting topic for philosophical exploration and involves analyses of concepts through which the data are organized. The topic is beyond the scope of this chapter however.

Qualitative researchers pushed back against KKV, disputing their central claims and arguing that they were turning qualitative research into quantitative research and thereby failing to acknowledge the unique contributions that qualitative research could make. Alexander George and Andrew Bennett questioned KKV's claim of one logic of inference noting its ambiguity. If KKV mean that there is a shared epistemological approach, that claim is probably true, but if the claim means that there is only one appropriate methodology suited to that approach, then they dispute the claim (George and Bennett 2005, 10–11).[12] *Rethinking Social Inquiry: Diverse Tools, Shared Standards* which appeared in 2004, with a second edition in 2010 (Brady and Collier 2004/2010) consolidated many of the criticisms of KKV, making the case that a variety of qualitative methods are valuable for political science research—the diverse tools of the Brady and Collier subtitle. The "shared standards" of the subtitle indicate their shared commitment to causal inference as the primary goal of (political) science. The debate consequently centers around methods—the diverse tools—and particularly on the question of whether, and if so, how, qualitative methods can serve the shared goal of causal inference.

Gary Goertz and James Mahoney, for example, argue that the way that quantitative methodologists think about what they are doing differs fundamentally from how qualitative methodologists do. In their view the two approaches constitute "two cultures" differing in values, beliefs, and norms that result in different research practices and ultimately in different understandings of causal inference (Goertz and Mahoney 2012). Goertz and Mahoney describe this as the difference between research emphasizing effects-of-causes or causes-of-effects. Quantitative researchers focus on effects-of-causes, "In particular, they have come down in favor of estimating the average effects of particular variables within populations or samples" (Goertz and Mahoney 2012, 41). The average effect is determined through the examination of a dependent variable (outcome) in relation to various independent variables in order to determine which make a difference to the dependent variable and so can be understood as causes. Qualitative researchers are more interested in the causes-of-effects, that is, they seek the causes of events, typically by tracing the causal pathway back from their outcomes. Case-based research might be within-case—the study of causality in a single case—or cross-case—the comparison of a number of cases in search of a causal regularity. For within-case causal analysis, a variety of techniques may be employed, but most commonly process-tracing and counterfactual analysis are used. Comparative, cross-case research may involve these methods as well, but also, as the number of cases explored increases, other techniques may be employed such as Qualitative Comparative Analysis (QCA).[13]

---

[12] George and Bennett suggest that the agreed upon epistemological approach might be characterized as deriving testable results from alternative hypotheses—a means of providing evidence for causal claims.

[13] For discussion of QCA see Zahle (2023) and Russo and Rihoux (2023).

14.3.2 Qualitative and Multi-method Research[14]

The organized section of APSA that is now the Qualitative and Multi-method Section began in 2002 as the Qualitative Methods Section. The announcement of the section in the Political Methodology newsletter specifically states as among its goals to "sponsor training and research focused on the several branches of methodology associated with the qualitative tradition. The section will also seek to promote an integrated understanding of these methods and their relationship to other branches of methodology, including quantitative methods" (Elman et al. 2002, 20).

The convergence on causal inference as an appropriate goal for political science raises concerns for qualitative research that are somewhat different than the worries raised for quantitative research. For the latter, the main consideration is how to distinguish mere correlations from causes. This is no small problem of course, but it is fundamentally different from the problem posed for qualitative research. First, consider the general methods that fall under that heading. The 2002 newsletter lists the following: "case study method, small-*N* analysis, the comparative method, the comparative-historical method, the ethnographic tradition of field research, constructivist and interpretive methods, concept analysis, and related areas of social and political theory" (Elman et al. 2002, 20). For a number of these (ethnographic, constructivist, and interpretive methods) causal inference does not seem to be the primary goal, but a number of the other methods listed—case study method, small-*N* analysis, and the comparative methods—are often used to support causal claims. Describing just how this is achieved became one of the main interests of the many of the members of the new section and it is these concerns that are reflected in *Rethinking Social Inquiry* and Goertz and Mahoney's *Two Cultures*.[15]

This description of the goals of the section also foreshadows interest in multi-method research that ultimately led to the renaming the section. It is now known as the Qualitative and Multi-method section. The interests of philosophers of social science have recently intersected with those of political science around multi-method research and how it works to support causal inference. Minimally using multiple methods might produce both more evidence for a claim, as well as evidence which is more diverse. Both philosophers and political methodologists have explored what this means.

Multi-method research can be understood as research that combines evidence produced by two or more methods with the goal of increasing inferential leverage in support of causal claims. In political science, the term has come to mean not just the use of multiple methods—case studies use a variety of methods within the

---

[14] I understand "mixed method" and "multi-method" research to refer to the same methodology. While some methodologists have argued for a more precise use of these terms and drawn distinctions that are marked by different terminology (Hesse-Biber 2015, xxxix), political methodologists appear to use the terms interchangeably and I adopt this usage.

[15] Although the section does support research and offer training in ethnographic and other traditions at its summer institute. Nonetheless the main focus remains causal inference, now often framed in relation to multi-method research.

case—but the use of evidence generated through different methods in order to support causal inference. This point is important because while it is clear that qualitative methods (case-based methods) may play a variety of heuristic roles—suggesting problems to be solved, providing background information, suggesting hypotheses to be tested, or simply providing an illustration for the causal claim argued for quantitatively—defenders of the importance of qualitative research are interested in showing that it can produce evidence that supports what is understood to be the primary goal of political science—causal inference.

If and, if so, how multi-method research designs accomplish this goal is debated and a variety of epistemological justifications have been offered. Two main justifications suggested are (1) that causal claims are better supported with evidence derived through a variety of sources and (2) since methods all have different strengths and weaknesses, combining methods will compensate for any weaknesses of individual methods. Both of these possible justifications appeal to the idea that the various methods employed should provide evidence that converges to support a causal claim. This is sometimes described as "triangulation". Jason Seawright offers the following characterization: "Simply put, triangulation designs involve asking the same question of causal inference using two different methods and checking that the same substantive conclusions are produced by both" (Seawright 2016, 4).

One worry about the triangulation metaphor is that it does not address the concern that Goertz and Mahoney raise about different conceptions of causality, different research practices, and different modes of inference that quantitative and qualitative approaches adopt—the two cultures. Beach and Pedersen also express this worry. "Debates about the nature of causation are important to understand because different conceptions of causation result in different types of causal claims, which then imply that different methodologies are needed to investigate them appropriately" (Beach and Pedersen 2016, 15). Seawright worries about triangulation as well, noting that it is often hard to determine in what sense the results might be said to agree given that the results are, in some sense, of different kinds. He illustrates his worries with an extended example of quantitative and qualitative (case study) evidence for the role of mountainous geography for civil wars. Looking at the logged percentage of Colombia's mountainous area differs in kind from looking that the geographical layout of the particular mountain ranges in Colombia. While it might be possible to compare the information that comes from case analysis—some of which shows potentially relevant differences between some mountainous areas and others when considered relative to conflict—to the logged percentage of mountainous area of Colombia in some way, it is not immediately obvious how to do so. These pieces of information appear to be answering quite different questions (Seawright 2016, 5–8).

One way of addressing this problem is suggested by a consideration of what kinds of evidence different methodological approach offer. We might use the language of effects-of-causes and so describe the evidence offered through quantitative evidence as evidence of effects-of-causes correlation (association or variance). Generally, it is evidence of average effects. In contrast, since a case study involves close examination of a single unit it does not offer this sort of evidence. Case study research is used to offer evidence of causes-of-effects (evidence of mechanism).

To better understand how a case may provide such evidence, consider one qualitative method—process tracing. Though there is some disagreement on what this method is and how it should be conducted, it is generally agreed that one use of process tracing is to produce evidence supporting within-case causal inference. "Process tracing is perhaps the tool of causal inference that first comes to mind when one thinks of qualitative methodology in political science" (Mahoney 2010, 123).

The basic idea of process tracing can be understood through Goertz and Mahoney's claim that qualitative research seeks causes-of-effects rather than effects-of-causes. Causes-of-effects are found by tracing the causal pathway backwards from the effect to the cause. Beach and Pedersen describe four variants of process tracing that are differentiated by their research goal. They are theory-testing, theory-building, theoretical-revision, and explaining outcome (Beach and Pedersen 2019, 9). These may all occur within the same research project. For theory-testing process tracing researchers start with a hypothesized mechanism that guides the search for entities, activities, or events that one ought to observe if that mechanism is operating in the case under investigation. Good process-tracing arguments compare competing hypotheses, showing one to be a better fit for the hypothesized mechanism than the other. A new or revised research problem may arise as a result of a case or cases that appear to provide counterexamples to previously proposed accounts and thus call for theory revision. Revising theory may also lead to theory-building or theory-building might happen directly from observations made within the case.

Theory-testing process tracing is the focus of the following discussion since it is most directly related to causal inference. Process tracing proceeds by postulating a causal mechanism (theory) and then seeking pieces of evidence that either indicate that the mechanism is or is not operating in the case under examination. The causal mechanism is either suggested by background knowledge—current political theory or game theoretic models—or by observed regularities in other cases thought to be similar to the case under investigation (this might involve theory-building process tracing). Qualitative data—e.g., information from archives, interviews, participant observations—as well as quantitative data (descriptive statistics)—e.g., demographic information, the number of particular kinds of events (e.g., civil unrest) that occur, increases or decreases in voter turnout—all might marshalled within the case to support that the hypothesized mechanism is present (or not) in this case.

The value of process tracing, a qualitative method, is thus thought to rest in how it provides evidence for causal mechanisms. Goertz and Mahoney argue that neither statistical nor experimental methods can do this (Goertz and Mahoney 2012, 101). Statistical and experimental methods may be able to establish a link between a dependent and independent variable, but they cannot reveal what is in the "black box"—the mechanism through which the cause brings about the effect. This analysis suggests a common use for multi-method research. Statistical techniques could be used to establish that there is a correlation between a dependent and independent variable and process tracing can provide evidence of the mechanism that sustains that correlation. Alternatively, the correlational evidence could bolster the process tracing analysis.

Quantitative (or experimental) methods that establish effects-of-causes (average effects) provide evidence of correlation and so suggest that there is a mechanism operating, but do not provide evidence for the specific mechanism that is responsible for the effect. In order to provide evidence of the specific mechanism (to open the black-box) a qualitative research method—process tracing—is necessary. Mechanistic explanations are the realm of the qualitative researcher—a different realm than that of the quantitative researcher.

14.3.3 Evidential Pluralism

Philosophers Yafeng Shan and Jon Williamson (2021, 2023) have recently proposed Evidential Pluralism and as a philosophical foundation for multi-method research in the social sciences.[16] They do so by extending the Russo-Williamson account of integrating evidence of correlation (association) and evidence of mechanism in the biomedical science (Russo and Williamson 2007). They offer the following characterization:

> Evidential Pluralism. In order to establish a causal claim, one normally needs to establish two propositions: that the putative cause and effect are appropriately correlated, and that there is some mechanism complex involving the putative cause which is responsible for the putative effect and which can account for the extent of the correlation. So, in order to assess a causal claim, one normally needs to assess relevant association studies and mechanistic studies where available. (Shan and Williamson, 2023, 3)

As they note, association studies are usually produced through quantitative methods (statistical or experimental methods) and mechanistic studies are usually produced through qualitative methods.[17] Evidential pluralism motivates the use of multi-method research since it calls for studies that produce both types of evidence (when available). They describe why this is desirable using language similar to that of Goertz and Mahoney: "Association studies confirm the existence of an appropriate correlation and can provide indirect evidence of the existence of an appropriate mechanism …. Mechanistic studies, on the other hand, can use either quantitative or

[16] Nancy Cartwright (2021b) argues that this view should be called evidential dualism rather than Evidential Pluralism, given that it does not embrace the wide variety of evidence that she believes should be taken into consideration for policy. I distinguish Shan and Williamson's Evidential Pluralism from evidential diversity for this reason.

[17] Although Shan and Williamson allow that quantitative studies produce mechanistic evidence such as providing evidence that an intervening variable mediates between some putative cause A and some putative effect B (Shan and Williamson 2023, 75). The question of whether mechanisms can be understood as a series of intervening variables is disputed and rejected by Goertz and Mahoney.

qualitative methods to test specific mechanism hypotheses …" (Shan and Williamson 2023, 75). Note that since they allow that quantitative methods may support mechanistic studies, this is not an explicit call for always using both quantitative and qualitative methods, but rather prescribes using both correlational and mechanistic studies. However, there are further reasons that they offer for the value of qualitative research. "Qualitative methods, meanwhile, can help to fill in richer details of the mechanism complex. For example, qualitative methods can help to identify and verify potential confounders, as well as key activities and entities on mechanism pathways, and the spatiotemporal organisation of a mechanism" (Shan and Williamson 2023, 75).

Shan and Williamson do not share Goertz and Mahoney's conception of these different methodological approaches constituting two cultures and indeed they reject causal pluralism. For Goertz and Mahoney, there are two different ways of thinking of causality—two concepts of cause—each of which is supported by evidence produced through different methods. Shan and Williamson argue contra Goertz and Mahoney that causal pluralism does not follow from either methodological or evidential diversity. As they frame it, Evidential Pluralism is stating that different methods (association studies and mechanistic studies) provide different sorts of evidence for the *same* causal claim—causal monism. "We would argue that although there is indeed methodological diversity in the social sciences, this is perfectly well explained by Evidential Pluralism, which appeals to a single concept of cause" (Shan and Williamson 2023, 84).

The causal pluralism question is not trivial for multi-method research. If indeed, different concepts of cause are supported by evidence of different types then it becomes more difficult to understand how different methods might be integrated to support the same causal claim. This is the problem of integration, a worry alluded to in the discussion of Seawright's concern about triangulation in the previous section. As Seawright suggests in that context, the two sorts of evidence are gathered in response to different research questions. Evidence of average effects does not support a causal claim that will answer questions posed about a particular case since it is evidence of the wrong type. At best it can support a probabilistic conclusion about cases of this type. Shan and Williamson argue that causal pluralism does not follow from methodological diversity however. On the contrary, they claim that causal monism is a better fit with Evidential Pluralism, in part, because the problem of integration disappears. Each method provides a different sort of evidence for the same causal claim.[18] Causal monism appears to provide a simple solution and so arguably is a better explanation of what is going on in multi-method research. IfEvidential Pluralism were combined with causal pluralism it would need to be supplemented by a more robust account of the integration of methods.

I have argued elsewhere that if one is a causal pluralist, then the integration of methods in multi-method research calls for an account of the relationship between

[18] This response assumes that the same question is being addressed. Seawright's suggestion that different research questions might be up for grabs is hard to assess in the abstract. There may be times when the research question addressed is the same one and other times when it is not.

different conceptions of causality that would support such an integration. Association studies provide evidence of average effects whereas mechanistic studies can provide evidence of singular causation and so one requirement for integration would be an account of the relationship between singular causation (actual causation) and causation understood as average effects or general causation (Crasnow 2012).

While such an account would allow for the integration of evidence from studies of different types, even if such integration is possible, it may not always be necessary or desirable to carry it out. Causal knowledge is sought for different purposes and some types of evidence may support one of these purposes better than others. This would be consistent with the idea that there are different research questions that need to be addressed reflecting different goals of research. To clarify, consider prediction, explanation, and control to be the general purposes for which causal knowledge is sought.[19] Mechanistic studies provide evidence that better supports the explanatory goal of causal knowledge than association studies do. Prediction is well-supported by association studies. Control might be supported by either sort of study although precisely what we are trying to control is probably relevant to making this determination.

An argument in favor of Evidential Pluralism might be that it aims at causal knowledge that can achieve all three of these goals—all-purpose causal knowledge. Such an argument also requires a monist understanding of "cause". If one abandons monism in favor of causal pluralism, the motivations for Evidential Pluralism are less clear. To sort out the relative merits of causal monism versus causal pluralism, we first take a detour to consider what appears to be an underlying presupposition about causal inference.

14.3.4 The "Detachability Ideal"

The idea that causal claims should serve all three goals—that causal claims should stand free from the aims of research or other contextual features surrounding research—is closely related to what I call the "detachability ideal". The detachability ideal is the idea that once warranted the beliefs that we count as knowledge are independent of the contexts in which they were produced. Context includes the purposes for which the knowledge was sought, the background conditions that make data relevant as evidence, as well as the evidence that warrants the causal claim. This ideal takes as the primary goal of (political) science producing evidence that allows for the causal claim to be detached from that evidence in much the way a rule of detachment in logic allows the detachment of the conclusion from the premises in a sound argument.[20]

---

[19] Illari (2011) and Williamson (2013), mentionthese as aims of causal knowledge. Shan, Taylor, & Williamson refer to the prediction, explanation, and control characteristics of causal claims as PECs (2024, 388).

[20] This is analogy, not a claim that the reasoning is deductive.

That this ideal is achieved through experimental methods—that they are the gold standard—of evidence is rightfully challenged by Shan and Williamson through Evidential Pluralism which requires both evidence of association and evidence of mechanism in order to detach the causal claim. The detachability ideal is intertwined with an understanding of knowledge as universal and general—lawlike. The transportability of the causal claim—where else it might be applicable—is thought to depend on these characteristics. But although Evidential Pluralism is a correction to evidential monism (particularly monism that only seeks evidence produced through association studies), it does not challenge the detachability ideal.

What are the costs and benefits of detachability? A benefit is that if causal claims are detachable from the evidential context then they can be used for all three goals—predication, explanation, and control. But we know that this is not entirely straightforward. A causal claim warranted in one setting can only be transported to other settings through a detailed comparisons of the contextual features of the new setting. When we detach a claim from its evidential argument, we risk losing the detail of the original context, which, in turn, makes it more difficult to carry out this comparison. Nancy Cartwright and Jeremy Hardie use the metaphor of a "causal cake" to emphasis that all the ingredients must be present in order for a causal claim about policy to be transported (Cartwright and Hardie 2012). The causal cake includes everything necessary for the cause to succeed in producing the effect. They argue as well that the sorts of evidence needed to show what goes into that cake (what are the support factors) as well as the evidence needed to assess the target context, is varied (evidential diversity). Shan and Williamson are not insensitive to the fact that extrapolating (transporting) causal knowledge in social settings is particularly difficult because there is so much variation in social contexts (Shan and Williamson 2023, 44–45). However, they maintain that all varieties of required evidence will be one of two types—evidence of association or evidence of mechanism. It is an open question whether all types of evidence needed for causal claims do fall into these two categories. Cartwright has argued that a finer parsing of evidence is needed (Cartwright 2021a, b).

14.3.5 Causal Monism Vs. Causal Pluralism

It is not surprising that Shan and Williamson embrace causal monism since they understand a causal claim to be fit for all purposes—that it be detachable. In order, to meet that goal, it must be supported by evidence relevant to all purposes. Claims supported by both evidence of association and evidence of mechanism fit the bill. Such claims should be detachable and useful in all contexts. One the other hand, causal pluralists, like Cartwright, more often support a broader evidential diversity (and consequently methodological pluralism) since causal claims ought to be assessed for relevance of their evidence to their purpose. Additonally, some argue that evidential diversity should lead one to embrace causal pluralism as we have seen with Goertz and Mahoney (2012).

Shan and Williamson argue that evidential diversity need not lead to causal pluralism and indeed I have also expressed this view (Crasnow 2011, 47). However, that one is not (logically) compelled to be a causal pluralist, does not mean that one should not be. If the view is neither compelled nor impermissible then other considerations come into play. For the present debate, the main considerations are whether causal pluralism or monism better comports with the practices of scientists and whether causal pluralism or monism provides a more fruitful understanding of science. For the first, Shan and Williamson argue that the way that scientist use "cause" when conducting multi-method research is unambiguous and thus supportive of monism (Shan and Williamson 2023, 124). While I agree with them that it is often clear that evidence from different sources is adduced to support a monistic causal claim, I am less convinced that this is always the case. Nonetheless, I accept their claim for the purpose of making a normative argument for causal pluralism, evidential diversity, and methodological pluralism. Political scientist may be operating with one concept of cause, but should they be?

As the history of the discipline of political science discussed in Part I reveals, calls from within for the discipline for reform have often cited its lack of relevance, that is, both its inability to explain current events and its inability to meaningfully inform policy. Thinking of causal knowledge as identifying detachable causes is supported by an understanding of knowledge as abstract and non-contextual. This is a conception of science that much current philosophy of science challenges, but it is something like this idea of science that motivated behavioralists and their successors as they conceived of political science *as a science*. The sorts of research methods they advocated were ones that promised to offer clear identification of causal relations and were thus be favored over methods that were messier. The urge to make qualitative methods more like quantitative—more scientifically rigorous—stems from this way of thinking.. Arguments in support of qualitative research designs aim to justify qualitative methods as potentially able to achieve such detachability as well. But the challenges of the complexity of the political world (or the social world in general) force a narrowing of scope in order to achieve such results. This reduction in scope also means that the context of research is highly specific and lowers the likelihood of transportability. In addition, methods used do not encourage the production of the sort of detailed evidence one would need to make the needed comparisons, since they frequently strip this information in the process of coding in order to construct data sets that can be used with quantitative techniques.

Political issues take place in complex contexts and knowledge of the evidential contexts of causal claims and the preferred methods of political science are in tension with each other. This is not to say that causal knowledge cannot be transported. Under the right circumstances it can but to do so requires understanding the context of origin, including what the evidence adduced for the claim is, and the extent to which the target of exportation is similar to or different from the original context. As Cartwright has argued, this goal is best served by evidential diversity and open to the use of multiple methods, both within and across research projects (methodological pluralism). Achieving these goals does not depend on causal pluralism, but the spirit of causal pluralism allows for the recognition that there may not be one understanding of causality that can serve all the goals of research and consequently reminds researcher to remain attentive to the context of their evidence—the background assumptions through which its relevance as evidence is established, the purposes for which knowledge is sought.

Relevance requires awareness of the context in which the evidence is produced, including the purposes for which it is produced. Nothing about this needs to commit researchers to causal pluralism, but if one is a causal pluralist, the different senses in which we employ the causal concepts will be a focus of interest. The cluster of attitudes and approaches—causal pluralism, evidential diversity, and a plurality of methods better support this way of thinking about scientific research. They encourage an awareness of the different goals for which evidence might be sought and the different purposes to which causal knowledge can be put. If we are concerned about relevance, as political scientist are, at least from time to time, then the particular importance of understanding the role of evidence for policy is encouraged by such pluralisms.

14.4 Conclusions

The history of political science during the first part of the twentieth century centered on concerns about in what sense political science should be understood to be a science. Although there was some overlap with philosophy of science, the specifics of the discipline had their own internal dynamic. What emerged was an understanding of the goal of science as the production of causal knowledge. Because experiments were the gold standard for providing evidence for causal claims, political science gravitated to statistical methods that emulated experimental reasoning. Qualitative researchers, following this understanding of the goals of research, sought to make qualitative reasoning more rigorous—more like quantitative reasoning. The result was a deeper analysis of the inferential role of qualitative research that resulted in a call for making use of the strengths of both types of methods in order to improve causal inference—multi-method research.

Philosophers have since weighed in on the question of the philosophical foundations of multi-method research (Shan and Williamson 2023). Part 2 explores the questions raised by Evidential Pluralism, which has been proposed as one analysis of multi-method research. I have argued that there is an epistemological presupposition, the detachment ideal, that shapes the discussion through offering an understanding of causal inference that treats causal claims as decontextualized. This ideal is part of an understanding of science that much contemporary philosophy of science challenges. I have argued that an alternative understanding of causal inference in which the evidential context is explicitly acknowledged would better serve a political science that seeks to be relevant.

References

Adcock, Robert. 2007. Interpreting Behavioralism. In *Modern Political Science: Anglo-American Exchanges Since 1880*, ed. Robert Adcock, Mark Bevir, and S. Stimson, 181–208. Princeton, NJ: Princeton University Press.

Adcock, Robert A., Mark Bevir, and Shannon C. Stimson. 2007. A History of Political Science: How? What? Why? In *Modern Political Science: Anglo-American Exchanges since 1880*, ed. Robert Adcock, Mark Bevir, and Shannon C. Stimson, 1–17. Princeton, NJ: Princeton Universrity Press.

Beach, Derek, and Rasmus Brun Pedersen. 2016. *Causal Case Study Methods: Foundations and Guidelines for Comparing, Matching, and Tracing*. Ann Arbor, MI: University of Michigan Press.

Beach, Derek, and Rasmus Brun Pedersen. 2019. *Process-Tracing Methods: Foundations and Guidelines*. Ann Arbor, MI: University of Michigan Press.

Bennett, Andrew, Aharon Barth, and Kenneth R. Rutherford. 2003. Do We Preach What We Practice? A Survey of Methods in Political Science Journals and Curricula. *PS: Political Science and Politics* 36 (3): 373–378.

Bevir, Mark. 2008. Meta-methodology: Clearing the Underbrush. In *The Oxford Handbook of Political Methodology*, ed. Janet M. Steffensmeier, Henry E. Brady, and David Collier, 48–79. Oxford: Oxford University Press.

Bevir, Mark. 2022. *A History of Political Science*. Cambridge, UK: Cambridge University Press.

Brady, H. E., and David Collier, eds. 2004/2010. *Rethinking Social Inquiry: Diverse Tools, Shared Standards*. Lanham, MD: Rowman & Littlefield.

Cartwright, Nancy. 2021a. Rigour Versus the Need for Evidential Diversity. *Synthese* 199:13095–13119.

Cartwright, Nancy. 2021b. What Counts as Evidence for a Mechanism? And Why? *The Reasoner* 15 (6): 53–54.

Cartwright, Nancy, and Jeremy Hardie. 2012. *Evidence-Based Policy: A Pratical Guide to Doing It Better*. Oxford: Oxford University Press.

Crasnow, Sharon. 2011. Evidence for Use: Causal Pluralism and the Role of Case Studies in Political Science Research. *Philosophy of Social Science* 41 (1): 26–49.

Crasnow, Sharon. 2012. The Role of Case Study Research in Political Science: Evidence for Causal Claims. *Philosophy of Science* 79:655–666.

Dowding, Keith. 1994. The Compatibility of Behaviouralism, Rational Choice and the 'New Institutionalism. *Journal of Theoretical Politics* 6 (1): 105–117.

Dowding, Keith. 2016. *The Philosophy and Methods of Political Science*. London: Palgrave.

Easton, David. 1969. The New Revolution in Political Science. *The American Political Science Review* LXIII (4): 1051–1061.

Easton, David. 1990. Political Science in the United States: Past and Present. In *The Development of Political Science: A Comparative Study*, ed. David Easton, John Gunnell, and Luigi Graziano, 272–291. London: Routledge.

Elman, Colin, David Collier, and Henry E. Brady. 2002. A New APSA Organized Section for Qualitative Methods. *The Political Methodologist* 11 (1): 20–21.

George, Alexander, and Andrew Bennett. 2005. *Case Studies and Theory Development in the Social Sciences*. Cambridge, MA: Cambridge University Press.

Goertz, Gary, and James Mahoney. 2012. *A Tale of Two Cultures: Qualitative and Quantitative Research in the Social Sciences*. Princeton, NJ: Princeton University Press.

Goodnow, Frank J. 1904. The Work of the American Political Science Association. *Proceedings of the American Political Science Association* 35–46.

Gunnell, John G. 2006. The Founding of the American Political Science Association: Discipline, Profession, Political Theory, and Politics. *American Political Science Review* 100 (4): 479–486.

Gunnell, John G. 1990. The Historiography of American Political Science. In *The Development of Political Science: A Comparative Study*, ed. David Easton, John Gunnell, and Luigi Graziano, 13–33. London: Routledge.

Gunnell, John G., and David Easton. 1990. Introduction. In *The Development of Political Science: A Comparative Survey*, ed. David Easton, John G. Gunnell, and Luigi Graziano, 1–12. London: Routledge.

Hesse-Biber, Sharlene. 2015. Introduction: Navigating a Turbulent Research Landscape: Working the Boundaries, Tensions, Diversity, and Contradictions of Multimethod and Mixed Methods Inquiry. In *The Oxford Handbook of Multimethod and Mixed Methods Inquiry*, ed. Sharlene Hesse-Biber and R. Burke Johnson, xxxiii–liii. Oxford: Oxford University Press.

Illari, Phyllis. 2011. Disambiguiting the Russo-Williamson Thesis. *International Studies in the Philosophy of Science* 25 (2): 139–157.

Kincaid, Harold, and Jereon van Bouwel. 2023. Putting Philosophy of Political Science on the Map. In *The Oxford Handbook of Philosophy of Political Science*, ed. Harold Kincaid and Jereon van Bouwel, 1–14. Oxford: Oxford University Press.

King, Gary, Robert O. Keohane, and Sidney Verba. 1994. *Designing Social Inquiry: Scientific Inference in Qualitative Research*. Princeton, NJ: Princeton University Press.

Kuhn, Thomas S. 1962. *The Structure of Scientific Revolutions*. Chicago, IL: Chicago University Press.

Lawler, Janet, and David Waldner. 2023. Intepretivism Versus Positivism in an Age of Causal Inference. In *The Oxford Handbook of Philosophy of Political Science*, ed. Harold Kincaid and Jereon van Bouwel, 221–242. Oxford: Oxford University Press.

Mahoney, James. 2010. After KKV: The New Methodology of Qualitative Research. *World Politics* 62 (1): 120–147.

Popper, Karl R. 1962. *Conjectures and Refutations: The Growth of Scientific Knowledge*. London: Routledge and Kegan Paul.

Quack, Alexandra, and Catherine Herfeld. 2023. The Role of Narratives in Transferring Rational Choice Models into Political Science. *History of Political Economy* 55 (3): 549–576.

Riker, William H. 1962. *The Theory of Political Coalitions*. New Haven: Yale University Press.

Russo, Federica, and Benoît Rehoux. 2023. Comparative Qualitative Analysis (QCA): A Pluralistic Approach to Causal Inference. In *The Oxford Handbook of Philosophy of Political Science*, ed. Harold Kincaid and Jeroen van Bouwel, 243–265. Oxford: Oxford University Press.

Russo, Federica, and Jonathan Williamson. 2007. Interpreting Causality in the Health Sciences. *International Studies in the Philosophy of Science* 21 (2): 157–170.

Seawright, Jason. 2016. *Multimethod Social Science: Combining Quantitative and Qualitative Tools*. Cambridge: Cambridge University Press.

Shan, Yafeng, and Jonathan Williamson. 2021. Applying Evidential Pluralism to the Social Sciences. *European Journal for the Philosophy of Science* 11 (96): 1–27.

Shan, Yafeng, and Jon Williamson. 2023. *Evidentail Pluralism in the Social Sciences*. New York: Routledge.

Shan, Y., S. D. Taylor, and J. Williamson. 2024. Epistemic Causality and Its Application to the Social and Cognitive Sciences. In *Alternative Approaches to Causation*, ed. Y. Shan, 241–277. Oxford: Oxford University Press.

Verbeek, Bruno, and Lee McIntyre. 2017. Why Is There No Philosophy of Political Science? In *The Routledge Companion to Philosophy of Social Science*, ed. Lee McIntyre and Alexander Rosenberg, 433–447. New York: Routledge.

Williamson, Jonathan. 2013. How Can Causal Explanations Explain? *Erkenntnis* 78:257–275.

Willoughby, W. W. 1904. The American Political Science Association. *Political Science Quarterly* 19 (1): 107–111.

Zahle, Julie. 2023. Qualitative Research in Political Science. In *The Oxford Handbook of Philosophy of Political Science*, ed. Harold Kincaid and Jeroen van Bouwel, 204–220. Oxford: Oxford University Press.

Chapter 15
Early Theorizing About Mechanisms in the Social Sciences

Petri Ylikoski

Abstract The idea of mechanism-based explanation has a long history in the social sciences. This article examines two influential early sources of this idea that have become almost invisible in current debates about social mechanisms. The first is Rom Harré's realist philosophy of natural science, which articulated an account of mechanisms that provided the foundation for later critical realist ideas about causal explanation. Harré's metaphysical theory about causation consists of a tight conceptual package that involves ideas about powers, natures, essences, and generative processes. After describing its core ideas, the article shows how Roy Bhaskar adapted them to the social sciences and discusses why Harré himself did not make the same move. The second source of mechanical inspiration in the social sciences is Jon Elster, whose approach is quite different from that of Harré and Bhaskar. The article describes how Elster's definition of mechanism has changed over time to reflect his theoretical concerns and how he introduces the important idea of a toolbox of mechanisms.

Keywords Explanation · Mechanism · Social sciences · Jon Elster · Rom Harré · Roy Bhaskar

15.1 Introduction

Talk about causal mechanisms and mechanism-based explanations has become mainstream in the present-day philosophy of science. They are now regarded as substantive and legitimate philosophical concepts (Glennan and Illari 2018). This is quite different from earlier times (pre-1990s) when philosophers writing about causation or explanation occasionally used the concept but did not regard it as a

P. Ylikoski (✉)
Sociology, University of Helsinki, Helsinki, Finland
e-mail: petri.ylikoski@helsinki.fi

Y. Shan (ed.), *History and Philosophy of the Social Sciences*,
https://doi.org/10.1007/978-3-032-08167-4_15

philosophically profound or illuminating concept. Instead, it was regarded either as an innocent metaphor or as one of the dead-ends of old natural philosophy (Suppes 1970).

The mechanisms have also become prominent in the philosophy of the social sciences. Ever since Peter Hedström and Richard Swedberg published their edited volume *Social Mechanisms* (1998), the idea of mechanism-based explanation has become increasingly prominent in the social sciences. Another important source is the critical realist tradition inspired by the early work of Roy Bhaskar (1975, 1979), which has long inspired social scientists to think about social processes in terms of mechanisms. In this paper, I will look at the long—and diverse—roots of the mechanism-based philosophy of social sciences. Contrary to the impression given by recent literature, the idea of mechanism-based is not an import from the recent philosophy of science inspired by the biological sciences (Bechtel and Richardson 1993; Machamer et al. 2000; Wimsatt 2007; Glennan 2017).

The word mechanism belongs to the non-technical causal vocabulary of many social scientists. They talk about mechanisms when they wish to refer to a cause, a causal pathway, or an explanation without explicit theorizing about the nature of the mechanism. Usually, this talk about mechanisms is casual and occasional, and the speakers do not attempt to define it explicitly. Many philosophers of science talk about mechanisms in this same informal way. In this paper, I will not attempt to discover when the informal concept (or concepts) of mechanism came to social sciences. Nor will I look for what we could regard as exemplars of mechanism-based theorizing. Instead, I will focus on *explicit theorizing about mechanisms*.

The origins of thinking about mechanisms are diverse and cannot be discussed in one paper. I will concentrate on two origins that I regard as the most important. The first is Rom Harré's realist philosophy of natural science, which is the source of critical realist ideas about mechanisms. I will show that Harré provided an articulated account of mechanisms anticipating many later developments. However, his primary focus was not on articulating an account of mechanism-based explanation but on developing a metaphysical theory about causation. For this reason, his concept of mechanism is part of a tight conceptual package that involves ideas about powers, natures, essences, and generative processes. This theory provided the foundation for Roy Bhaskar's formulation of critical realism. However, in Harré's later work on the philosophy of social sciences (Harré and Secord 1972; Harré et al. 1985; Van Langenhove 2010), mechanisms played a much less prominent role.

Harré ideas were brought to social sciences by Roy Bhaskar, whose two first books are still foundational for the critical realism movement. I argue that while Bhaskar's philosophical style and many of his ideas are quite different from Harré's, his discussion of mechanisms, causation, and explanation is based on Harre´s ideas with very little difference. The key differences are Bhaskar's use of the concept of tendency and his discussion of social structures. While the former is related to the conceptualization of causal power, the latter is directly connected to the idea of social mechanism. Bhaskar often talks about structures as generative mechanisms while defining the former in terms of internal relations. This implies quite a significant change in the concept of mechanism that set him apart from Harré and, more generally, has made many philosophers of science suspicious of his ideas.

The second source of mechanical inspiration is Jon Elster, the most influential early advocate of mechanism-based thinking outside critical realism. He has advocated mechanisms since the early 1980s. I will trace how his ideas about mechanisms change over time and show how they arise from concerns that differ significantly from Harré's. While the latter started with theoretical explanations in the social sciences, Elster's primary interest has been the causal explanation of particular events in the social sciences. His definition of mechanism has changed over time to reflect his theoretical concerns. Elster also introduces the important idea of a toolbox of mechanisms as a basis of general knowledge in the social sciences. The toolbox idea replaces the law-based account of social scientific generalizations and has influenced much of later thinking about mechanism-based theorizing in the social sciences.

15.2 Harré on Causal Powers and Generative Mechanisms

Principles of Scientific Thinking (PST 1970) is a systematic presentation of Rom Harré realist position in the philosophy of science. It brings together ideas Harré had developed in his earlier works (Harré 1960, 1961, 1964, 1965) or would later expand (Harré and Secord 1972; Harré and Madden 1975; Harré 2005).[1] The book advocates for a Copernican revolution in the philosophy of science against positivism and conventionalism. Instead of thinking of theories as deductive systems of laws and regarding ideas about mechanisms and permanent objects as mere heuristic devices, Harré's realist position turns things upside down. According to it, modeling (unobservable) mechanisms and entities is essential for theorizing, while developing deductive systems of laws is just a useful heuristic device. In the following, I will leave aside Harré's criticisms of "deductivism" (Harré 1970: 1–29) and focus on his positive proposal for a realist account of science. However, it is important to remember that his account of mechanisms was initially developed as an alternative to the prominent empiricist account of science.

Harré summarizes theoretical thinking in science as follows:

> Scientists, in much of their theoretical activity are trying to form a picture of the mechanisms of nature which are responsible for the phenomena we observe. The chief means by which this is done is by the making or imagining of models. Since enduring structures are at least as important a feature of nature as the flux of events, there is always the chance that some models can be supposed to be hypothetical mechanisms, and that these hypothetical mechanisms are identical with real natural structures. (Harré 1970: 34–35)

In other words, theories are trying to answer the question: Why do phenomena behave the way they do? by "supplying an account of the constitution and behaviour

[1] For an overview of Harré's thinking, see van Langenhove (2010).

of those things whose interactions with each other are responsible for the manifested patterns of behaviour." (Harré 1970: 35). Often this requires conceiving models of presently unknown "structures and constitutions of things" (Harré 1970: 35), that is, mechanisms of nature. In contrast to an empiricist, a scientific realist takes these postulates seriously and organizes the scientific research process to find evidence of whether these models are analogical to real mechanisms in the world. Importantly, Harré does not explicitly define mechanisms. The concept is a placeholder for whatever processes and entities are responsible for explaining phenomena:

> I do not in the least intend anything specifically mechanical by the word 'mechanisms'. Clockwork is a mechanism, Faraday's strained space is a mechanism, electron quantum jumps is a mechanism, and so on. Some mechanisms are mechanical, others are not. I choose the word 'mechanism' for this use largely because it is the word most usually used for this purpose. We talk of 'the mechanism of a chemical reaction', 'the mechanism of bodily temperature control', 'the mechanism of star creation', and so on. (Harré 1970: 36)

Thus, a mechanism is implicitly defined in the context of a theoretical explanation of phenomena. Mechanisms are constellations of entities, structures, or processes needed to understand the phenomena under study. However, mechanisms are also crucial for how Harré defines causation and causal laws:

> … a causal relation obtains when one state of things or substances generates or produces another, and some generative or productive mechanism can be found, or imagined under the licence of science, by which the production or generation is achieved. (Harré 1970: 85)

> Causal laws are those which state the permanent or enduring conditions under which a certain kind of phenomenon will occur, in short they describe the modes of generation or mechanisms of production of phenomena. (Harré 1970: 102)

> This is a generative account of causation: the cause produces, generates, necessitates, makes, brings about, or leads to the effect. Causal laws include not only appropriate regularity but also an idea of underlying generative mechanisms. In contrast, accidental generalizations characteristically lack the underlying generative mechanism. An empiricist might regard generative mechanisms as a mere heuristic idea, but for Harré, it is the source of our belief in natural necessity (Harré 1970: 104–122).

Harré is not proposing a mechanical account of causation. Instead, the fundamental concept is that of causal power.[2] He understands causality to be active production of effects by powerful particulars.[3] The bases for their powers are found in the workings of generative mechanisms. A working of a generative mechanism involves

[2] For a recent reassessment of causal powers ontology, see Hill et al. (2021).

[3] The typical examples of dispositions in philosophical debates are not good examples of causal powers. Properties like fragility, solubility, brittleness, inflammability, etc., are, in Harré terminology, liabilities: dispositions to suffer change in virtue of essential nature due to an external stimulus (Harré 1970: 272). Most causal powers do not involve such change, nor do external stimuli necessarily activate them. Instead, there is a continuum from powers to liabilities. Harré distinguishes between intrinsic and internal conditions and extrinsic and external conditions. While the first distinction deals with proper parts or features of the thing under consideration, the latter distinction deals with things within the spatial envelope of the thing. Assuming that intrinsic conditions are always internal conditions constitutes a category mistake and a fallacy.

organized activity of powerful particulars constituting it. Thus, the attribution of causal powers is stratified:

> To say of a thing or material that it has a power to do, or to be, or to effect something or other is to say specifically what would happen under appropriate conditions, and to say that these effects occur in virtue of the nature of the thing or material, whatever that may be. It is to say specifically what a thing can do but only unspecifically what nature it has, because it is to say only that it is in virtue of whatever nature it has that it can affect things, materials and observers the way it does. (Harré 1970: 122)

In other words, to say that X has the power to A means that "if X is subject to stimuli or conditions of an appropriate kind, then X will do A, *in virtue of its intrinsic nature*." (Harré 1970: 270). By implication, there is a modal difference between having a power and exercising it. A power can be exercised over and over again or not exercised at all. The basis for attributing a power, even when it is not exercised, is having an appropriate nature—the causal basis—for the power. When a power is ascribed to a thing, the existence of its causal basis is only a promissory note. According to Harré, whether we can ever cash it is an open question, but we have to believe that it is logically possible to do so (Harré 1970: 276, 283).

There are two kinds of explanations involved here. The explanation for *having a power* appeals to the nature or constitution of the thing or a material. This nature consists of a mix of powerful particulars organized in some particular way so that it accounts for the power. To explain *exercising power, in addition to its intrinsic nature, the explanation also has to* refer to (internal or external) triggering conditions that stimulate the power into action. The first explanation is an example of a constitutive explanation, while the latter is a case of causal explanation (Ylikoski 2013).

The relation between the powers and the underlying nature is not simple. The same nature can provide a basis for multiple distinct powers. Thus,

> the same electronic constitution confers two distinct powers upon chemical atoms: the power of entering into certain chemical combinations only, and the power to radiate a particular spectrum. (Harré 1970: 281)

Similarly, two things might have the same power despite having different natures. In other words, the powers can be multiply realized. (If the concept of supervenience had been invented, Harré might have characterized the relation between the powers and nature in terms of it.) These complications further justify the conceptual distinction between powers and their natures.

Harré account of the constitution of powers involves a regress of causal powers: an entity has its causal powers due to causal powers of constitutive parts.[4] However, there is no regress of explanations: it is an empirically verifiable fact that an entity

[4] According to Harré, the regress of explanations can be either to the direction of micro or macro. In micro explanation the behavior and properties of wholes are to be explained by the interacting behavior and properties of their parts. In macro explanation, the explanatory work is done by the system in which behaviors to be explained are embedded. Such explanations are found biology, but also in cosmology (1970: 260–262).

has certain powers; it is an entirely different question of *why* it has those powers. An analogical account is provided for the logical form of laws:

> To accord a statement the status of a law is to treat it as describing a process having an associated generative mechanism, but not to describe what that mechanism is. Thus we have the same specific and explicit, tacit and unspecific structure we have just identified as the logical form of a powers statement. (Harré 1970: 123)

Here is the role of mechanisms in the big picture. While the metaphysics of causation is rooted in powers, for scientific explanations, we need accounts of what powerful entities are involved and how they interact. This is what mechanisms do: they describe how a collection of powerful entities interact in a specific way to produce the law-like behavior we can observe. In a typical causal explanation, the mechanisms play a central role:

> Let us call the cause event 'a', the effect event 'b' and the mechanism 'M'. Now the mechanism M consists of certain components which interact according to certain laws, and we find or propose certain conjunctions of phenomena, say p, q and r, within M, which account for the way in which when it is stimulated by a-type events it produces or generates b-type events. Causal relations between these are filled out by describing other mechanisms, M1′, M2′, ..., which are such that p stimulates M1′ to produce or generate q, and q stimulates M2′ and so on until r is produced or generated; and a produces or is identical with p, and r produces or is identical with b. To such questions as, 'How does p produce or generate q?' we answer by describing yet more mechanisms M1″, M2″, ..., until we come to a pair of components, in a mechanism, which are related by the basic interactions of the science we are using. (Harré 1970: 109)

The dual structure of power ascriptions allows for a division of scientific labor:

> The components and mode of operation of the causal mechanisms are, in general, different in kind from the phenomena that they explain, which are either things in various relations, or things undergoing qualitative change. So explanatory mechanisms become a new subject for scientific study and the explanation of their principles of operation calls for the hypothesis of further explanatory mechanisms, new model building and so on. (Harré 1970: 261)

The stratified structure of science is crucial for Harré (and later for Bhaskar). Explanatory mechanisms form a hierarchy. Chemistry has observable chemical reactions that are explained by atomic hypothesis and valency (Mechanism I), which are, in turn, explained by the electrical theory of atomic structure (Mechanism II), which is a target for a still-to-be-discovered mechanism.

According to Harré, there is an ultimate or final mechanism for any given historical period that does not call for an explanation. (Harré 1964: 25–27). Harré discusses the nature of ultimate causal powers quite extensively (Harré 1970: Chapters 10 & 11; Harré and Madden 1975: Chapter 9). While we might not ever be in an epistemic position to know the ultimate powerful entities or even in a position to know whether we have reached the ultimate powerful entities, they pose a conceptual challenge for his theory as they will not, by definition, have underlying mechanisms and the standard account of ascription of powers will not work for them. However, as this paper focuses on mechanisms of the social world, I will leave these considerations aside.

An important aspect of Harré's theory is the central role of natural kind concepts. The causal powers of entities are rooted in their natures. According to Harré and Madden

> the nature of a kind, or a material substance or an individual is its essence, that is, those of its qualities which are inseparable from its being that kind, that material or that individual. (Harré and Madden 1975: 101)

This essence is the Lockean real essence that we come to know through *a posteriori* scientific inquiry. Do we know any real essences? Probably not, Harré and Madden state:

> The real essence is only finally discovered when the analysis is complete. So far as we can tell, no analyses of things and substances have yet been completed, so our knowledge of the natures of things is as yet an approximation to knowledge of their real essences, but can stand in for that knowledge in all relevant contexts. (Harré and Madden 1975: 102)

The connection of mechanisms with natures, essences, and real definitions informs us about what kinds of *explananda* Harré's theory is intended to deal with. Most of his examples are from physics and chemistry, where it is easy to understand how theoretical explanations are concerned with explaining the causal properties of substances and materials. This pattern is then presumed to apply also to biological and social sciences. Although his theory also applies to the explanations of events, or happenings as Harré calls them, the primary focus is on general theoretical explanations addressing kinds. These explanations address *constitutive* questions: why or how does a thing have its causal capacities and dispositions? Harré focuses on general principles of scientific metaphysics, so he does not pay much attention to the details of explanations of particular facts. In other words, he spends more time discussing the nature of possible fundamental entities than the pragmatics of explaining particular events.

PST is mainly focused on natural sciences but contains only brief examples of attribution powers to humans. *Causal Powers* (CP) is focused on causation and natural necessity and does not contain any ideas about explanation or social sciences. Thus, the primary source for applying Harré's realist philosophy to the social sciences is *The Explanation of Social Behaviour* (EoSB), published in 1972. Interestingly, EoSB was designed to be an argument *against the Mechanistic Model of Man*, so the rhetorical setting is quite different. The Mechanistic Model of Man explicates behavior in terms of external stimuli, aims to minimize the contribution of factors internal to the organism, and expects to find uniform behavioral regularities through laboratory experiments. Harré and Secord present *the Anthropocentric Model of Man* as an alternative foundation for social psychology. This alternative perspective starts with the idea that science should treat humans as agents acting according to rules. Thus, human behavior should be conceived as action mediated by meanings rather than responses caused by stimuli. The new model takes lay explanations of behavior to be the best model for a psychological theory.[5] According

[5] In Harré's stratified view of science, the psychological theory is supported by physiology which provides the proper scientific foundation for study of social behavior.

to Harré and Secord, their new model is more in accordance with actual methodologies of real natural sciences than the Mechanistic model that draws its inspiration from the Humean conception of cause and logical empiricist philosophy.

The rhetorical choice of campaigning against the Mechanistic Model explains why mechanisms do not play a very prominent role in Harré's writings about social sciences. However, there is also a more principled reason that derives from the core idea of the new *ethogenic* research program. The idea is that all social behavior has a purpose for the person performing it and a meaning for others observing it. The behavior has to be understood in terms of local standards of correctness, that is, as monitored rule-following. In this setting, the human ability to meaningful agency—including perceptual and cognitive abilities to learn and follow rules, to monitor rule-following of oneself and others, etc.—constitutes the central cluster of causal powers and mechanisms for the social sciences. How humans, as biological organisms, can acquire and have these abilities is a question for the natural sciences, like evolutionary biology and physiology. The social sciences and psychology can take these abilities for granted and focus on explaining the behaviors in terms of relevant rules and social interactions. "A man then is a mechanism, but one which monitors and controls the way he performs" (1972: 97). The central challenge is to figure out how relevant meanings guide the core explanatory mechanism of human behavior.

> … we must have some idea of the causal mechanism at work, or, in the case of the social behaviour of people, of the meanings they perceive, of the plans they conceive, and of the reasons they have for their actions. (Harré and Secord 1972: 57)

This makes much of mechanism talk obsolete. The basic mechanisms of social interaction are already familiar, and the disciplinary division of labor makes looking under the hood unnecessary. The real challenge is figuring out the relevant meanings, rules, and roles; for them, the traditional causal language is mostly inept or does not add any additional value.

There is an additional reason why the conceptual apparatus of PST is not used extensively in EoSB. Harré's account of mechanisms is heavily embedded in the essentialist conceptual scheme of natures, essences, and natural kinds. These concepts might be well-adapted to disciplines like chemistry, but their application becomes increasingly tricky when expanded to biology and human sciences. When Harré faces the difficulty of applying essentialist thinking to the social world, he gives it up. The concept of mechanism is implicitly defined within this conceptual scheme and is given up along with the package.

In his later work, Harré changed his understanding of rules and rule-following, but the basic set-up in terms of mechanisms remains the same: they do not play a prominent role in his discussion of the social sciences. His social psychology primarily focuses on explaining individual social behavior and local interactions, in other words, micro-social phenomena. Thus, there is no need to coin special notions like *social mechanism*. The word mechanism appears in the title of Harré et al. (1985)—*Motives and Mechanisms*—but it does not do any substantial work. The book basically provides an updated version of the EoSB argumentation.

15.3 Bhaskar's Tendencies and Structures

To bring mechanisms properly to the social sciences, we need the work of Roy Bhaskar.[6] I will first discuss his general views about causation and explanation presented in *A Realist Theory of Science* (RTS 1975) and then address their application to the social sciences in *The Possibility of Naturalism* (PoN 1979).[7]

Bhaskar's philosophical goals and style of argumentation in A Realist Theory of Science (RTS) are pretty different from Harré's. Bhaskar favors transcendental arguments and is keen to present broad claims about the fallacies of competing philosophical schools. However, regarding causation, Bhaskar accepts all of Harré's basic ideas. Bhaskar has the same ontology of causal powers, a similar analysis of causal powers in terms of natures, essences, natural necessity, and generative mechanisms, and an identical view of hierarchies of causal powers and mechanisms. Furthermore, like Harré, his notion of mechanism is implicitly defined in terms of its relation to other causal notions and its role in explaining powers. The only difference seems to be that Bhaskar routinely talks about "generative mechanisms and structures" rather than just "generative mechanisms."[8]

Bhaskar's novelty is taking causal laws as statements of tendencies.[9] Causal powers of things are only expressed as regularities in closed systems, where the experimenter has causally shielded the system from interference from outside causal mechanisms. However, these causal laws are not supposed to hold only in these exceptional and artificial circumstances. Instead, the idea is to learn from the experiment about the causal powers of enduring mechanisms that are active also outside of the experimental system. According to Bhaskar, the condition for the intelligibility of experiments in the sciences is that

> … the world is an open system consisting of things possessing causal powers or potentialities and liabilities in virtue of their intrinsic structures (essential natures), which may or may not be exercised, and when exercised may or may not be manifest in a particular outcome, hence may be exercised unactualised and/or unmanifest to people. (Bhaskar 1975: xvii)

Causal laws do not hold only counterfactually, but transfactually, influencing things even in circumstances where they are not displayed as observable regularities. They operate in open and closed systems, expressing the activity of generative mechanisms and structures that exist independently of the experimenter. Generative

[6] Harré's formulation of realism also influenced philosophically oriented social scientists, especially in the UK (*e.g.* Keat and Urry 1975), but they were overshadowed by the emerging critical realism movement.

[7] For the development of Bhaskar's thinking, see Bhaskar and Hartwig (2010).

[8] In a Postscript to PoN, Bhaskar notes that he: "… often use the term 'structure' and 'generative mechanism' as if they were synonyms." (Bhaskar 1979: 187).

[9] Another novel thing is the appeal to *manipulation* in characterization of causality in experimental systems. Here Bhaskar's inspiration is von Wright (1971). However, Bhaskar does not end up presenting a full-blown interventionist account of causation.

mechanisms are nothing but ways of acting of things, and they provide the basis for causal laws. Because the activities of generative mechanisms and structures are independent of the experimentally or accidentally produced regularities, the causal laws must be interpreted as tendencies.

The concept of tendency is familiar from J. S. Mill (1906, Book III, 10, §5), although Bhaskar does not fully accept Mill's discussion.[10] Instead, his main inspiration comes from Peter Geach's (Anscombe and Geach 1961: 101–108) discussion of Augustine's concept of inclinatio. Tendencies are not de facto regularities but claims about activities of powerful particulars that also cover conditions where possibly many interfering factors affect the observable outcomes. Tendencies are neither claims about what happens usually: all heavy bodies tend to fall, not just most. Even balloons that do not fall are subject to gravity's influence. Tendency is more than mere potentiality. In open systems where multiple, possibly countervailing, powers are at work, the observable outcomes are the resultants of their activities. Thus, there is a crucial difference:

> … whereas powers are potentialities which may or may not be exercised, tendencies are potentialities which may be exercised or as it were 'in play' without being realized or manifest in any particular outcome. (Bhaskar 1975: 40)

> Whereas a powers statement says A would ψ in appropriate circumstances, a normic statement says that A really is ψ'ing, whether or not its actual (or perceivable) effects are counteracted. (Bhaskar 1975: 41)

In Bhaskar's metaphysics, there is a crucial distinction between the real structures and mechanisms of the world and the actual patterns of events they generate. Against the empiricists, he holds that the world consists of mechanisms, not events. These mechanisms combine to generate the flux of phenomena we can observe, that is, the domain of actual. According to Bhaskar, there is nothing esoteric in generative mechanism mechanisms:

> … a generative mechanism is nothing other than a way of acting of a thing. It endures, and under appropriate circumstances is exercised, as long as the properties that account for it persist. (Bhaskar 1975: 42)

What does the exercise of a tendency involve? Are tendencies like forces that can cancel each other out and more generally add up? While some tendencies might work like this, this is not the general case. According to Bhaskar:

> when a tendency is exercised unfulfilled two things are not in doubt: (a) that something actually happens, towards explaining which the exercise of the tendency goes some way; and (b) that something is really going on, i.e. there is a real generative mechanism at work, which accounts for the influence of the factor the tendency represents in the generation of the event. (Bhaskar 1975: 89)

The requirement seems relatively weak: the things have to be acting; that is, their mechanisms have to be at work and contributing *somehow* to the outcome. The

[10] There is much overlap between Bhaskar and Cartwright (1989), this probably due to Mill's influence on both.

contrast between open and closed systems creates an impression that Bhaskar is concerned with the problem of extrapolating from experiments. However, he does not provide a solution to the epistemological problem of extrapolation—what does the experiment tell us about the world outside it?—but points to a metaphysical assumption of extrapolative inferences. Bhaskar's discussion is confusing as he mostly talks about tendencies as if they were analogical to forces that could be directly extrapolated. However, on closer reading, tendencies are more like mechanisms than forces. Bhaskar often writes about mechanisms and tendencies in the same way. For example, he states:

> … four questions can be asked of any generative mechanism G:
>
> (i) is the mechanism enduring?
> (ii) is the mechanism operating?
> (iii) are the results of the activity of the mechanism unaffected by the operations of others (of either the same or different types)?
> (iv) are the results of the activity of the mechanism perceived or otherwise detected by men? (Bhaskar 1975: 238)

Is Bhaskar's account different from the one presented by Harré and Madden? The latter are less interested in providing an account of causal laws. Also, in their account, the activities of powerful particulars are crucial for understanding how observable happenings are generated. Furthermore, they do make a distinction between powers and liabilities and state that apart from some verbal matters, they would approve the essentials of Aquina's theory as interpreted by Geach (Harré and Madden 1975: 100). The only differences seem to be that Bhaskar employs the word mechanism more liberally and that he highlights the importance of between tendencies and ordinary powers.

In sum, Bhaskar's thinking about powers and mechanisms is in accordance with Harré theory. He also follows Harré in the essentialist conceptualization of the theoretical objects in science:

> Scientists attempt to discover what kinds of things there are, as well as how the things there are behave; to capture the real essences of things in real definitions and to describe the ways they act in statements of causal laws. The real essences of things are their intrinsic structures, atomic constitutions and so on which constitute the real basis of their natural tendencies and causal powers. (Bhaskar 1975: 165)

In contrast to Harré, Bhaskar is ready to use this schema to make sense of the social sciences in PN. His overall argument begins by challenging methodological individualism. According to Bhaskar, the individualist program will not work because many individual attributes presuppose social context. Furthermore, individualists are misguided by their assumption that sociology is mainly concerned with large-scale mass or group behavior. However, according to Bhaskar, the subject matter of sociology is relational (Bhaskar 1979: 29–34). Second, against the social constructivist model, he argues that social structures cannot be understood as created by humans, nor are they moments of the same dialectical process. Instead, people and society are two radically different kinds of things. Rather than creating society, the people reproduce or transform it. Society always pre-exists people and is a necessary condition for their activity as all activity presupposes the prior existence of social forms (Bhaskar 1979: 37). These two claims are the starting point for Bhaskar's transformational model of social activity. It builds on a duality of structure: society is both the ever-present condition and the continually reproduced outcome of human agency. Thus,

> ... society must be regarded as an ensemble of structures, practices and conventions which individuals reproduce or transform, but which would not exist unless they did so. (Bhaskar 1979: 39)

Bhaskar argues that there is a categorical distinction between people and societies, and human agency and social structures (Bhaskar 1979: 38). To connect them, we need a mediating system of

> positions (places, functions, rules, tasks, duties, rights, etc.) occupied (filled, assumed, enacted, etc.) by individuals, and of the practices (activities, etc.) in which, in virtue of their occupancy of these positions (and vice versa), they engage. I shall call this mediating system the position-practice system. Now such positions and practices, if they are to be individuated at all, can only be done so relationally. (Bhaskar 1979: 44)

This is the background for Bhaskar's claim that social structures are the appropriate analog for mechanisms in the natural sciences. Only social relations endure in social life; they hold between positions and practices, not between the individuals occupying them. However, there are essential ontological differences:

> 1. Social structures, unlike natural structures, do not exist independently of the activities they govern.
> 2. Social structures, unlike natural structures, do not exist independently of the agents' conceptions of what they are doing in their activity.
> 3. Social structures, unlike natural structures, may be only relatively enduring (so that the tendencies they ground may not be universal in the sense of space-time invariant). (Bhaskar 1979: 42)

Social structures are crucial for addressing proper sociological *explananda*:

> ... the initial conditions in any concrete social explanation must always include or tacitly presuppose reference to some or other social relation (however, the generative structures invoked are themselves best conceived). And it is, I suggest, in the (explanation of the) differentiation and stratification, production and reproduction, mutation and transformation, continual remoulding and incessant shifting, of the relatively enduring relations presupposed by particular social forms and structures that sociology's distinctive theoretical interest lies. (Bhaskar 1979: 44)

Bhaskar's argumentation is very dense, and his terminology is quite loose as he discusses social forms, relations, and structures interchangeably. However, he seems to regard social structures as entities that have causal powers, essential natures, and generative mechanisms. Furthermore, social structures can involve internal and external relations (Bhaskar 1979: 46–47). This is a considerable change from Harré's account, where the carriers of powers are always particulars. Harré's account allows that the way particulars are organized into a whole affects their causal powers, but this causal relevance does not imply that they could have causal powers. This issue has been a bone of contention between Harréan and Bhaskarian realists since the publication of PN.

For example, Harré (2009) argues that critical realists following Bhaskar have reified social structures. According to Harré, the natural sciences explain by presenting models of things, processes, etc, which are possible generators of the phenomena. In the social sciences, a similar role is occupied by schemata, conventions, rules, and so on, which the people who live this kind of life use to guide their meaning-making and management, consciously or unconsciously. The social world is Heraclitean, not Parmenidean (Harré 2009: 134–135). The structural concepts in the social sciences are only heuristic devices, not proper explanatory entities. The mistake of critical realists is to assume that the social world has other fundamental causal agents than human beings acting in accordance with the grammar of behavior. A realist interpretation of macro-social and structural concepts is a fool-errant and an "unicorn project":

> Structural hypotheses in social sciences are heuristic devices sourced from models which could not secrete existential hypotheses, so searching for social structures as beings on a par with actual human interactions is a unicorn project. By failing to ground the social world as an evolving flux, a process driven by the actions of human beings as active agents, despite the overt horror with which old style Critical Realists regard positivism, they tacitly endorse it. Structural concepts in human sciences are heuristic models only—there are no structures. (Harré 2009: 138)

It is an open question whether Harré manages to demonstrate that "… the concept of "social mechanism" either picks out shared grammars or serves a taxonomic role" (2009: 139). Still, it is clear that Bhaskar's adaptation of Harré's causal ideas to the social sciences is contentious (Lewis 2000). Furthermore, the definition of social structure—and associated concepts—have been a constant theoretical challenge for the critical realist tradition. Thinking structures in terms of internal relations has led to theorizing employing transcendental arguments and a priori conceptual analyses that have made more empirically oriented social scientists cautious (Kaidesoja 2013). I think it is fair to say that the challenges with the conceptualization of social structures and the general difficulties in applying essentialist concepts (Pratten 2012) have been the main reasons for the limited acceptance of critical realism among practicing social scientists. While many are eager to adopt the general outlook of realism and accept the criticism of empiricist accounts of explanation and causation, they rarely fully accept Bhaskar's program. Most people inspired by critical realism have picked up the heuristic idea of generative mechanisms and dropped the metaphysical luggage—causal powers, natural necessity, natural kinds, etc.

15.4 Elster's Two Theories About Mechanisms

Jon Elster is one of the most significant early advocates of mechanism-based social science. It is hard to say when he started to think in terms of causal mechanisms. In *Logic and Society* (L&S 1978), he already uses the notion of causal mechanism but does not discuss it explicitly. For example, he describes natural selection as a causal mechanism. He also talks about mechanisms for social change, consequences of action, counterfinality, etc. At this phase, it is clear that he does not treat it as a philosophical concept but as a part of scientific vernacular. However, he presents his famous critical analysis of functional explanations in the social sciences (1978: 121–122). The crux of his argument is that an acceptable functional explanation requires a causal loop by which the beneficial consequences maintain the practice or institution to be explained. He observes that this requirement is rarely satisfied and that only processes that seem to work are analogical to the *mechanism* of natural selection. It is natural to think about the feedback process and the whole functional explanation in terms of mechanisms. In both cases, some organized processes are required, and merely embedding the facts under some regularities is insufficient. Here, he is quite close to presenting a more general challenge to the covering law account of explanation.

Elster explicitly addresses mechanisms in Explaining Technical Change (ETC 1983), his most systematic discussion of social scientific explanation. Here, he associates mechanisms with the search for micro-foundations and employs the metaphor of the black box.[11]

> To explain is to provide a mechanism, to open up the black box and show the nuts and bolts, the cogs and wheels of the internal machinery. (Elster 1983: 24)

According to Elster, mechanisms have two roles. First, they enable going from the larger to the smaller: from molecules to atoms, societies to individuals. Second, they reduce the time lag between the alleged cause and its effect. Thus, a social scientific explanation provides a continuous and contiguous chain of causal or intentional links, which opens the black box. In other words, the preference for mechanisms is grounded in the preference for *fine-grained explanations*.

Elster says that causal explanation subsumes events under causal laws. However, he notes that Hempel's covering law account does not provide an adequate account of explanation as it is open to counterexamples such as causal epiphenomena (i.e., the correlation between A and B caused by a common cause) and pre-emption (Elster 1983: 26). Elster highlights the importance of the locality of causation: a cause always acts on something that is contiguous in space and time. This feature of causation is connected to the idea of explanation with mechanisms. The belief that the world is governed by local causality compels searching for mechanisms of ever-finer grain. The local nature of causality provides an escape from problems with Hempelian theory: epiphenomena and pre-emption can be identified by focusing on continuous and contiguous causal chains (Elster 1983: 29). Elster also challenges the idea of statistical explanation in the social science in principle on the same grounds. Without assumptions about causal mechanisms, statistical correlations are no "more than indications that 'there is something going on' worth looking at in

[11] Philippe van Parijs is a possible source for the black box metaphor. He argues that explanatory understanding requires causal mechanisms and that "As long as the nature of the underlying mechanism is left unspecified, our explanation remains a (clearly not fully satisfactory) black-box explanation." (van Parijs 1981: 12–15).

more detail" (Elster 1983: 47) and "no amount of mere manipulation or controlling of the variables" (Elster 1983: 48) can substitute for causal mechanisms. The idea of *a continuous causal process* is thus a fundamental building block of Elster's thinking about causal mechanisms.

Mechanisms are more central in *Nuts and Bolts for the Social Sciences* (N&B 1989). Elster now launches the idea of *a toolbox of mechanisms* as a way of thinking about the growth of social scientific knowledge. Consequently, the book is now organized around mechanism. According to Elster, mechanisms are the basic units of social sciences, nuts and bolts, or cogs and wheels, that can be used to explain complex social phenomena. However, the articulation of this idea is relatively sparse in N&B, and I will return to it when discussing *Political Psychology* (PS 1993), where it is articulated more fully.

Elster's discussion of mechanisms in N&B mainly repeats the position presented in ETC and consists of a compact discussion of selected observations about causal explanation. First, he argues that the primary *explananda* for causal explanation are events, not facts. He goes on to distinguish between simple causal statements and causal explanations. The latter requires providing, or at least suggesting, a causal mechanism. Elster argues that while the explanatory mechanisms are not explicitly stated in much of historiography and social sciences, they are tacitly presupposed. After distinguishing between causal explanations and statements of correlation, Elster argues that causal explanations are different from assertions about necessitation. A causal explanation provides an account of why the event happened *as* it actually happened. The cause is the event that caused the effect, not a possible cause whose action it might have pre-empted. He also distinguishes between proper causal explanation and storytelling. The explanation is not just how something could have happened but how it actually happened. Finally, causal explanations are not predictions. Explanation without prediction is possible, as is explanatory understanding without an ability to explain. (Elster 1989: 4–8). The same points will reoccur in *Explaining Social Behavior* (ESB 2007, 2015), showing continuity in his thinking about causal explanation.

In ETC, Elster still assumed that causal explanation presupposes causal laws; now, he damps down expectations that the social sciences could, in practice, provide explanations by laws. He observes that the social sciences:

> can isolate tendencies, propensities and mechanisms and show that they have implications for behavior that are often surprising and counterintuitive. What they are more rarely able to do is to state necessary and sufficient conditions under which the various mechanisms are switched on. (Elster 1989: 9)

This is an additional reason to think of explanation in terms of mechanisms rather than laws. While laws are general and exception-less, mechanisms do not make such claims about generality. Instead, when a new mechanism is identified, knowledge progresses by adding a new item to the repertoire of possible causes. The generality of the toolbox is based on its potential usability for novel cases, not on the incorporation of increasingly general empirical claims.

Elster expands the toolbox idea in *Political Psychology* (1993, the French original 1990), where he argues against the idea that social sciences should aim to construct general theories such as “historical materialism, Parsonian sociology, or the theory of economic equilibrium” (Elster 1993: 2). Citing Paul Veyne (1984) and Raymond Boudon (1986), he argues that the aim of establishing general and invariable propositions is, and will always remain an illusory dream. However, the alternative is not to be satisfied with descriptive narratives about particular events. Instead, the solution is to focus on mechanisms. According to Elster:

> The distinctive feature of a mechanism is not that it can be universally applied to predict and control social events, but that it embodies a causal chain that is sufficiently general and precise to enable us to locate it in widely different settings. It is less than a theory, but a great deal more than a description, since it can serve as a model for understanding other cases not yet encountered. (Elster 1993: 5)

He further argues that moving from a plurality of mechanisms to a unified theory would require an ability to identify in advance the conditions in which one or the other mechanism would be triggered. However, current social sciences cannot identify such conditions, and it is unlikely that the situation will improve in the future. In contrast to simplified experimental situations, in the real world, there are too many possible permutations of conditions for us to figure out the characteristic mechanism or the combination of mechanisms operating in each of them.

The idea of mechanisms as midway between theories and narratives is quite similar to Robert Merton’s idea of theories of middle range (Merton 1968). Elster—Robert K. Merton Professor of the Social Sciences at Columbia University—does not make this connection explicit, but Peter Hedström (Hedström and Swedberg 1998; Hedström and Udéhn 2009) has.

In PS, Elster defines mechanisms as “a special causal pattern that can be recognized after the event but rarely foreseen” (Elster 1993: 3). The central theoretical thread of the volume is woven around pairs of contrary mechanisms. For example, the first pair consists of the “sour grapes syndrome,” in which desires are adjusted in accordance with the means of achieving them, and its opposite mechanism in which people want what they cannot have precisely because they cannot have it. This is a significant change from his earlier thinking about mechanisms. Elster describes the change as follows:

> In earlier work (Elster 1983), reductionism was also characterized as a search for mechanisms, but not in the sense in which that term is used here. According to this earlier argument, the difference between theory and mechanism is one fineness of grain. A theory is lawlike “If, then" statement relating an antecedent state to a subsequent one. A mechanism provides the causal chain that mediates between the two states. In the absence of a mechanism, the law is a mere black box. […] In the present exposition, theories and mechanisms differ in level of generality rather than in fineness of grain. In my earlier terminology, going from theory to mechanism is to go from "If A, then always B" to "if A, then always C, D, E, F and B." On the view set out here going from theory to mechanism is to go from "If A, then always B" to “If A, then sometimes B.” However, I also urge the further move to "If A, then sometimes C, D, E, F and B.” (Elster 1993: 7–8, note 13)

The same line of ideas is further developed in *Alchemies of the Mind* (AM 1999), where Elster defines mechanisms as "*frequently occurring and easily recognizable causal patterns that are triggered under generally unknown conditions or with indeterminate consequences*" (Elster 1999: 1). Elster now distinguishes between two types of mechanisms characterized by indeterminacy.[12] In one set of cases (type A), the indeterminacy concerns which of the possible mechanisms will be triggered, while in another set of cases (type B), it concerns the net effect of opposite mechanisms that are triggered simultaneously (Elster 1999: 3; Elster 2011).

In accordance with PS, Elster argues against explanatory laws in the social sciences. He clarifies that in ETC, the antonym of mechanisms was a black box, while now the proper antonym is a scientific law. In his earlier account, mechanisms were associated with the reductionist strategy and methodological individualism. He still considered micro-foundational explanations in terms of laws, although at a lower level of aggregation. Mechanisms were good because they provided finer grain explanations and reduced the risk of spurious explanations. Now he states that "[T]he mechanism provides an explanation because it is *more general* than the phenomenon that it subsumes." (Elster 1999: 6). Mechanisms are good because they enable explanation when generalizations break down. Instead of "if A, then always B," we now have "if A, then sometimes B," or preferably "if A, then sometimes C, D, and B." However, he still says that the call for mechanisms is only *faute de mieux*; they are not desirable in themselves. Explanation by laws would still be ideal, but unfortunately, it is not achievable in the social sciences (Elster 1999: 10).

This discussion is interesting for multiple reasons. First, Elster has not given up methodological individualism; he still wishes to avoid spurious explanations and regards explanation as opening black boxes. So, he is not really changing the meaning of 'mechanism'; instead, he is providing it with an additional meaning that concerns the format of causal explanations. Second, he discusses a mechanism as a fragile generalization, a causal pattern with unspecified exceptions and background conditions. "if A, then sometimes B" is compatible with both definitions provided in PS ("a special causal pattern that can be recognized after the event but rarely foreseen") and in AM ("frequently occurring and easily recognizable causal patterns that are triggered under generally unknown conditions or with or with indeterminate consequences"). The word *causal* is essential in both definitions since these are not simple subjunctive conditionals. There should be a proper causal link rather than a mere regularity. Interestingly, Elster characterizes the change in the meaning of mechanism slightly differently some 10 years later. A footnote in *Explaining Social Behavior* (ESB 2007) signals some remorse for the terminological confusion:

> In some of my earlier writings I used "mechanism" to denote what I now call "causal chains." In more recent work I began to use "mechanism" in the sense defined later in this chapter. I should probably have chosen a different terminology, but it is too late now. (Elster 2007: 32)

[12] Indeterminacy here refers to epistemic indeterminacy. It is indeterminate for us which mechanisms will be active, or what would be the net effect. Elster explicitly denies spooky connections to quantum theory (Elster 2011: 50).

The crucial difference is that the causal chain is singular, while the causal pattern is generic. Here, Elster may have finally given up the final connection to the covering law account. In any case, the idea of causal chain captures some of the meaning of causal mechanism, but not all. It suggests a chain of events leading to the *explanandum*; it is unclear if descriptions of such chains always produce an understanding of *how* the outcome was generated. This would probably require a more technical definition of a causal chain. Also, it makes sense to develop theories about causal mechanisms, not so much about mere causal chains. In the same sense, the idea of a growing toolbox of causal chains is not as appealing as a toolbox of mechanisms.

How about Elster's theory about A- and B-type mechanisms discussed extensively in AM? It can be assumed that any combination of mechanisms is also a mechanism, so there is no principal problem with calling them mechanisms, even if one would like to stick to a broader notion of mechanism. In fact, in AM, Elster introduces the idea of molecular mechanisms.[13] However, for some curious reason, A- and B-type mechanisms are not described as molecular mechanisms even though both are combinations of simpler mechanisms and, clearly, special cases of a more general class of mechanisms. So, calling them *indeterminate molecular mechanisms* or *molecular mechanisms with indeterminate outcomes* would have made sense. Given their prominent role in the discussion, the reader gets the impression that they are the main categories of mechanisms. However, most of the mechanisms discussed in N&B or ESB are not examples of *indeterminate molecular mechanisms*, so the more general notion of mechanism still has plenty of uses.

Significantly, all main examples of mechanisms in PS and AM are *psychological*. This reflects Elster's research interests and his methodological individualism. I think this focus also explains his new worries about mechanisms. We don't have proper theories about the workings of the mind/brain, so we must be satisfied with fragile folk psychological generalizations. The mind/brain, as a causal configuration, is mostly inaccessible to us. Furthermore, the minute differences between individual minds and their histories are also inaccessible to us, which also makes generalizations more difficult. All this makes the epistemic indeterminacy of these causal configurations highly salient and unavoidable. The critical question from the point of view of social sciences is, in what measure are social mechanisms similar to psychological mechanisms? The workings of the social world are not as inaccessible to us as those of the mind. We could have more than just observed "causal patterns" because we can observe social processes. This would imply that the situation is not as desperate with social mechanisms as it is with psychological mechanisms. However, social scientists would still be constantly facing situations where multiple mechanisms could be working together, and many initial and background conditions would be unknown, thus raising the issue of indeterminacy analogical to psychological mechanisms.

[13] Later, in the second edition of ESB (2015), they will called macro mechanisms.

In any case, it is accurate to say that psychological *indeterminate molecular mechanisms* are the kinds of mechanisms Elster finds especially interesting.[14] In AM, Elster aims to show that the idea of mechanism has deep roots in common sense psychology and the writings of some classical social thinkers. One of the unique features of his thinking about mechanisms is the connection he sees between proverbs and mechanisms. According to Elster, proverbs are culturally accumulated knowledge that has proven illuminating. Typically, proverbs are simple, robust, and hedged with qualifications. They do not pretend to be law-like statements. Furthermore, almost all proverbs have other proverbs that point in the opposite direction. This fits nicely with the idea of *indeterminate molecular mechanisms.* Another source of insight highlighted by Elster is classic authors. In AM, he especially mentions Montaigne and Tocqueville, but the book illustrates his use of classic sources for insights about emotions and their mechanisms. Elster tells us that reading Tocqueville's *Democracy in America* helped him formulate his idea of a mechanism while reading Montaigne's *Essays* helped him understand Tocqueville and refine the concept of mechanism. In particular, Montaigne teaches us to be wary of arbitrary story-telling and demonstrates the centrality of indeterminacy in mechanisms. In contrast, Tocqueville provides generalizations that may look like law-like but turn out to be contradictory. Elster argues that to make sense of these apparently conflicting claims, we must read them as statements about mechanisms. Here again, the idea of *indeterminate molecular mechanisms* helps to save the deep insights from too hasty falsification.

Embedded in Elster's discussion of *indeterminate molecular mechanisms* is a diagnosis of the lack of laws in social science. As noted, Elster would be happy with causal laws if they were available for the social sciences. However, we don't have them. We only have pretty fragile generalizations that need to be supported by mechanistic considerations. If the social world were full of *indeterminate molecular mechanisms*, this would explain the fragility of causal generalizations and the difficulties in predicting events. And if this were the main reason for the fragility of social science generalizations, calling *indeterminate molecular mechanisms* mechanisms could make some sense. However, this would still be confusing. *Indeterminate molecular mechanism* pairs are hardly the only or the most crucial explanation for the fragility of social science generalizations. Furthermore, the fragility of generalizations is a separate issue from the nature of causal explanation.

After AM, Elster's thinking about mechanisms has not changed. The editions of ESB combine materials from N&B and AM. As the passage quoted above shows, Elster has shown some remorse for changing the way in which he has talked about mechanisms. However, it is fair to notice that Elster's terminological changes

[14] Interestingly, Elster frequently talks about tendencies in PS and AM. He is not using the notion as a theoretical concept, but as an everyday concept that describes typical outcomes of psychological processes. Definitely, he is not equating mechanisms and tendencies like Bhaskar, although he probably thinks that there is a causal mechanism behind the tendency. Maybe we could coin one more monster concept, *asymmetric indeterminate molecular mechanism,* for cases where one of the outcomes is most typical.

occurred before the idea of mechanism-based explanation became popular.[15] Maybe he felt free to change the concept as it is not yet in wide use. Despite this hassle, his underlying ideas about causal explanation have remained surprisingly similar over the years. However, his other ideas, for example, about the explanatory power of rational choice theory, have changed substantially over the years.

15.5 Discussion

I have discussed two early sources for mechanism-based thinking in the social sciences. They are critical reactions to the covering law model of explanation and regularity theory of causation. While both take causal explanations seriously, their perspectives are drastically different. The starting point for Harré and Bhaskar is a focus on theoretical explanations in the natural sciences. For them, mechanisms are part of a tight conceptual package that involves powers, natures, essences, and generative processes. Maybe this is why they never analyzed mechanism-based explanations in detail or provided detailed examples. Jon Elster's primary concern is singular explanations of events in the social sciences, and an intuitive idea of the continuity of causal processes drives his thinking. He never mentions causal powers, natures, or other essentialist ideas. Although Elster is a realist about causes and explanations, he is not in the business of presenting a general argument for scientific realism. He is not developing philosophical theory but articulating intuitive ideas about adequate social scientific explanations. Elster's books are full of examples of possible mechanisms, although their presentation is usually quite sketchy. Furthermore, Harré and Bhaskar have nothing that corresponds to Elster's idea of a toolbox of mechanisms.

Apart from being different from each other, these two sources for thinking about mechanisms have been independent of the developments in the philosophy of science in general. Both Harré and Elster predate the new mechanism movement (Machamer et al. 2000; Glennan and Illari 2018) and have very different sources of inspiration. While the new mechanism movement originates in the work of philosophers of biological sciences inspired by Herbert Simon (for example, Bechtel and Richardson 1993; Wimsatt 2007), the sources for Harré and Elster are in fields like chemistry and history. In the recent literature, these different streams of thinking have joined. However, it is good to remember the heterogeneous origins of the mechanistic ideas as it helps to explain some of the confusion in the current literature: the idea of a causal mechanism can be associated with many sorts of philosophical projects.

While the authors I have discussed have been hugely influential in the later development of mechanism-based thinking in the social sciences (Hedström and

[15] A version of Chapter 1 of AM was published in Hedström and Swedberg (1998), a volume that made the idea of mechanism-based social explanation widely known among social scientists.

Ylikoski 2010; Ylikoski 2018), some things have not been transmitted to later accounts. For example, while essentialism is a crucial element for both Harré and Bhaskar, it has not been accepted outside the critical realist tradition. Later mechanistic thinkers in the social sciences have been much less oriented towards general metaphysical issues and have separated the concept of mechanism from the essentialist package. In this sense, they have not followed either Harré's or Bhaskar's visions for mechanisms in the social sciences. Wittgensteinian rule-following considerations (Harré) nor transcendental argumentation (Bhaskar) have found much following in the later literature on social mechanisms, no matter how central they are for these pioneers.

Similarly, Elster's later work on indeterminate psychological mechanisms has not generated much following, and very few have found theoretical inspiration in proverbs or classic authors. More importantly, the later discussions have been much more focused on social mechanisms proper, with much less attention to subtle psychological mechanisms. While Elster is indisputably among the founders of the analytical sociology tradition, these features make him an outlier compared to more recent accounts.

References

Anscombe, G. E. M., and P. T. Geach. 1961. *3 Philosophers: Aristotle, Aquinas & Frege*. Ithaca: Cornell University Press.

Bechtel, William, and Robert C. Richardson. 1993. *Discovering Complexity. Decomposition and Localization as Strategies in Scientific Research*. Princeton: Princeton University Press.

Bhaskar, R. 1975/2008. *A Realist Theory of Science*. 2nd ed. London: Routledge.

Bhaskar, R. 1979/2005. *The Possibility of Naturalism. Philosophical Critique of the Contemporary Human Sciences*. 3rd ed. London: Routledge.

Bhaskar, R., and M. Hartwig. 2010. *The Formation of Critical Realism. A Personal Perspective*. London: Routledge.

Boudon, R. 1986. *Theories of Social Change*. Oxford: Blackwell.

Cartwright, N. 1989. *Nature's Capacities and Their Measurement*. Cambridge: Cambridge University Press.

Elster, J. 1978. *Logic and Society*. Chichester: Wiley.

Elster, J. 1983. *Explaining Technical Change*. Cambridge: Cambridge University Press.

Elster, J. 1989. *Nuts and Bolts for the Social Sciences*. Cambridge: Cambridge University Press.

Elster, J. 1993. *Political Psychology*. Cambridge: Cambridge University Press.

Elster, J. 1999. *Alchemies of the Mind. Rationality and the Emotions*. Cambridge: Cambridge University Press.

Elster, J. 2007. *Explaining Social Behavior. More Nuts and Bolts for the Social Sciences*. Cambridge: Cambridge University Press.

Elster, J. 2011. Indeterminacy of Emotional Mechanisms. In *Analytical Sociology and Social Mechanisms*, ed. P. Demeulenaere, 50–63. Cambridge: Cambridge University Press.

Elster, J. (2015). Explaining social behavior: More nuts and bolts for the social sciences (Rev. ed.). Cambridge: Cambridge University Press.

Glennan, S. 2017. *The New Mechanical Philosophy*. Oxford: Oxford University Press.

Glennan, S., and P. Illari, eds. 2018. *The Routledge Handbook of Mechanisms and Mechanical Philosophy*. London: Routledge.

Harré, R. 1960/1983. *Introduction to the Logic of the Sciences*. 2nd ed. London: Macmillan.
Harré, R. 1961. *Theories and Things*. London: Sheed and Ward.
Harré, R. 1964. *Matter and Method*. London: Macmillan.
Harré, R. 1965. *The Anticipation of Nature*. London: Hutchinson.
Harré, R. 1970. *The Principles of Scientific Thinking*. London: Palgrave Macmillan.
Harré, R. & E. H. Madden 1975. *Causal Powers. A Theory of Natural Necessity*. Oxford. Basil Blackwell.
Harré, R. 2005. *Modeling: Gateway to the Unknown*. Amsterdam: Elsevier.
Harré, R. 2009. Saving Critical Realism. *Journal for the Theory of Social Behaviour* 39:129–143.
Harré, R., and P. F. Secord. 1972. *The Explanation of Social Behaviour*. Oxford: Basil Blackwell.
Harré, R., D. Clarke, and N. De Carlo. 1985. *Motives and Mechanism. An Introduction to the Psychology of Action*. London: Methuen.
Hedström, P., and R. Swedberg. 1998. *Social Mechanisms: An Analytical Approach to Social Theory*. Cambridge: Cambridge University Press.
Hedström, P., and L. Udéhn. 2009. Analytical Sociology and Theories of the Middle Range. In *The Oxford Handbook of Analytical Sociology*, ed. Hedström and Bearman, 25–49. Oxford: Oxford University Press.
Hedström, P., and P. Ylikoski. 2010. Causal Mechanisms in the Social Sciences. *Annual Review of Sociology* 36:49–67.
Hill, B., H. Lagerlund, and S. Psillos. 2021. *Reconsidering Causal Powers. Historical and Conceptual Perspectives*. Oxford: Oxford University Press.
Kaidesoja, T. 2013. *Naturalizing Critical Realist Social Ontology*. London: Routledge.
Keat, R., and J. Urry. 1975. *Social Theory as Science*. London: Routledge.
Lewis, P. 2000. Realism, Causality and the Problem of Social Structure. *Journal for the Theory of Social Behaviour* 30:249–268.
Machamer, P., L. Darden, and C. F. Craver. 2000. Thinking About Mechanisms. *Philosophy of Science* 67:1–25.
Merton, R. K. 1968. *Social Theory and Social Structure*. New York: Free Press.
Mill, J. S. 1906. *System of Logic*. 8th ed. London: Longmans, Green & Co.
Pratten, S. 2012. Essentialism and the Social. *The Sociological Review* 60:241–266.
Suppes, P. 1970. *A Probabilistic Theory of Causality. Acta Philosophica Fennica XXIV*. Amsterdam: North-Holland.
Van Langenhove, L., ed. 2010. *People and Societies. Rom Harré and Designing the Social Sciences*. London: Routledge.
Van Parijs, P. 1981. *Evolutionary Explanation in the Social Sciences. An Emerging Paradigm*. Totowa: Rowman and Littlefield.
Veyne, P. 1984. *Writing History*. Middletown, CT: Wesleyan Press.
Von Wright, G. H. 1971. *Explanation and Understanding*. London: Routledge & Kegan Paul.
Wimsatt, W. 2007. *Re-engineering Philosophy for Limited Beings: Piecewise Approximations to Reality*. Cambridge: Harvard University Press.
Ylikoski, P. 2013. Causal and Constitutive Explanation Compared. *Erkenntnis* 78:277–297.
Ylikoski, P. 2018. Social Mechanisms. In *The Routledge Handbook of Mechanisms and Mechanical Philosophy*, ed. S. Glennan and P. Illari, 401–412. Routledge.

Chapter 16
Dimensions of the Methodological Individualism/Holism Debate

Jeroen Van Bouwel

Abstract Analyzing the doctrine of methodological individualism and its opposition to methodological holism, I start by briefly reviewing three historical periods in which the discussion around it was very lively (i.e., the turn of the century around 1900, the 1950s, and the 1980s–90s) explicating the variety of characterizations of methodological individualism. To highlight the connection of these philosophical discussions to social scientific practice, intradisciplinary as well as interdisciplinary dynamics, I then look into the debates around microfoundations (in the 1980s) and so-called economics imperialism in the social sciences (and its impact within political science in the 1990s). While both the microfoundations project and economics imperialism are driven by methodological individualists, here too it seems important to pay attention to the variety of understandings of methodological individualism as well as the exact ambition of their undertakings (evaluated along different dimensions, i.e., ontological, epistemological, axiological, and institutional). Overall, I conclude that when looking into the history of the social sciences, the best philosophers can do when discussing methodological individualism and holism is to carefully distinguish the broad variety of positions that can be evaluated along multiple dimensions, to demonstrate the benefits of methodological pluralism as well as expose the downsides of methodological monocultures.

Keywords Methodological individualism · Microfoundations · Economics imperialism · Unification · Methodological holism · Explanatory pluralism · Scientific pluralism

---

J. Van Bouwel (✉)
Ghent University, Ghent, Belgium
e-mail: jeroen.vanbouwel@ugent.be

Y. Shan (ed.), *History and Philosophy of the Social Sciences*,
https://doi.org/10.1007/978-3-032-08167-4_16

16.1 Introduction

The debate between methodological individualists and methodological holists deals with one of those classic problems in social science where defining the problem itself amounts to entering a minefield, there are so many aspects to the problem that it is hard to cover all of them all at once. On top of that, the problem has been refocused time and again throughout the history of the social sciences. In Sect. 16.2, I briefly review three historical periods in which the discussion around methodological individualism and its opponents was very lively which allows us to explicate a variety of definitions of and motivations for methodological individualism. In Sect. 16.3, I introduce a couple of caveats and distinctions to attend to in the debate. To highlight the connection of these philosophical discussions to social scientific practice, intradisciplinary as well as interdisciplinary dynamics, I look first into the debate about microfoundations, in Sect. 16.4, and then into the so-called *economics imperialism* in the social sciences, prominent in the 1980s–90s, in Sect. 16.5. In Sect. 16.6, I conclude that when looking into the history of the social sciences, the best philosophers can do when discussing methodological individualism and holism is to carefully distinguish the broad variety of positions (considering multiple dimensions along which these positions can be evaluated), demonstrate the benefits of methodological pluralism as well as expose the downsides of methodological monocultures.

16.2 Some Historical Milestones

With respect to methodological individualism, Lars Udehn (2002: 479) identifies three periods of intense debate in the history of the social sciences as professionalized since the nineteenth century, i.e., the turn of the century around 1900, the 1950s, and the 1980–90s. This debate was not only animating the social sciences at times, it could also be considered foundational for the philosophy of the social sciences as the long-time editor of the *Philosophy of the Social Sciences* journal, Ian Jarvie, writes: "Emerging from Weber and the Austrian school before there was a philosophy of the social sciences, this [debate over methodological individualism] in a way was the founding debate of the subject in the 1950s" (Jarvie 2011: 24).

While the popularization of the wording *methodological individualism* is traditionally linked to the economist Joseph Schumpeter in 1908 (in German), the first exposition of the doctrine of methodological individualism is usually ascribed to Carl Menger who in the 1870s developed Austrian methodological individualism in economics in opposition to the German Historical School—this dispute is characterized as the *Methodenstreit*.[1] Economics—following the *marginalist revolution* of

[1] Udehn (2002: 484) writes: "Already in his *Principles of Economics* (1871), Menger tried to reduce the complex phenomena of the economy to their simplest elements, that is, the actions of

the 1870s of which Menger was part—then further develops into the methodological individualist discipline *par excellence*, while sociology and anthropology are more predominantly methodologically holistic—with explanations invoking social 'wholes', their structure, culture, and/or function—much influenced by Emile Durkheim's anti-individualism. Durkheim was one of the leading advocates of methodological holism at the turn of the century (see, e.g., his *The Rules of Sociological Method*), famously stating that "the determining cause of a social fact must be sought among the antecedent social facts and not among the states of the individual consciousness" (Durkheim [1895] 1982: 134). A sociologist who was not defending methodological holism though, and typically contrasted with Durkheim, was Max Weber.

Weber is often considered the main proponent of methodological individualism during this period. In a letter to economist Robert Liefmann, Weber writes: "… if I now happen to be a sociologist according to my accreditation papers, then I became one in order to put an end to the mischievous enterprise which still operates with collectivist notions (*Kollektivbegriffe*). In other words, sociology, too, can only be practiced by proceeding from the action of one or more, few or many, individuals, that means, by employing a strictly 'individualist' method" (Weber quoted by Guenther Roth 1976: 306). Interpretive understanding is central to sociology, according to Weber: sociology is "a science concerning itself with the interpretive understanding of social action and thereby with the causal explanation of its course and consequences" (Weber quoted by Roth 1976: 315) and sociology "treats the single individual and his action as the basic unit, as its 'atom,' if a questionable analogy is allowed here" (Weber quoted by Udehn 2002: 485). It might indeed be a questionable analogy, as an atom does not attach any particular motives or meaning to its actions, while Weber's plea for methodological individualism is to be understood as an explanatory guideline according to which we want social phenomena to be explained in terms of the actions of individual agents enabling us to subjectively understand the meaning or motives for action of those individual agents. These phenomena cannot be solely explained by collectivist notions, by a form of sociocultural determinism which Weber perceived in methodological holism.

The second period of intense debate around methodological individualism and holism happens in the 1950s. Let us look at John Watkins, a student of Karl Popper, who argues for methodological individualism as follows: "All social phenomena are, directly or indirectly, *human* creations. A lump of matter may exist which no one has perceived, but not a price which no one has charged, or a disciplinary code to which no one refers, or a tool which no one would dream of using. From this truism I infer the methodological principle which underlies this paper, namely, that the

individual human beings (Menger [1871] 1976: 46ff). In his next work, *Problems of Economics and Sociology* (1883), he went from practice to principle and formulated the first ambitious program of methodological individualism, or "atomism," in the history of the social sciences. In order to understand economic phenomena we must go back to their true elements, individual human beings, and try to find "the laws by which the former are built up from the latter" (Menger [1883] 1963: 93)."

social scientist can continue searching for explanations of a social phenomenon until he has reduced it to psychological terms" (Watkins 1952a: 28–29).[2] Next, he specifies what sort of knowledge we are after: "An understanding of a complex social situation is always derived from a knowledge of the dispositions, beliefs, and relationships of individuals. Its overt characteristics may be *established* empirically, but they are only *explained* by being shown to be the resultants of individual activities" (Watkins 1952a: 29).

Notice how Watkins emphasizes explanation and the logical derivations from psychological regularities and laws. In a follow-up article, Watkins writes about methodological individualism: "This principle states that social processes should be explained by being deduced from principles governing the behaviour of participating individuals and from analyses of their situations, and not from super-individual, 'holistic,' sociological laws" (1952b: 186). The language is very much aligned with the Hempel and Oppenheim (1948) paper in which they proposed that the logic of explanation requires, at least implicitly, the use of general laws. Watkins also reasons within what would later be called the layer-cake model (cf., Oppenheim and Putnam 1958), with a level of psychological laws and a level of sociological laws, as well as a relation of reduction from the sociological to the psychological level.

Watkins, however, stops at the psychological level in the layer-cake, which is according to him a 'natural stopping-place': "But the ultimate premisses of social science are human dispositions, i.e. something familiar and understandable. They 'are so much the stuff of our everyday experience that they have only to be stated to be recognized as obvious.' And while psychology may try to explain these dispositions, they do provide social science with a natural stopping-place in the search for explanations of overt social phenomena" (Watkins 1952a: 32–33). Similar to Max Weber, Watkins refers to familiarity and understanding as central aspects of social science. Different from Weber though, as Udehn highlights, Watkins and his contemporaries were advancing methodological individualism as a principle about social explanation: "Methodological individualism was advanced by Max Weber and the Austrians in particular, as a principle about the definition of collective concepts. It has been advanced, especially by Popper and his followers, as a principle about the explanation of social phenomena" (Udehn 2002: 497).

The critics of Watkins' methodological individualism also focus on explanation and laws, for instance, Leon Goldstein: "And this can be provided only by a system of socio-cultural laws. There is nothing absurd about the possibility of discovering such laws, and the fact that the principle of methodological individualism is not compatible with the search for them is hardly ground for the rejection of a perfectly tenable program" (Goldstein 1956: 811). "Such laws would be concerned with the kinds of conditions under which determinate change may be expected in

[2] Watkins also explicitly uses the terminology "methodological holism", e.g.: "I am arguing against methodological holism, and for methodological individualism. Our arguments [i.e., Watkins' and Walter Eucken's] tend to coincide because 'historicism' is closely related to 'holism': the belief in laws of development presupposes a 'whole' which undergoes the development" (Watkins 1952a: 27).

sociocultural systems, and if the theoretically necessary and sufficient conditions for some determinate kind of change were to obtain, we may reasonably expect that the change would take place. Who the people are whose institutions are changing is not relevant to the problem. The view of the methodological individualist seems to be that social change results from conscious effort, though the results are often more than people bargained for" (Goldstein 1956: 812). Without those laws, and with only methodological individualism, many questions cannot be answered: "To the theoretical questions of institutional development and change, of major interest to sociologists and anthropologists, they seem to have paid little attention. And it is precisely upon the analysis of such questions that the principle of methodological individualism breaks down" (Goldstein 1956: 802–803).[3]

The third period of intense debate I would like to highlight starts in the late 1970s and goes on in the 1980–90s, with, *inter alia*, the appearance of Analytical Marxism. Three of the most prominent scholars of this variety of Marxism, combining Marxism with analytical methods, were G.A. Cohen, John Roemer, and Jon Elster. As Marxism is arguably one of the most methodologically holist research approaches in the social sciences, it is interesting to see how (most) Analytical Marxists advocate methodological individualism.

Let us briefly look at Elster's advocacy of methodological individualism. The language has now evolved from laws to mechanisms and Elster is interested in laying bare the microfoundations (micro-mechanisms if you prefer) of Marx's claims about social causalities. Elster writes that: "History is made by individuals and must be explained in terms of individual action" (Elster 1983: 32), and how methodological individualism is 'trivially true': "To explain social institutions and social change is to show how they arise as the result of the actions and interaction of individuals. This view, often referred to as methodological individualism, is in my view trivially true" (Elster 1989: 13). To explain is "to provide a mechanism, to open up the black box and show the nuts and bolts, the cogs and wheels" (Elster 1985: 5). Thus, Elster defends looking for mechanisms, microfoundations for Marxism, as nothing but individual properties, goals, beliefs, actions, and interactions, to explain social structure and social change: "It is the belief that the world is governed by local causality that compels us to search for mechanisms of ever-finer grain" (Elster 1983: 29).

While the Watkins-era might be pictured in the layer-cake model—with level-specific laws and the idea of micro-reduction—the Elster-era is best captured in the Coleman boat with intra- and inter-level mechanisms. Many defenders of methodological individualism in this period are also looking at rational choice theory as the best option to deliver the microfoundations, something which coincides with the export of rational choice theory from economics to other social sciences. I will return to the Coleman boat and the microfoundations requirement (and its critics) in Sect. 16.4, and, rational choice individualism and imperialism in Sect. 16.5. Let us

[3] For readers interested in further exploring these 1950s debates between methodological individualists like Hayek, Popper, and Watkins, on the one hand, and their critics, like Ernest Gellner, Maurice Mandelbaum, and, Goldstein, on the other hand, I recommend the volume by O'Neill (1973) which contains many of the important contributions from this period.

first sort out some important distinctions and limitations of the debate between methodological individualists and holists.

16.3 Some Important Distinctions and Limitations

As the distinction between Max Weber—focusing on defining collective concepts—and John Watkins—focusing on the explanation of social phenomena—already shows, self-declared methodological individualists might have diverging ideas about what individualism is mainly about. A variety of questions seem to be addressed: (1) Can collectivist concepts like *class* or *state* be reduced to individualist concepts? (2) Are social-level entities and events reducible to individual-level entities and events? Is society just an aggregation of individuals? (3) Can social entities be agents? (4) Are individuals socially constrained? (5) Can social phenomena always be explained in terms of individuals, their actions, and beliefs? Depending on whether you are an individualist or a holist your answer to these questions—some ontological, others semantic, or methodological—will vary. I will not discuss these different questions here, I just want to stress that this chapter deals with *methodological* individualism and holism, which should neither be mixed up with *ontological* individualism and holism, nor with *political* versions thereof.[4]

The first important distinction not to lose out of sight is the one between ontological and methodological claims. The former considers whether macro-level social entities are built exclusively out of individuals and their properties ("social entities are composed of individuals"), or whether social entities or their properties are *sui generis* ("social entities act independently of individuals"). The latter, the methodological one, is not a claim about what there is but about how to gain knowledge of what there is, and this is often understood in terms of explanation, i.e., whether all macro-level social phenomena are ultimately explicable in individual terms, or whether some social phenomena are better explained in social terms. Taking an individualist position on the ontological question does not necessarily imply an individualist position on the methodological one, or one could be a methodological holist without subscribing to ontological holism. In the history of social science and philosophy, we can find quite some defenses where the distinction between ontological and methodological claims is not being made. Take Watkins (1952a), quoted above, who infers methodological individualism from the truism (according to him) that all social phenomena are, directly or indirectly, human creations. (I must add that he qualifies that argument in another article soon afterward (1952b): "I no longer believe that methodological individualism is entailed either by the truism that social objects are created by personal attitudes or by the 'invisibility'

[4] Those interested in *ontological* individualism might turn to Epstein (2015), for *political* individualism one could start with Kymlicka (2001).

of social structures (though I believe that these two considerations support methodological individualism).")

Although the distinction between ontological and methodological issues is being emphasised by philosophers time and again, this way of reasoning keeps on popping up; a recent example is Pierre Demeulenaere in his introduction to *Analytical Sociology and Social Mechanisms* where he writes that methodological individualism, "can be expressed very simply: Social life exists only by virtue of actors who live it; Consequently a social fact of any kind must be explained by direct reference to the actions of its constituents" (2011: 4). Like many have done before him, Demeulenaere makes an inference from composition to explanation; one starts with certain *a priori* or necessary truths concerning social ontology, the nature of social reality, thus deciding on the locus of causation, justified by a 'metaphysical commonplace', doubtful transcendental arguments, or even without further argument ... and, then, the methodological consequences, e.g., the best level of explanation, seem to follow 'automatically' from the ontological stance. This could be called the *ontological fallacy* (cf., Van Bouwel 2003).[5]

Secondly, pleas for methodological individualism are often confused with claims for political individualism (and politics more generally), or motivated by the idea that the use of methodological holism would necessarily result in a threat to political individualism. Schumpeter already emphasized a long time ago the importance of making a sharp distinction between political and methodological individualism: "Both concepts have nothing in common. The first refers to general statements like the freedom of people to develop themselves and to take part in well-being and to follow practical rules. The second does not include any proposition and does not involve a specific starting point. It just means that one starts from the individual in order to describe certain economic relationships" (Schumpeter [1908] 1998: 90–91).

One could respond by pointing at some form of affinity between methodological individualism or methodological holism and some political positions. For instance, the sociologist applying methodological holism might picture the individual as rather determined or steered by social structures or cultural influences, while the economist working within methodological individualism might emphasize the opportunities and choices for the individual to flourish. First, these perspectives are not necessarily incompatible (although they might be perceived as such by the public, cf. Nettle et al. 2023). Second, as we have seen above with Analytical Marxism, there are examples of social scientists who might share similar political values but have a very different methodological position (on the individualism/holism debate).

[5] The ontological fallacy refers to problematic ways in which consequences regarding the best form of explanation are being drawn from ontological arguments. Notwithstanding the critical questions I raise concerning the ontological approach, let me emphasize that ontological debates could play a legitimate role in considering methodological possibilities of a particular theory, model or approach. The ontological moves I criticize are different in that they are often made *a priori* (not on the basis of a thorough study of social scientific practice) and that the results of the ontological statements are to be generalized across the social sciences (not limiting them to the particular theory, model or approach).

Some Marxists (e.g., Mandel 1989) completely disagreed with Elster's methodological interventions and Elster's conviction that they would uphold Marxist values (or would help to explain and change social reality).

The distinction is important to stress as there are still so often political arguments being formulated for or against certain methodological positions, while the possible combinations between the methodological positions and different political positions are manifold. It is unfortunately not uncommon in the debate that critics of a methodological position try to associate it with an unpopular political position, a form of *guilt by association*. Let us hope we can get beyond those caricatures.

Now, let us turn to what the methodological debate is about, namely what position provides the better social explanations. We can distinguish at least five different positions in the literature (as seen from contemporary philosophy, which is of course historically situated too)[6]:

Strong MI: Social phenomena can/must always be explained in terms of individuals, their properties, and actions, i.e., by individualist explanations.

Strong MH: Social phenomena can/must always be explained in non-individualist terms, by social structures, institutions, social functions, or, culture, i.e., by holist explanations.

Weaker MI: Individualist explanations of social phenomena are always better than holist explanations, even if there would be satisfactory holist explanations (there is a fine-grain preference). The individualist explanations also might assign a role to social institutions or social structure (*inter alia*, in combination with a microfoundations requirement).[7]

Weaker MH: Holist explanations of social phenomena are always better than individualist explanations, even if there would be satisfactory individualist explanations.[8]

---

[6] A reviewer recommended me to mention that some might disagree with the characterization of these positions.

[7] Among this weaker versions of methodological individualism, one could also count the versions that are less restrictively individual and allow social elements (without giving up the label of methodological individualism), i.e. *structural individualism, social individualism, institutional individualism,* etc., they are including *explanantia* that are not as strictly individualist as the strong version of MI would require, relaxing the requirement of exclusivity. Udehn (2002: 498) writes: "The doctrine of methodological individualism, then, ranges from versions requiring that social phenomena be fully explained in terms of individuals, to versions requiring only that they be partly explained in terms of individuals. How large this individualistic part must be is not stated, and cannot be stated, at least not precisely, but it is possible to conceive of a version of methodological individualism that assigns virtually all explanatory power to social institutions and social structure, and only a small fraction of it to individuals."

[8] One could also add holist explanations supplemented with individualist parts. In the end, one might wonder to what extent *Weaker MH* is then any different from *Weaker MI*. Udehn writes: "With the occurrence of institutional and structural individualism on the scene, important holistic elements were included in methodological individualism. The result is that the previous line separating methodological individualism and holism has become blurred and the two doctrines no longer appear as clear-cut opposites" (2002: 502).

MP or Explanatory Pluralism: Both individualist and holist explanations are indispensable in the social sciences. What is the better explanation will be dependent on the explanation-seeking question at stake (and explanation-seeking questions vary depending on the epistemic interests that have to be addressed).

In Sect. 16.4, I will explore the debate about microfoundations in the social sciences, as one instance in which 'what is the better explanation' is up for discussion. Before that, let me add one last reflection about some "givens" in the debate in particular with respect to the notion of stratified levels and the idea of a well-defined individual level.

Watkins (1952a) wrote about the 'natural stopping-place', the level of individual dispositions, when providing explanations of social phenomena. As Petri Ylikoski (2012) has convincingly argued, this idea of a natural stopping-place is very present in the debate, with the individual level being understood in an absolute way. The individual level, according to Ylikoski, is understood as (1) *unique*, in all social explanations the individual level would always be the same; (2) *comprehensive*, there is a consistent and well-defined individual level that is sufficient to cover all social phenomena and serve as a reduction basis for all non-individual social notions; and, (3) *privileged*, explanations in terms of this individual level have some special explanatory qualities that set them apart from explanations at other levels (cf., Ylikoski 2012: 26). Ylikoski argues against this mainstream view of the individual level and defends that levels should be understood perspectivally, in the sense that the levels are dependent on the explanatory target (Ibid., 25). The exact contrast between a lower or micro-level and a higher or macro-level depends on one's explanatory interests, not on a priori considerations, according to Ylikoski; there is no predetermined individual and/or social level. There is also no reason to stop opening the black box at the individual level, while it is often assumed (like Watkins did) that the individualist can stop opening the black box once we are at an individual level, the 'natural stopping-place', where contrary to explanations invoking supra-individual social structures or macro-mechanisms, no further microfoundations have to be provided. Furthermore, Ylikoski (2012) also argues that the idea of *levels* better be replaced by *scales*.

The problematic notion of levels is not limited to the social sciences, of course. Potochnik and McGill, among other philosophers of science, question the concept of universal and discrete hierarchical levels in the sciences more generally: "Indeed, the very notion of stratified levels depends on not only the ubiquity, but also the uniformity, of part-whole composition. For strata to emerge, atoms must always compose molecules, populations must always compose communities, and so forth. But the uniformity of composition needed for stratified levels simply does not exist" (2012: 126).

The take-home lesson of much of this literature in philosophy of science seems to be to avoid turning the levels into something absolute. Nevertheless, they could be of analytical usage when one focuses on particular levels that are relative to and specified in relation to one's research questions. So, exactly what level is relevant or how it is being specified might shift depending on what one wants to study.

Explanations and theories might then exhibit a mono-level structure or rather include multiple levels simultaneously. Let us see how that played out in the history of the social sciences.

16.4 The 1980s Microfoundations Debate

In this section, I return to the Analytical Marxists and further explore how they give form to methodological individualism, with a particular focus on the work of Jon Elster. Elster provides a systematic criticism of functionalism and teleological thinking in Marx's work and insists on the need for microfoundations in social science. He defines 'methodological individualism' in the following terms: "Social science explanations are seen as three-tiered. First, there is a causal explanation of mental states, such as desires and beliefs … Next, there is intentional explanation of individual action in terms of the underlying beliefs and desires … Finally, there is causal explanation of aggregated phenomena in terms of the individual actions that go into them. The last form is the specifically Marxist contribution to the methodology of the social sciences" (Elster 1985: 4). The three steps of these explanations are visualized in the so-called Coleman boat—sometimes called the Coleman bathtub. This boat, a diagram taken from James Coleman, pictures a mechanism model of explanation in which a social phenomenon is, via underlying mechanisms, brought about by another social phenomenon; as such the black box of functionalism and teleological thinking in Marx's work is opened, and "the nuts and bolts, the cogs and wheels" (Elster 1985: 5) are shown. The purely holist explanations of Marx actually do not constitute satisfactory explanations unless they are supplemented by microfoundations, by accounts of the underlying mechanisms on the individual level, or so the reasoning of Analytical Marxists goes.

To make this more concrete, let us look at a diagram taken from Coleman (1986) that clarifies how social mechanisms bring about social macro-phenomena or social outcomes, discussing the effect of religious doctrine on the economic system (based on Max Weber's book *The Protestant Ethic and the Spirit of Capitalism*). The diagram has, on the one hand, a macro-to-macro relation (which Coleman labels as methodological holism) that connects large macro-level variables. Coleman (1986: 1321) writes about macro-macro: "Some social theory and some social research are based entirely on relations of this sort. They exhibit a methodological holism that contrasts to the methodological individualism that grounds sociology in a theory of action." On the other hand, Coleman speaks of macro-micro-macro relations which he links to methodological individualism. This involves arrows 1, 2, and 3 in the diagram. Arrow 1, macro-to-micro represents how social situations and cultural environments influence and constrain individuals' actions, desires, and beliefs. Next, arrow 2 represents how individuals with their beliefs, desires, and action opportunities choose their preferred action among the feasible alternatives. Finally, arrow 3 represents how individual actions are transformed into various intended and unintended social outcomes, the way actions combine and generate a collective outcome, thus producing a new social situation. Coleman aims to have a theory of purposive action as a foundation for social theory, entailing the acceptance of a form of methodological individualism and the rejection of methodological holism (related to the macro-to-macro arrow) (Fig. 16.1).[9]

Coleman (1986: 1327) then calls for contributions to the knowledge of relations of types 1, 2, and ("in particular to the most elusive of these three relations, the micro-to-macro relation") the relation of type 3. He seeks microfoundations.

Thus, the idea of microfoundations basically stipulates that: "all social facts, social structures, and social causal properties depend ultimately on facts about individuals within socially defined circumstances. Social ascriptions require microfoundations at the level of individuals in concrete social relationships" (Little 2012: 138). Next, "an explanation of a macro-social phenomenon must be accompanied by a sketch of plausible microfoundations for the causal linkages it postulates" (Little 2013: 605).

What does the microfoundations requirement imply for an explanation to be the better one or to be satisfactory? According to scholars that support this microfoundations requirement, the validity of social explanations depends on whether a credible sketch of the microfoundations underlying a macro-to-macro relation can be given. Reading this requirement, a lot depends on how one understands "sketch", how fully specified do we want the microfoundations to be, and what does the microfoundations requirement exactly imply for a social explanation to be satisfactory? Will the satisfactory explanation (1) only contain the lower part of the boat excluding the macro-to-macro relation, or (2) is the satisfactory explanation a

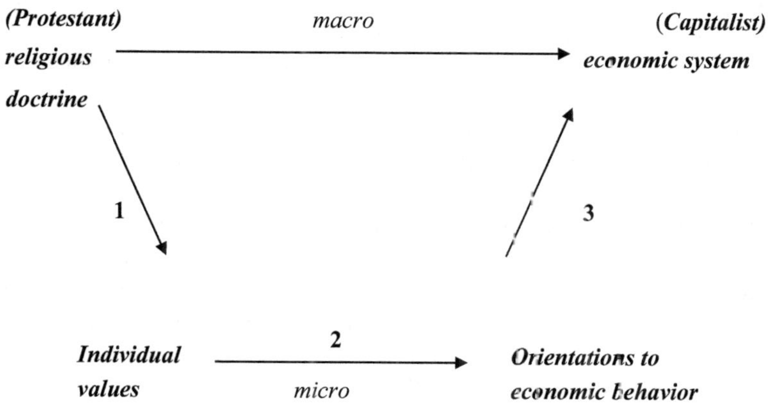

Fig. 16.1 Coleman's boat

[9] Coleman's methodological individualism can be characterized as a weak version of MI. He writes, e.g., that the "theoretical position of methodological individualism is fully compatible with recognition of the constraints on action that social structure creates" (Coleman 1986: 1310). Udehn (2002: 495) calls Coleman's MI "peculiar" and classifies it as "structural individualism" which "is distinct from institutional individualism because it adds something to the latter. It is also more holistic, since it implies a set of interrelated positions that determine the interaction between individuals occupying these positions, an idea that is central to most versions of social holism" (Ibid.).

combination of the macro and the micro level, or (3) is it just the macro-to-macro relation as a satisfactory explanation with the condition that a plausible account of the micro-level can be provided (without the latter having to be part of the explanation)? Advocates of the microfoundations requirement have been formulating different answers to these questions (cf., Van Bouwel 2019):

1. For some, a macro-explanation (i.e., the upper part of the Coleman boat relating macro-factors like the Protestant religious doctrine and a capitalist economic system) will never be satisfactory. Hedström and Swedberg (1998), for instance, state in their presentation of the social mechanisms approach: "In the social sciences, however, the elementary 'causal agents' are always individual actors, and intelligible social science explanations should always include explicit references to the causes and consequences of their actions" (Hedström and Swedberg 1998: 11–12). For them, this does not only imply that we need a micro-level part in every explanation, but also: "that there exist no macro-level mechanisms; macro-level entities or events are linked to one another via combinations of situational, individual action, and transformation mechanisms, i.e., all macro-level change should be conceptualized in terms of three separate transitions (macro-micro, micro-micro, and micro-macro)" (Hedström and Swedberg 1996: 299). As becomes clear after reading this quote, excluding (macro-to-macro) mechanisms on a macro-level does not mean that the defenders of social mechanisms want to exclude all references to entities on the macro-level from social explanations; they just consider a reference to (individual actions on) the individual, micro-level as a condition *sine qua non* of a satisfactory explanation. Underlying this claim seems to be an ontological conviction, a conviction concerning causation, namely that causal agents are always individual actors.
2. Where some see only the lower part of the Coleman boat as a satisfactory explanation (macro-to-micro, micro-to-micro, micro-to-macro), others see the satisfactory explanation as integrating both macro-level and the micro-level in the explanation. Some weaker versions of methodological individualism might recognize that there are indispensable holist (macro-to-macro) explanations, but that they do not stand on their own, they should be supplemented with individual-level microfoundations. Daniel Little (1994) writes, for instance: "social explanation must be explicitly grounded on an account of the microfoundations that produce them. [...] A putative explanation couched at the level of high-level social factors whose underlying individual-level mechanisms are entirely unknown is no explanation at all" (Little 1994: 484).
3. Interestingly, Daniel Little revised his position a bit in his later work: "The requirement of microfoundations is not a requirement on explanation; it does not require that our explanations proceed through the microfoundational level. Rather, it is a condition that must be satisfied on *prima facie* grounds, prior to offering the explanation (…) In short, we are not obliged to trace out the struts of Coleman's boat in order to provide a satisfactory macro- or meso-level explanation or mechanism" (Little 2012: 143). So, in this third option, the macro-level explanation is being considered satisfactory with the condition that some account

of the lower part of the boat can be provided (without the latter having to be part of the explanation); microfoundations would play a justificatory rather than an explanatory role. Little adds that one argument for his position is scientific practice itself, "the fact that good sociologists do in fact make credible use of such claims" (Little 2012: 145).[10]

Thus, in sociology, this search for microfoundations driven by methodological individualists, engendered an intradisciplinary dynamic in which competing theories and explanations were confronted with one another. The microfoundations project ended up being epistemically beneficial along at least three different epistemic dimensions, I think:

(a) As mentioned above, among methodological individualists there is quite some variation about what the better explanation would be even if they agree that searching for microfoundations is important. The visualization of microfoundations making use of the Coleman boat is a great tool to spell out this variety of positions among methodological individualists. (In practice, some of these *Weaker MI* positions might be very difficult to distinguish from *Weaker MH* positions. One important difference here might be that most holists would never subscribe to a microfoundations requirement in all situations; they would rather wonder whether we are not all of the time providing all kinds of explanations and causal claims, without knowing the underlying mechanisms or foundations (cf., Kincaid 2015: 1129)? Why should we have this requirement specifically for social explanations?)

(b) A second epistemic dimension of the microfoundations project is that it could also be considered as developing a heuristic; aiming to establish microfoundations might help us to integrate the causal information of different levels with one another, stimulate the interaction and integration of pieces of explanatory knowledge. (A critic of the microfoundations requirement might welcome the integrative spirit, but why should we limit it to microfoundations at the individual level, why not extend it to identifying *macrofoundations* or a *macroroof*, aiming to integrate or to check the compatibility of the information contained in the social explanation with pieces of explanatory knowledge at a higher level, how the macro makes the micro work? (cf. Van Bouwel 2019) Or, why not require microfoundations of microfoundations (in this case that would be the sub-individual level), for instance, is this another instance of the 'natural stopping-place'?[11])

---

[10] Let me add here that Daniel Little labels his current approach "methodological localism" and distinguishes it from "classical methodological individualism" (cf., Little 2014: 56).

[11] That *both* higher levels and lower levels might help to integrate causal information is a point that has been made more than once in the mechanisms literature, for instance, in relation to biology: "Higher-level entities and activities are thus essential to the intelligibility of those at lower levels, just as much as those at lower levels are essential for understanding those at higher levels. It is the integration of different levels into productive relations that renders the phenomena intelligible and thereby explains it" (Machamer et al. 2000: 23).

(c) The third epistemic aspect I would like to highlight is the confirming or justifying role of microfoundations. One could, for instance, understand Coleman's project as an intent to *justify* Max Weber's causal claim concerning the Protestant ethic and the rise of modern capitalism by providing micro-mechanisms; knowing the causal mechanisms by which individual actions are formed and influence social outcomes might play an important role in justifying causal claims. This does not imply that Coleman sees the micro-mechanisms as a *sine qua non* of every social explanation (i.e., that it has to be *part of* the explanation), or that Coleman would aim for reductionistic accounts, reducing the macro to the individual micro-level. Quite the contrary, Coleman's boat requires us to pay attention to the mechanisms by which social macro-facts condition the decision-making processes of individual agents, as such highlighting and attributing causal importance to the social macro level. (The critic might acknowledge that analyzing the interaction, cooperation, or mutual influence between micro- and macro-levels, definitely is epistemically beneficial, however, some questions do remain. What does this microfoundations requirement mean in practice; when are microfoundations satisfactorily stipulated in order for a macro-explanation to be justified?)

Overall, the quest for microfoundations in the social sciences is a positive and fruitful project of which I tried to highlight the different epistemic benefits (beyond being a mere attempt to prove methodological individualism correct). Dan Little, an important contributor to the microfoundations project, links his own account of explanation to pluralism, "this implies the legitimacy of a fairly broad conception of methodological pluralism in the social sciences, constrained always by the requirement of microfoundations" (Little 2012: 146). It remains to be seen to what extent the microfoundations requirement leads to pluralism, or to what kind of pluralism. I have been defending methodological and explanatory pluralism myself and I disagree with Little about the microfoundations requirement always having to be fulfilled. As I illustrated in Van Bouwel (2006, 2014), in response to some explanation-seeking questions, holist explanations do not have to be complemented by accounts of microfoundations. We can have satisfactory and complete answers to some explanation-seeking questions without the underlying individual-level mechanisms, adding them would not make the explanation any better.

Moreover, there are some situations in which the microfoundations requirement might be counterproductive and just macro-explanations at the social level are to be preferred:

First, a macro-explanation without any microfoundations might provide the required explanatory information –answering the explanation-seeking question– in a very effective and non-redundant way. One could ask why it would be a problem to have a bit more explanatory information by providing microfoundations at the individual level (even if it is not strictly necessary to have a satisfactory explanation). It is not obvious that microfoundations always add useful explanatory information, just having the macro might sometimes be epistemically more advantageous or, given our limited cognitive systems, just the optimal level of explanatory information we can deal with –more microfoundational details might make the explanatory story more difficult to comprehend or process. (cf. Van Bouwel 2019)

Second, macro-explanations are easier to construct: we do not need specific causal information about individuals, nor to address the heterogeneity of underlying individual details. Sometimes microfoundations might increase intelligibility or provide more effective interventions, but in other cases, they might just provide too much and unnecessary information, decrease intelligibility—for instance, in cases of multiple realization—and blur what interventions would be more efficacious.

Third, macro-explanations might be more stable (insensitive to disturbance). Micro-explanations are sometimes misleading when it comes to answering contrastive explanation-seeking questions: each micro-explanation specifies one way to remove the contrast. The numerous other possibilities, involving other individual accounts, are neglected. The micro-explanation does not always give information about the sensitivity of the macro-state to changes at the micro-state; it sometimes picks one specific set of micro-changes that is sufficient to provoke a change at the macro-level. It does not tell us which other perturbations at the micro-level would produce the same change in the macro-state, and which perturbations would produce no change or a different change in the macro-state. (cf., Van Bouwel 2006)

Fourth, in practice, there is also the risk that the microfoundations requirement narrows the range of acceptable macro-explanations, i.e., the range of phenomena macro-accounts can address (something which happened in mainstream economics, for instance, according to Wren-Lewis (2007), I will return to that below). Thus, putting too much emphasis on plausibly connecting the macro to the micro as a way to assure ontological compatibility, or on having to be confident that microfoundations exist, carries the risk of missing out on satisfactory macro-explanations.

That said, for some explanation-seeking questions to be answered in the best way possible, we need micro-explanations, and given its epistemic benefits, the quest for microfoundations should definitely not be stopped. Both individualist and holist explanations are indispensable in the social sciences in order to address our different epistemic interests as well as possible, or so I have argued in my defenses of methodological and explanatory pluralism.

16.5 The Dimensions of Economics Imperialism

Historically speaking, the so-called *economics imperialism* in the second half of the twentieth century offers us a good illustration of how methodological individualism, and the microfoundations requirement, impacted the social sciences.[12] The central idea of economics imperialism is that conceptual and methodological tools from economics are being used to improve knowledge in the social sciences outside of economics, simultaneously expanding the explanatory scope of economics into other disciplines. Uskali Mäki describes it as follows: "Economics imperialism is a matter of persistent pursuit to increase the degree of unification provided by rational choice theory by way of applying it to new types of explanandum phenomena that are located in territories that are occupied by disciplines other than economics" (2002: 238).

It should be emphasized that it is the *dominant* economic theory and methodology that is being applied far beyond its original home, in particular the technical apparatus of microeconomics. This apparatus includes the rational individual, maximizing utility in all decision-making, whose economic behavior is being coordinated through the markets (characterized by supply and demand functions). Typical non-market environments are then being treated as if a market were present, for instance, families, or political and governmental institutions (cf. public choice theory). In terms of social scientific disciplines, it implies that economic theory and methods are being applied in disciplines like political science, sociology, history, anthropology, jurisprudence, and so on; topics and questions traditionally pertaining to these disciplines are then being addressed by the use of rational choice theory or game theory. The application in—some would say the invasion of—disciplines outside of economics is then understood as economics *imperialism*, the *colonization* of other social sciences—some would also point out how the particular microeconomics approach colonized economics internally.[13] In the late twentieth century, it accompanies a move from a market economy to a market society.[14]

We have already discussed how methodological individualism is present in the microfoundations project, and the individualist rational choice imperialism pushes the same line, simultaneously defending its universal applicability, see, e.g., Hirshleifer: "What gives economics its imperialist invasive power is that our

---

[12] This is one historical incarnation of MI, a very influential one within the social sciences of the second half of the twentieth century. There existed of course also other projects driven by MI in the social sciences during that period, but I chose this one as it had a big impact (I am mainly focussing on its impact in political science here).

[13] The label *imperialism* might sound too strong, but Gary Becker—author of the *locus classicus* (his 1976) of economics imperialism—actually agrees with the label: "This definition of "economic imperialism" is probably a good description of what I do" (Swedberg 1990: 39). The use of *imperialism* in the title by Lazear (2000), a supporter, is also telling.

[14] There is an interesting externalist account to be told here too, but that is beyond the scope of this chapter. Just one illustration, the 1980s Analytical Marxism and rational choice imperialism played an important role in the universal basic income debate, see Jäger and Zamora Vargas (2023: 118ff).

analytic categories—scarcity, cost, preferences, opportunities, etc.—are truly universal in applicability. Even more important is our structured organization of these concepts into the distinct yet intertwined processes of optimization on the individual decision level and equilibrium on the social level of analysis" (Hirshleifer 1985: 53).[15]

A notable exponent of economics imperialism is Gary Becker, who was a professor at the University of Chicago and advocated for applying utility-maximizing behavior, market equilibrium, stable preferences, rational choice, and other core economic tenets to the domains of sociology, psychology, law, and other fields, see e.g., Becker's *A Treatise on the Family* (1981) as a paradigmatic work. Becker (1976) contended "that the economic approach is uniquely powerful because it can integrate a wide range of human behavior" (1976: 5) and it is "not restricted in application to material goods or even markets" (Becker 1976: 5); it goes beyond economics' traditional domain of market phenomena, to what had been considered outside of it, non-market settings, cf., his work on household behavior, discrimination, crime, addiction, etc. In his 1981, Becker writes: "The economic approach … assumes that individuals maximize their utility from basic preferences that do not change rapidly over time, and that the behavior of different individuals is coordinated by explicit and implicit markets" (1981: ix). He continues, "this volume uses the assumptions of maximizing behavior, stable preferences, and equilibrium in implicit or explicit markets to provide a systematic analysis of the family" (1981: ix). And in his Nobel speech he concludes that: "The rational choice model provides the most promising basis presently available for *a unified approach* to the analysis of the social world by scholars from the social sciences" (Becker 1992: 52, my italics).

The reactions to this economics imperialism within the social sciences have been wide-ranging, from wild appraisal (e.g., Lazear 2000), to considering it just a hegemonic practice with zero scientific surplus and radically rejecting it (e.g., Fine 2019). How appropriate is such imperialism? Here I will review some criteria developed by Mäki (2009) that can be used in evaluating economics imperialism. Mäki starts from the observation that economics imperialism can be understood as striving for unification, and that unification is considered a respectable achievement amongst most scientists and philosophers of science; it boils down to expanding the domain of phenomena explained by a more parsimonious theory. Becker stated in his Nobel Lecture that the rational choice model is "the most promising basis currently available for a unified approach" in social science. (Becker 1992: 52) However, we cannot just conclude that every attempt at unification and explanatory expansion is to be considered an unconditional success. Certain criteria can help us evaluate the unification and expansion. We will consider (Sect. 16.5.1) *ontological*,

[15] Let me just remark that in rational choice models the agent does not necessarily have be a human individual, sometimes it is a nation-state, firm, etc. Remember the perspectival nature of levels mentioned above. Some philosophers have also argued that rational choice theory is not a 'pure' form of methodological individualism as it uses prior social information, information that originates from some prior social processes or phenomena.

(Sect. 16.5.2) *epistemological,* (Sect. 16.5.3) *axiological*, and, (Sect. 16.5.4) *institutional* criteria. I will spend some pages on those criteria, as I think they explicate different dimensions which are also helpful in evaluating the different positions within the methodological individualism/holism debate.

16.5.1 Ontological Dimension: Does Economics Imperialism Involve Genuine Ontological Unification?

A first question to consider is whether the instance of economics imperialism that one is evaluating actually leads to genuine ontological unification. As quotes from Becker and Hirshleifer mentioned above already illustrated, among supporters of economics imperialism ("truly universal in applicability") there is the idea that it leads to unification ("a unified approach to the analysis of the social world"). Mäki (2009) teaches us, however, that it is important to distinguish derivational unification from genuine ontological unification. To explain the difference, Mäki writes: "One of my favorite examples of derivational unification has been given by game theorist Robert Aumann (1985). Aumann is explicit about unification being a major virtue of theories (all the italics are added in the following). "Part of the greatness of theories like gravitation or evolution, or the atomic theory of matter, is *that they cover so much ground, that they 'explain' so many different things. …* The idea of gravitation itself, in the abstract, is rather mysterious; it is important because it enables us to relate the tides to the motion of the planets and to the trajectories of shells and missiles" (30). The accompanying feature is "spareness; as few as possible exogenous parameters should be used to account for any particular phenomenon. … In addition … one would like spareness in the basic structure of the theory…" (31)" (Mäki 2009: 363).

He continues: "Further textual evidence suggests that what Aumann has in mind is derivational unification. As a general point, he says: "In constructing such a theory, *we are not trying to get at the truth, or even to approximate to it: rather, we are trying to organize our thoughts and observations in a useful manner*" (31–32). In relation to game theory, the principle applies… *a solution notion is the scientists' way of organizing in a single framework many disparate phenomena and many disparate ideas*" (34–35). Economics is no exception to the general principle of derivational unification: Aumann says, "the validity of utility maximization does not depend on its being an accurate description of the behavior of individuals. Rather, it derives from its being the underlying postulate that *pulls together most of economic theory* … Alternatives such as satisficing have proved next to useless in this respect. While attractive as hypotheses, there is little theory built on them*; they pull together almost nothing*; they have few interesting consequences. In judging utility maximization, we must ask not 'Is it plausible?' but '*What does it tie together, where does it lead?*'" (35)"(Mäki 2009: 364).

The second kind of unification Mäki distinguishes is the more genuine ontological unification: "Ontological unification is a matter of redescribing large classes of apparently independent explanandum phenomena as forms or manifestations of a common system of entities, causes, and mechanisms. It is based on the representational capacities of theories in depicting such underlying systems" (Mäki 2009: 364). It is then "a matter of generating a unification of a theory across a wider range of phenomena, not simply in a derivational way showing how the logical formulae of the theory could be applied to new phenomena, but in an ontological way where the theory is shown to increasingly represent the 'the simplest mechanisms and processes of the world's workings'"(Mäki, 2009: 364). If the contribution of a unifying theory is only derivational, little is gained in scientific terms, i.e., a theory might provide the tools to derive a lot of conclusions analytically, deriving a set of conclusions from a set of premises, however not empirically verified, achieving an expansion that does not increase the explanatory power of science. It might be considered a form of scientific expansion, but, if not accompanied by verification with empirical tests, it does not necessarily advance scientific understanding.

I already mentioned the *ontological fallacy* in Sect. 16.3. The derivational unification is a similar strategy, as it is a way of unifying the social sciences that starts with some *a priori* assumptions concerning social ontology (often based on questionable arguments, and raising doubts on whether the ontological assumptions could at all be revised). Next, this *a priori* ontological stance reduces epistemological matters into an ontological one—it is a failure to adequately sustain the distinction between ontology and epistemology, that is, a failure to deal with both ontology and epistemology in a non-reductive way. This fallacy is not limited to defenders of methodological individualism as I illustrated in Van Bouwel (2003).

Besides questions about how ontological unification is brought about, questions might also be raised about whether ontological unification should be an aim. The recent literature on scientific pluralism (e.g., Kellert et al. 2006, Chang 2012) shows that in scientific practice there are cases of theoretical and explanatory pluralism that will never converge in a unified account, being irreducible to one another. Moreover, those defenders of scientific pluralism argue that it is actually an epistemically preferable situation to have this plurality, and that aiming for ontological unification does not necessarily contribute to a better understanding or better explanations of the phenomena studied. These are all aspects to consider when evaluating along the ontological dimension.

16.5.2 Epistemological Dimensions of Unification and Interdisciplinarity Driven by Economics Imperialism

When evaluating the epistemological dimension, if one theory more successfully explains the phenomena to which it initially was applied as well as those previously explained by another theory, then the encompassing theory seems to be preferable; scientists generally prefer parsimony in theory choice, preferring theories that make the fewest assumptions. When we look at rational choice theory being applied outside of economics, and the promise that this could be a parsimonious encompassing theory for the social sciences, it seems important to consider some reactions (in terms of epistemic benefits or adequacy) to rational choice imperialism.

Political scientists, for instance, pointed out that economics imperialism is theory- and method-driven, not problem- or question-driven. The method dictates the problem, not vice versa. The focus is on the development of techniques, not on substantial political questions. Ian Shapiro (2002) elaborated on this distinction between theory- and method-driven on the one hand, and problem- and question-driven on the other. Problem- or question-driven "is understood to require specification of the problem under study in ways that are not mere artefacts of the theories and methods that are deployed to study it" (2002: 590). Method-driven research leads "to self-serving construction of problems, misuse of data in various ways, and related pathologies summed up in the old adage that if the only tool you have is a hammer everything around you starts to look like a nail" (2002: 598). For example, "making a fetish of prediction can undermine problem-driven research via wag-the-dog scenarios in which we elect to study phenomena because they seem to admit the possibility of prediction rather than because we have independent reasons for thinking it worthwhile to study them" (2002: 609).

Economics imperialism is thus perceived by these political scientists as eager to prove that the dominant economic theory and method is superior by the self-serving construction of political problems and by substituting existing theories and/or methods in political science. This raises questions about the epistemic adequacy of the theory; is it apt to answer unaddressed political questions or rather driven to get the social world into one theory, with one preferred form of explanation? If one wants the social sciences to be as well-equipped as possible to answer a broad range of explanation-seeking questions, addressing different epistemic interests, then it will have to evaluate theories in terms of how adequate they answer our explanation-seeking questions. Falling back on one theory and one form of explanation might result in a suboptimal situation as concerns the amount of explanation-seeking questions that can be dealt with in the best way possible (cf., Van Bouwel 2014).

Green and Shapiro (1994) also elaborated on the "Methodological Pathologies" of rational choice applications in political science, pointing at "the syndrome of fundamental and recurrent methodological failings rooted in the universalist aspirations that motivate so much rational choice theorizing. These concern the ways hypotheses are conceptualized, the manner in which they are transformed into testable propositions, and the interpretation of empirical results when tests are conducted" (Green and Shapiro 1994: 33). The analytical elegance and often sophisticated mathematics might help to mask uncertainties, but Green and Shapiro illustrate convincingly the often sloppy empirical testing, the biased fashion of selecting evidence, selective interpretation of data, arbitrary domain restrictions, and so on (Green and Shapiro 1994: ch.3).

Moreover, *commitment* or *confidence* might be filling in gaps left by a lack of evidential support, as Gary Becker himself stated: "So I start with the assumption that behavior is rational, and ask, "As I apply this to a particular problem, is there behavior that I *cannot* explain with the rationality model?" Since rationality can be pretty flexible and the data are often limited, I don't frequently encounter decisive evidence against rationality. Anyway, that is my way of doing things. Others are more agnostic about the scope of rationality, so they will approach a problem by asking, "Does this look like rational behavior or is it better interpreted in a different way?" Part of the difference, therefore, is the degree of *commitment* or *confidence* one has of finding rational behavior when investigating a particular set of phenomena" (Becker in Swedberg 1990: 41).

Epistemologically speaking, the uncertainty regarding evidence is papered over by commitment. That does not make the uncertainty disappear, of course, what it mainly shows is the excessive confidence typical of some economics imperialism proponents. One could ask for some epistemic humility. A first step could be to not consider applications of rational choice imperialism as *substitutes* for the existing theories in a colonized field, but rather consider them as *complements* to the existing theories. This would open up for critical interaction, for assessing the relative strengths and weaknesses of different theories, acknowledging that they answer different explanation-seeking questions, considering some division of labor, and so on. The economics imperialist might contemplate replacing their excessive confidence with some epistemic modesty[16] (cf., Mäki 2020).

As I already mentioned above, when we talk about economics imperialism, we talk about the dominant or mainstream microeconomics. Even within economics, many feel colonized by this dominant theory. It imposes an idea of microfoundations within economics too, i.e., that we can only accept something at the macro-level when we are confident that there are plausible microfoundations following the dominant theory. Simon Wren-Lewis (2007) warns us, based on his analysis of

[16] It might not be easy for mainstream economists to consider, for instance, some form of methodological pluralism. Thinking beyond methodological individualism is hard for mainstream economics according to Fine: "mainstream economics is extremely uncomfortable with the non-individualistic or what can be derived from it, not least with respect to issues of power and conflict, and equally ill at ease with the critical examination of the meaning and reconstruction of concepts and their normative content" (Fine 2019: 140). "[the concept of financialisation] whilst understandably prominent as a deficiency of the mainstream, brought to light by the GFC, this is merely the tip of the iceberg of a legion of deficiencies driven even by a suspended methodological individualism: how we deal with innovation, distribution, monopolisation, globalisation, neoliberalisation, the exercise of power, conflicts, their meanings, contextualisation and determinants" (Fine 2019: 141).

macroeconomics, that the microfoundations requirement in mainstream economics narrows the range of acceptable macro-explanations, i.e., the range of phenomena macro-accounts can address. He argues that the microfoundations requirement changed the way macroeconomics is done, with there being a much greater emphasis on the internal consistency of models nowadays and a greater tolerance of external inconsistency (i.a., inconsistency with the data): "Features of real economies may not be incorporated into models because their rationale in terms of microeconomic theory has yet to be established. Taken literally, the microfoundations methodology implies that the pace of development of macromodels is governed by the speed of theoretical innovation, rather than empirical discovery" (ibid., 58). Thus, we learn from Wren-Lewis's analysis that requiring macro-explanations and macro-accounts to be plausibly microfounded could have as a consequence that explanations with empirical backing but lacking a microfoundations rationale would not be allowed/published. In those cases, a microfoundations requirement ends up hampering explanatory progress.[17] In short, when considering the epistemological dimension of economics imperialism applications, we must make sure the claims are empirically evidenced, compare whether they answer substantial explanation-seeking questions (more) adequately (than existing theories), and check their ambition (substitutive vs complementary).

16.5.3 Axiological Dimension: Are the Contributions of Economics Imperialism Valuable?

The next question to raise is whether the contributions of economics imperialism are at all valuable, significant, or, relevant. Some political scientists criticized the growing body of formal rational choice theory literature and labeled it irrelevant. According to them, this literature does not answer important societal questions, 'real-world problems'. See, for instance, Ian Shapiro's criticism already mentioned above, but also the analysis of the influential international scholar Stephen Walt: "In this sense, much of the recent formal work in security studies reflects the "cult of irrelevance" that pervades much of contemporary social science. Instead of using their expertise to address important real-world problems, academics often focus on narrow and trivial problems that may impress their colleagues but are of little practical value. If formal theory were to dominate security studies as it has other areas of

---

[17] Along the same lines, Ben Fine writes: "With the New Classical Economics (NCE), and the emergence of rational expectations, representative individuals, and so on, macroeconomics was increasingly driven to the extremes of microeconomics, and assumptions to suit the theory rather than the subject matter." (…) "With real business cycles, and technical change treated as random shocks, the NCE was macro treated as micro, and not only taken to extremes but also in extreme ways. Lucas (2003) famously declares that micro has the prospect of making macro superfluous, possibly by defining as economics only what can be done by microeconomics" (Fine 2019: 136–137).

political science, much of the scholarship in the field would likely be produced by people with impressive technical skills but little or no substantive knowledge of history, politics, or strategy" (Walt 1999: 46).

And Walt continues: "Recent formal work in security studies has little to say about contemporary security issues. Formal rational choice theorists have been largely absent from the major international security debates of the past decade (such as the nature of the post-Cold War world; the character, causes, and strength of the democratic peace; the potential contribution of security institutions; the causes of ethnic conflict; the future role of nuclear weapons; or the impact of ideas and culture on strategy and conflict). These debates have been launched and driven primarily by scholars using nonformal methods, and formal theorists have joined in only after the central parameters were established by others. Thus one of the main strengths of the subfield of security studies—namely, its close connection to real-world issues—could be lost if the narrow tendencies of the modeling community took control of its research agenda" (Walt 1999: 47).

One reply you could give to Walt's analysis is that policymakers working in international relations might tell what is *irrelevant* or what is *valuable* to them, but this is not necessarily generalizable to other segments of society. This raises some important questions about the relevance or significance of certain research lines and how to decide on which lines to develop (cf., Kitcher 2001). Are these decisions to be taken by the social scientist herself, or some segment of the social scientific discipline or the community one is working in, or funding organizations, or citizens through some deliberative or representative democratic format?

A second, related aspect is the question of the *availability* of different research lines. One could make a case that the social sciences should be valuable for society at large, and that relevant knowledge for addressing the public's significant problems should be provided. (cf. Van Bouwel forthcoming) This obliges one to avoid an economics imperialism approach in substitutive mode or a unified approach in which some epistemic interests are not being addressed and are just unaddressable.

This latter aspect also shone through in some other critiques by political scientists on rational choice imperialism. Robert D. Putnam, for instance, warned us of the danger of *policy research* migrating toward *schools of public administration,* just as happened with practical economic studies that changed the economics department for *business schools*. "If one compares the size of economics departments and business schools in today's academy, the cost of reducing a social science to sterile theoretical endeavors is obvious" (Quoted on *www.paecon.net*). So, the attempt at explanatory unification under rational choice imperialism risks decreasing the relevance of academic political science for (parts of) the public. One way to ensure that the social scientific disciplines remain relevant is to look into the institutional (social-epistemological) set-up, establishing a division of cognitive labor that addresses a variety of epistemic interests and values, including the public's significant problems.[18]

16.5.4 Institutional Dimension: How to Ensure That Economics Imperialism Contributes Valuable, Relevant, and Sound Knowledge?

Let me again start with some reactions of the political science discipline to rational choice imperialism in the 1990s. The increasing 'colonization' of political science by economists led to so much discontent that an anti-imperialist movement, *Mr. Perestroika*, was created in 2000.[19] The Perestroikans' main focus was that major journals of the field had been preoccupied with publishing research that conforms to the *economics imperialism* features. A lot of attention was paid to the institutional factors that favored economics imperialism. One main target was the *American Political Science Review* which, according to its critics, did not reflect the breadth and diversity of high-quality research done within political science. Kurt Jacobsen wrote: "One might imagine that the American Political Science Association preaches that the best governing system, despite all its faults, is a democratic one, but APSA luminaries obviously display grave doubts as to how far democracy ought to be allowed to go. From inception, the Association never entertained the wildly radical notion of conducting internal elections. What rules is a cozy arrangement whereby a committee chosen by the president nominates its successor members who picks the next governing council who pick the next president, and so on. Disgruntled political scientists link the absence of democracy in the organization to the suffocating disciplinary dominance, especially in the last decade, of formal models and rational choice theory." And, "the APSA of late has been run by rational

[18] Notice that in social-epistemological debates there is also some form of economics imperialism present, in that ESK (the economics of scientific knowledge) develops alternatives to SSK (the sociology of scientific knowledge), cf., Fernandez Pinto (2016). Fernandez Pinto writes: "The crucial point here is to acknowledge that both ESK's thin conception of the social and its narrow understanding of diversity come (at least in part) from the conceptual and methodological framework imported from economics. For instance, economics' asocial understanding of rationality as utility maximization, simply ignores the ways in which different social and political values might influence positively or negatively the epistemic fruitfulness of scientific practice (…). Moreover, the strong methodological individualism of the economic models used in ESK impedes modeling a variety of nonhomogeneous agents, which in turn limits ESK's understanding of diversity in science" (2016: 465).

[19] The first steps of the movement were taken with a mass e-mailing by "Mr. Perestroika", as described in The New York Times on Nov. 4, 2000, "THINK TANK; Political Scientists Leading a Revolt, Not Studying One." Schram (2003) discusses the Mr. Perestroika movement in political science extensively, e.g.: "the ways in which contemporary social science all too often fails to produce the kind of knowledge that can meaningfully inform social life" (Schram 2003: 836). For more about "Mr. Perestroika" and the institutional dimensions at play, see Van Bouwel (2004).

choice exponents or sympathizers and its flagship journal, the American Political Science Review, reflects their unbending bent" (Jacobsen 2001).

Here the attention goes to the professional associations of the discipline (and how they are organized) as well as the flagship journals (and how they select papers), both *institutional* dimensions. In a more general way, the critic emphasizes how the increasing weight and dominance of the mainstream limit the space available for alternative approaches. In evaluating imperializing approaches—if we are in favor of keeping the social sciences relevant for the public—we do not want them to impose their theories and methodologies without them following some set of social-epistemic norms to ensure good scientific practice rather than using their institutional weight to exclude other approaches (neither do we want the mainstream to be completely replaced by the current alternative approaches). What the exact set of social-epistemic norms should be is not the topic of discussion here, we are just considering possible institutional dimensions of the challenges posed by economics imperialism here.[20] Stipulating a set of social-epistemic norms might help avoid an imperialist imposition of theories and methodologies as well as the substitution of alternative theories in the colonized field, and, on the opposite, contribute to a fair but critical interaction as well as the availability of a plurality of approaches to study the social phenomena concerned. Economics imperialism does not take place in a social vacuum, it affects what goes on in the disciplines being colonized, so these are all elements to be checked along the institutional dimension.[21]

[20] One possible set could be Helen Longino's (2002: 128–134) four norms for effective critical interaction between scientific approaches: (1) there should be venues for criticism (including e.g. journals being open for alternative approaches); (2) uptake of criticism; (3) publicly recognized standards for criticism; and, (4) communities must be characterized by tempered equality of intellectual authority. (cf., Van Bouwel 2008) An alternative approach to Longino's social-epistemic norms is to appeal to the economist's virtues like, e.g., intellectual integrity, cf., Fine: "Nonetheless, there is much to be said both for intellectual integrity—saying how it is as opposed to what is acceptable to the mainstream—and never forgetting that pluralism is a stance of equal voice in very unequal circumstances, given the mainstream's dominance" (Fine 2019: 144).

[21] Spelling out how each of these dimensions could play out in the different instances of the individualism/holism debate would lead us far beyond my word limit. For those interested in the third and fourth dimension (which traditionally get less attention in the debate), I could refer you to Van Bouwel (2014) for discussion of the axiological dimension and how to address the *relevance* question in the individualism/holism debate. For the institutional dimension and the importance of *availability* of different approaches (be them individualistic or holistic) see Van Bouwel (forthcoming).

16.6 Philosophers and the Multi-dimensional Individualism/Holism Debate in the Social Sciences

While the historical milestones reviewed in Sect. 16.2 were perhaps announcing a long trench war between methodological individualists and holists, I hope that the subsequent sections made it clear that the debate is a lot richer once you look beyond two warring trenches. Considering the history of the social sciences, I think the best philosophers can do when discussing methodological individualism and holism is to carefully distinguish and spell out the broad variety of positions that can be evaluated along multiple dimensions, *inter alia*, the four dimensions specified in Sect. 16.5.

It is kind of surprising that still nowadays very straightforward defenses of methodological individualism are being written with great fervor. As discussed in Sect. 16.3, the passion might come from political or ontological convictions rather than the methodological intricacies of social scientific explanation (the kind of methodological individualism that is being defended is often a watered-down version, and it is not always entirely clear what the defenders want to save except for the label). On the other hand, the ones that formulate criticisms of some aspects of methodological individualism, seem to be too quickly put in the bag of methodological holism (e.g., Zahle 2023 labels me, an outspoken methodological pluralist as a methodological holist). The limitations of this thinking in terms of methodological monocultures should be exposed, I think. Therefore, I emphasized the upsides of some historical methodologically individualist projects like the microfoundations project in Sect. 16.4, just as much as I pointed out the downsides of an all too monistic approach of some MI-driven economics imperialism applications in Sect. 16.5.

The way in which I portrayed the debate is an invitation to give up all-or-nothing answers on philosophical grounds concerning the individualism/holism debate, answers that were often very a-historical and not very well-informed by social scientific practice. In this chapter, I looked for variety and nuance in historical examples, showing the positive contributions as well as the setbacks of particular methodological approaches. Rather than a winner-take-all approach that would favor one particular approach, I suggested shifting toward the development of criteria or dimensions along which the methodological approaches and their interaction can be assessed—assessments that can be very local and contextual. Hopefully, this might lead us to a further fine-tuning of philosophical positions, stimulate more dialogue and comparison among competing approaches, and perhaps engender a better intellectual division of labor in studying the social world.

Throughout, I emphasized the benefits of taking a pluralist stance, considering the complementarity of methodologies and approaches, keeping an eye on how the social sciences can best address the plurality of epistemic interests that exists among the public(s), and how we might need a wide range of forms of explanations to address those interests in the best way possible, searching for the most adequate explanatory information. Such a methodological pluralism is preferable to a theory-, method- or ontology-driven unification that would imply some (useful) forms of explanation being lost and wanted kinds of explanatory information becoming unavailable to us.

References

Aumann, R. 1985. What Is Game Theory Trying to Accomplish? In *Frontiers of Economics*, ed. K. Arrow and S. Honkapohja. Oxford: Basil Blackwell.
Becker, G. 1976. *The Economic Approach to Human Behavior*. Chicago, IL: University of Chicago Press.
Becker, G. 1981. *A Treatise on the Family*. Cambridge, MA: Harvard University Press.
Becker, G. 1992. The Economic Way of Looking at Life. *Nobel Lecture*, December 9, 1992. https://www.nobelprize.org/uploads/2018/06/becker-lecture.pdf
Chang, H. 2012. *Is Water H_2O? Evidence, Realism and Pluralism*. London: Springer.
Coleman, J. 1986. Social Theory, Social Research, and a Theory of Action. *American Journal of Sociology* 91 (6): 1309–1335.
Demeulenaere, P., ed. 2011. *Analytical Sociology and Social Mechanisms*. Cambridge: Cambridge University Press.
Durkheim, E. 1982 [1895]. *The Rules of Sociological Method*. Basingstoke: Macmillan.
Elster, J. 1983. *Explaining Technical Change. A Case Study in the Philosophy of Science*. Cambridge: Cambridge University Press.
Elster, J. 1985. *Making Sense of Marx*. Cambridge: Cambridge University Press.
Elster, J. 1989. *Nuts and Bolts for the Social Sciences*. Cambridge: Cambridge University Press.
Epstein, B. 2015. *The Ant Trap. Rebuilding the Foundations of the Social Sciences*. New York: Oxford University Press.
Fernandez Pinto, M. 2016. Economics Imperialism in Social Epistemology: A Critical Assessment. *Philosophy of the Social Sciences* 46 (5): 443–472.
Fine, B. 2019. Economics and Interdisciplinarity: One Step Forward, N Steps Back? *Revista Crítica de Ciências Sociais* 119:131–148.
Goldstein, L. 1956. The Inadequacy of the Principle of Methodological Individualism. *Journal of Philosophy* 53 (25): 801–813.
Green, D., and I. Shapiro. 1994. *Pathologies of Rational Choice Theory*. New Haven, CT: Yale University Press.
Hedström, P., and R. Swedberg. 1996. Social Mechanisms. *Acta Sociologica* 39 (3): 281–308.
Hedström, P., and R. Swedberg, eds. 1998. *Social Mechanisms. An Analytical Approach to Social Theory*. Cambridge: Cambridge University Press.
Hempel, C., and P. Oppenheim. 1948. Studies in the Logic of Explanation. *Philosophy of Science* 15:135–175.
Hirshleifer, Jack. 1985. The expanding domain of economics. *American Economic Review Special Issue* 75:53–68.
Jacobsen, K. 2001. Revolt in Political Science. *Post-autistic Economics Newsletter* 6:6.
Jäger, A., and D. Zamora-Vargas. 2023. *Welfare for Markets. A Global History of Basic Income*. Chicago, IL: University of Chicago Press.
Jarvie, I. 2011. Introduction: Philosophical Problems in the Social Sciences: Paradigms, Methodology and Ontology. In *The SAGE Handbook of the Philosophy of Social Sciences*, ed. I. Jarvie and J. Zamora-Bonilla, 1–36. Thousand Oaks, CA: Sage.
Kellert, S., H. Longino, and K. Waters, eds. 2006. *Scientific Pluralism*. Minneapolis: University of Minnesota Press.
Kincaid, H. 2015. Open Empirical and Methodological Issues in the Individualism-Holism Debate. *Philosophy of Science* 82:1127–1138.
Kitcher, P. 2001. *Science, Truth, and Democracy*. New York: Oxford University Press.

Kymlicka, W. 2001. *Contemporary Political Philosophy*. 2nd ed. New York: Oxford University Press.

Lazear, E. 2000. Economic Imperialism. *Quarterly Journal of Economics* 115:99–146.

Little, D. 1994. Microfoundations of Marxism. In *Readings in the Philosophy of Social Science*, ed. M. Martin and L. McIntyre, 479–496. Cambridge, MA: MIT Press.

Little, D. 2012. Explanatory Autonomy and Coleman's Boat. *THEORIA. An International Journal for Theory, History and Foundations of Science* 27 (2): 137–151.

Little, D. 2013. Microfoundationalism. In *Encyclopedia of Philosophy and the Social Sciences*, ed. B. Kaldis, 605–606. Thousand Oaks: Sage.

Little, D. 2014. Actor-Centered Sociology and the New Pragmatism. In *Rethinking the Individualism-Holism Debate*, ed. J. Zahle and F. Collin, 55–75. Springer.

Longino, H. 2002. *The Fate of Knowledge*. Princeton: Princeton University Press.

Lucas, R. 2003. Macroeconomic Priorities. *American Economic Review* 93 (1): 1–14.

Machamer, P., L. Darden, and C. Craver. 2000. Thinking About Mechanisms. *Philosophy of Science* 67 (1): 1–25.

Mäki, U. 2002. Explanatory Ecumenism and Economics Imperialism. *Economics and Philosophy* 18 (2): 237–259.

Mäki, U. 2009. Economics Imperialism: Concept and Constraints. *Philosophy of the Social Sciences* 39 (3): 351–380.

Mäki, U. 2020. Notes on Economics Imperialism and Norms of Scientific Inquiry. *Revue de Philosophie Économique* 21 (1): 95–127.

Mandel, E. 1989. How to Make No Sense of Marx. *Canadian Journal of Philosophy Supplementary* 15:105–132.

Menger, C. 1963 [1883]. *Problems of Economics and Sociology*. Champaign, IL: University of Illinois Press.

Menger, C. 1976 [1871]. *Principles of Economics*. New York: New York University Press.

Nettle, D., W. Frankenhuis, and K. Panchanathan. 2023. Biology, Society, or Choice: How Do Non-Experts Interpret Explanations of Behaviour? *Open Mind: Discoveries in Cognitive Science* 7:625–651.

O'Neill, J., ed. 1973. *Modes of Individualism and Collectivism*. London: Heinemann.

Oppenheim, P., and H. Putnam. 1958. The Unity of Science as a Working Hypothesis. In *Minnesota Studies in the Philosophy of Science*, ed. H. Feigl, G. Maxwell, and M. Scriven, vol. 2, 3–36. Minneapolis: Minnesota University Press.

Potochnik, A., and B. McGill. 2012. The Limitations of Hierarchical Organization. *Philosophy of Science* 79 (1): 120–140.

Roth, G. 1976. History and Sociology in the Work of Max Weber. *British Journal of Sociology* 27 (3): 306–318.

Schram, S. 2003. Return to Politics. Perestroika and Postparadigmatic Political Science. *Political Theory* 31 (6): 835–851.

Schumpeter, J. 1998 [1908]. *Das Wesen und der Hauptinhalt der theoretischen Nationalökonomie*, 3. Aufl. Berlin: Duncker & Humblot.

Shapiro, I. 2002. Problems, Methods, and Theories in the Study of Politics, or What's Wrong with Political Science and What to Do About It. *Political Theory* 30 (4): 596–619.

Swedberg, R. 1990. *Economics and Sociology. Redefining Their Boundaries: Conversations with Economists and Sociologists*. Princeton, NJ: Princeton University Press.

Udehn, L. 2002. The Changing Face of Methodological Individualism. *Annual Review of Sociology* 28:479–507.

Van Bouwel, J. 2003. When Unveiling the Epistemic Fallacy Ends with Committing the Ontological Fallacy. On the Contribution of Critical Realism to the Social Scientific Explanatory Practice. *Philosophica* 71:81–98.

Van Bouwel, J. 2004. Imperialistische tendensen in de economiebeoefening en politieke wetenschappen. Over heterodoxie en pluralisme in de sociale wetenschappen. *Ethiek & Maatschappij* 7 (1): 32–45.

Van Bouwel, J. 2006. The Idea of Social Mechanisms in Social Scientific Explanations. In *Advances in Social Psychology Research*, ed. J. Arlsdale, 83–95. NY: Nova Science Publishers.
Van Bouwel, J. 2008. Ways of dealing with scientific plurality: Inquiring Helen Longino's epistemic democracy. Paper presented at the *6th International Network for Economic Method Conference* in Madrid, Spain.
Van Bouwel, J. 2014. Explanatory Strategies Beyond the Individualism/Holism Debate. In *Rethinking the Individualism-Holism Debate*, ed. J. Zahle and F. Collin, 153–175. Springer.
Van Bouwel, J. 2019. Do Mechanism-Based Social Explanations Make a Case for Methodological Individualism? *Journal for General Philosophy of Science* 50 (2): 263–282.
Van Bouwel, J. Forthcoming. Democratizing Explanatory Practice in the Social Sciences (with John Dewey). In *Pragmatism and Philosophy of Science: Historical and Contemporary Perspectives*, ed. R. Gronda, M. Janack, and G. Marchetti. London: Routledge.
Walt, S. 1999. Rigor or Rigor Mortis? Rational Choice and Security Studies. *International Security* 23 (4): 5–48.
Watkins, J. 1952a. Ideal Types and Historical Explanation. *British Journal for the Philosophy of Science* 3:22–43.
Watkins, J. 1952b. The Principle of Methodological Individualism. *British Journal for the Philosophy of Science* 3:186–189.
Wren-Lewis, S. 2007. Are There Dangers in the Microfoundations Consensus? In *Is There a New Consensus in Macroeconomics?* ed. P. Arestis, 43–60. Basingstoke: Palgrave Macmillan.
Ylikoski, P. 2012. Micro, Macro, and Mechanism. In *The Oxford Handbook of Philosophy of Social Science*, ed. H. Kincaid, 21–45. Oxford: Oxford University Press.
Zahle, J. 2023. Methodological Holism in the Social Sciences. *Stanford Encyclopedia in Philosophy.*

Index

Y. Shan (ed.), *History and Philosophy of the Social Sciences*,
https://doi.org/10.1007/978-3-032-08167-4

MIX
Papier aus verantwortungsvollen Quellen
Paper from responsible sources
FSC® C105338

If you have any concerns about our products,
you can contact us on
ProductSafety@springernature.com

In case Publisher is established outside the EU,
the EU authorized representative is:
Springer Nature Customer Service Center GmbH
Europaplatz 3, 69115 Heidelberg, Germany

Printed by Libri Plureos GmbH
in Hamburg, Germany